RESEARCH TRENDS
IN FLUID DYNAMICS

RESEARCH TRENDS IN FLUID DYNAMICS

*Report from the United States National Committee
on Theoretical and Applied Mechanics*

EDITORS

J. L. Lumley
*Sibley School of Mechanical
and Aerospace Engineering
Cornell University
Ithaca, New York*

Andreas Acrivos
*The Benjamin Levich Institute for
Physico-Chemical Hydrodynamics
The City College of New York
New York, New York*

L. Gary Leal
*Department of Chemical Engineering
University of California
Santa Barbara, California*

Sidney Leibovich
*Sibley School of Mechanical
and Aerospace Engineering
Cornell University
Ithaca, New York*

American Institute of Physics   Woodbury, New York

AIP Press
American Institute of Physics
500 Sunnyside Boulevard
Woodbury, NY 11797-2999

Library of Congress Cataloging-in-Publication Data
Research trends in fluid dynamics : report from the United States
   National Committee on Theoretical and Applied Mechanics /
   editors, J. L. Lumley ... [et al.]
     p. cm.
   ISBN 1-56396-459-7
   1. Fluid dynamics. I. Lumley, John L. (John Leask), 1930- .
II. U.S. National Committee on Theoretical and Applied Mechanics.
QC145.2.R48 1996   95-25488
532'.05'072--dc20   CIP
[B]

10 9 8 7 6 5 4 3 2 1

CONTENTS

CONTRIBUTORS

Hassan Aref *Department of Theoretical and Applied Mechanics, University of Illinois, Urbana, Illinois 61801-2935*

Peter Bradshaw *Mechanical Engineering Department, Stanford University, Stanford, California 94305-3030*

Susan N. Brown *Department of Mathematics, University College, London WCIE 6BT England*

David A. Caughey *Sibley School of Mechanical and Aerospace Engineering, Cornell University, Ithaca, New York 14853-7501*

Robert H. Davis *Department of Chemical Engineering, University of Colorado, Boulder, Colorado 80309-0424*

Morton M. Denn *Department of Chemical Engineering, University of California, Berkeley, California 94720-9989*

Benjamin Gebhart *Department of Mechanical Engineering and Applied Mechanics, University of Pennsylvania, Philadelphia, Pennsylvania 19104-6315*

James T. Jenkins *Department of Theoretical and Applied Mechanics, Cornell University, Ithaca, New York 14853*

Ed J. Kerschen *Department of Aerospace and Mechanical Engineering, University of Arizona, Tucson, Arizona 85721*

Joel Koplik *Levich Institute and Department of Physics, City College of New York, New York, New York 10031*

Sidney Leibovich *Sibley School of Mechanical and Aerospace Engineering, Cornell University, Ithaca, New York 14853-7501*

Sanjiva K. Lele *Department of Aeronautics and Astronautics and Mechanical Engineering, Stanford University, Stanford, California 94305-4035*

E. J. List *Department of Environmental Engineering Sciences, California Institute of Technology, Pasadena, California 91125*

John L. Lumley *Sibley School of Mechanical and Aerospace Engineering, Cornell University, Ithaca, New York 14853-7501*

Chiang C. Mei *Department of Civil Engineering, Massachusetts Institute of Technology, Cambridge, Massachusetts 02139*

Parviz Moin *Department of Mechanical Engineering, Stanford University, Stanford, California 94305*

René Moreau $MADYLAM, N°1340, ruedelaPiscine-DomaineUniversitire-38400Saint-Martin-d'hères$

E. Phil Muntz *Department of Aerospace Engineering, University of Southern California, Los Angeles, California 90089-1191*

J. N. Newman *Department of Ocean Engineering, Massachusetts Institute of Technology, Cambridge, Massachusetts 02139*

Stephen B. Pope *Sibley School of Mechanical and Aerospace Engineering, Cornell University, Ithaca, New York 14853-7501*

Andrea Prosperetti *Department of Mechanical Engineering, Johns Hopkins University, Baltimore, Maryland 21218*

William C. Reynolds *Department of Mechanical Engineering, Stanford University, Stanford, California 94305*

Katepalli R. Sreenivasan *Mason Laboratory, Yale University, New Haven, Connecticut 06520-2159*

Paul H. Steen *School of Chemical Engineering, Cornell University, Ithaca, New York 14853*

Sheldon Weinbaum *Department of Mechanical Engineering, City College of New York, New York, New York 10031*

John A. Whitehead *Department of Physical Oceanography, Woods Hole Oceanographic Institution, Woods Hole, MA 02543*

Norman J. Zabusky *Department of Mechanical and Aerospace Engineering, Rutgers University. Piscataway, New Jersey 08855-0909*

FOREWORD

The U. S. National Committee for Theoretical and Applied Mechanics (USNC/TAM) has commissioned a series of reports on research directions in various areas. The first in the series was *Research Directions in Computational Mechanics,* published by the National Research Council. The present volume is the second in the series.

To prepare the volume, the Editorial Committee appointed by the USNC /TAM, under the chairmanship of John Lumley, identified a number of research areas in fluid mechanics which were felt to be either rapidly developing, or particularly critical to an area of commercial, or environmental significance, or in some other way exceptional. The Editorial Committee selected individuals from around the world who were felt to be uniquely qualified to comment on these research areas. With a very small number of exceptions the editors were able to obtain the participation of responsible experts. Most of the short sections written on each of these areas by these individuals were read by several other contributors, and underwent a certain evolution in consequence, much like the refereeing process to which technical papers are subject when they are published in a scientific journal. These sections are written for the non-specialist, but necessarily assume a scientific training. These sections will be most useful to specialists from other areas, trying to orient themselves in unfamiliar territory, to graduate students interested in a broad view of fluid mechanics today, and to Program Monitors looking for guidance in a rapidly changing situation.

Finally, the Editorial Committee prepared an executive Summary, intended for the interested general reader (perhaps a Congressional Staff Member, or indeed anyone concerned with public funding of scientific research). This section emphasizes the impact of each of the research areas on environmental issues, commercial questions, on anything that might be the subject of legislation or might otherwise influence public policy.

For financial reasons, this volume is not being published through the National Research Council, and hence has not met the editing and production requirements of an NRC report.

John L. Lumley

Sidney Leibovich

ACKNOWLEDGMENTS

The Editorial Committee wishes to thank the following agencies for partial support to the various members during the preparation of this volume: the Air Force Office of Scientific Research (Control and Aerospace Programs), the Office of Naval Research (Fluid Dynamics and Oceanography Programs), the National Science Foundation (Chemical & Thermal Systems and Physical Oceanography Programs), the National Aeronautics and Space Administration (Lewis, Langley and Ames Research Centers) and the Department of Energy (Fusion Energy Program). In addition, the Editorial Committee wishes to thank the following individuals who have helped bring this volume to completion in various ways: Peter Bradshaw, David Caughey, Paul Durbin, Parviz Moin, Joseph Marvin, and William C. Reynolds. The Committee wishes particularly to thank Gail Cotanch, who was responsible for the endless work involved in organizing this operation, including harassing the contributors to produce their manuscripts, retyping much of the material, and finally formatting the document.

1994-95 MEMBERSHIP OF THE U.S. NATIONAL COMMITTEE FOR THEORETICAL AND APPLIED MECHANICS

Officers (4), with end of term indicated

Chair, Gary Leal 10/96

Vice Chair, J. Tinsley Oden 1 10/96

Past Chair, Sidney Leibovich 1 10/96

Secretary, Philip G. Hodge, Jr. 10/96

Members representing Societies and Institutes (14)

American Institute of Chemical Engineers, Robert S. Brodkey 10/95

American Physical Society, Gary Leal 10/95

Society for Industrial & Applied Mathematics, Frederic Wan 10/95

American Mathematical Society, C. M. Dafermos 10/96

American Society of Mechanical Engineers, Thomas L. Geers 10/96

Society of Rheology, Andrew Kraynik 10/96

American Academy of Mechanics, Earl Dowell 10/96

American Society of Civil Engineers, Paul Spano 10/97

American Society for Testing & Materials, David McDowell 10/97

Society of Naval Architects & Marine Engineers, Michael Bernitsas 10/97

Acoustical Society of America, Allan D. Pierce 10/97

American Inst. of Aeronautics & Astronautics, Robert Jones 10/98

Society for Experimental Mechanics, Michael Fourney 10/98

Society of Engineering Science, George Dvorak 10/98

Members at Large (7)

John Lumley 10/95

Ron Adrian 10/96

Albert Kobayashi 10/95

David B. Bogy 10/96

L. B. Freund 10/95

Richard James 10/96

Dan Joseph 10/95

U. S. Members of IUTAM (7)

Bruno A. Boley 10/96

Jan Achenbach 10/96

Andreas Acrivos 10/96

N. J. Hoff 10/96

Hassan Aref 10/96

Y. H. Ku 10/96

Daniel C. Drucker 10/96

Ex-Officio Members (Non-voting)

Foreign Secretary, NAS, Gerald P. Dinneen

Chair, CPSMR, NRC, Richard N. Zare

Chair, Comm. on Engr. & Tech Sci., NRC, Albert R. C. Westwood

Chair, Board on Physics & Astronomy, NRC, Charles F. Kennel

NRC Staff (Manufacturing Studies Board)

Member of Congress Committee

Member at Large

Treasurer & Bureau Member

Member of Symposium Panel

EXECUTIVE SUMMARY

The purpose of this book is to illustrate some of the exciting activities currently underway in various areas of fluid mechanics, and to bring forth the broad range of ideas, challenges and applications which permeate the field. The greater part of the book, the individual chapters on various research topics, is written for specialists in fluid mechanics, including Program Monitors, and concentrates on the scientific questions that determine the research directions. The present section, however, is addressed to the general reader, who is more interested in the ways in which this research may influence public policy, or enhance the economy and US competitiveness in international markets, than in the technical details.

We might begin with a few general statements about fluid mechanics, the study of the motion of 'fluids', meaning liquids and gases, and the effects of such motion. Fluid motions are responsible for most of the transport and mixing (of materials or properties) that take place in the environment, in industrial processes, in vehicles, and in living organisms. Hence, they are responsible for most of the energy required to power aircraft, ships and automobiles, to pump oil through pipelines and so forth. In the environment, fluid motion is responsible for most of the transport of pollutants (thermal, particulate and chemical) from place to place, as well as for making life possible by transporting oxygen and carbon dioxide and heat from the places where they are produced to the places where they will be used or rejected. In industrial processes, it is largely responsible for the rates at which many processes proceed, and for the uniformity of the resulting product. Research in fluid mechanics has as its ultimate goal improvement in our ability to predict and control all of these situations, so as to improve our ability to design devices (for example, aircraft gas turbines, automobile engines) and to regulate (for example, industrial emissions). If fluid motions appear to be ubiquitous, one might recall that the ancient Greek philosophers postulated that there were but four elements, air, earth, fire, and water. Of the four, three are fluid states, and the fourth, Earth, is not only saturated with water in the thin continental skins on which we live, but is mostly liquid metal just below the continents.

It is a good idea to bear in mind that modern fluid mechanics, as a discipline, is comparatively old, having had its roots in the first half of the eighteenth century, although some initial work was done by the Greeks and Romans, beginning in the last few centuries BC. However, even after two hundred and fifty years, (or 2500, depending on the viewpoint) many unsolved problems remain, and our ability to predict many flows is limited.

1

Many reasons for this are possible. Examination of the record, however, suggests that it was not lack of federal funds or of military or commercial interest that was responsible. Indeed, military and commercial interest in the applications of fluid mechanics has nearly always been intense, beginning with that of Hierŏn the Tyrant of Syracuse (who employed Archimedes, but otherwise gave the title a bad name), who had an intense interest in the development of anti-siege weapons, and continuing to the present day. The slow progress has been due, rather, to the extraordinary difficulty of the subject itself. Many reasons for this, inherent to the subject and not of concern to us here, can be adduced, but the fact remains. Progress is difficult, and is likely to remain so, but the payoff can be considerable.

Let us turn now to specific areas. Compressible flows are those in which the changes in pressure from place to place in the flow are so large that the density of the fluid is changed. The flow around a commercial aircraft is compressible, as is the flow inside the engine. These flows present special difficulties: waves propagate in these flows at the speed of sound, and temperatures are high and non-uniform, causing a number of effects that are difficult to predict. Velocities in these flows are close to, or exceed, the speed of sound (supersonic), perhaps by a great deal (hypersonic). Compressible flows are most common in aeronautical applications involving high speed internal and external flows, but there is also a wide range of non-aeronautical applications such as laser technology, vacuum technology, gas-phase reactors, plasma processing of materials, manufacturing processes involving shock waves, and the rapidly developing field of micro-electronic flow sensors and actuators associated with control. The development of a new generation of high-speed military and civilian aircraft, the development of new aircraft engines using high pressure-ratio compressors and turbines and supersonic combustion ramjets for high altitude air-breathing propulsion, and the development of new helicopter concepts all require research on compressible flows. Applications involving high altitude flight or operation in earth orbit or space entail hypersonic flows. Some new materials (such as diamond films) are synthesized from gases so hot that many molecules come unstuck into their component atoms, and the atoms are stripped of many electrons; a fluid in this state is called a plasma. This is a compressible flow too, but a particularly difficult one. In this plasma synthesis, as well as in the development of high-power gas-dynamic lasers, things change so much and so rapidly that the fluid's internal state is always lagging seriously behind its surroundings, creating special problems of prediction. Models of processes occurring in nature such as solar convection, dynamics of cosmic gas clouds, interstellar jets, galactic evolution, and so forth, also involve compressible flows.

All these flows, as well as their lower-speed, relatively incompressible counterparts, can and must be calculated numerically, as part of the design process. This procedure is called computational fluid dynamics, or computational aerodynamics, with their subsets: direct and large eddy simulation of turbulence. The ability to calculate these various flows has in part replaced experiment, and has become an essential part of the design process, allowing rapid evaluation of changes in design parameters. This substantially shortens design cycle time, which results in corresponding reductions in the cost of new designs.

Most of these flows are turbulent, that is, unsteady and chaotic, not repeating in detail. The turbulent state is opposed to the laminar state, which is smoothly varying, organized, and not chaotic. The difference is significant, since the chaotic motions of the turbulent flow produce 1000 times the drag or heat transfer of the corresponding laminar flow. Turbulence is the last great unsolved problem of classical physics; there is no comprehensive theory of turbulence, although much partial qualitative understanding has been achieved. Even in the absence of complete understanding, we have been forced to develop (necessarily not completely satisfactory) ways of computing turbulent flows for design purposes. The inadequacy of the models used is the factor limiting further development of computational fluid dynamics. The use of dynamical systems theory and approaches such as fractal and multifractal measures (separate chapters of this book are devoted to these topics, where definitions can be found) are attempts to build models of various aspects of turbulent flows that will permit us to make more accurate calculations.

The possible payoffs are many, and we will mention only a few: reduction of drag (relative to lift) of aircraft, or increase of propulsive efficiency, would result in a commercial aircraft fleet with much reduced specific fuel consumption, and lower costs per passenger mile, improving competitiveness, and reducing dependence on foreign oil. More generally, development of aircraft having a broader performance envelope (higher altitude, longer range, higher speed, greater payload) would improve competitiveness. In that, as in many other areas, we currently face stiff competition from Europe and perhaps soon from the Pacific Rim. NASA feels that in order to remain competitive in the next two decades, we will have to improve our lift/drag ratio by a factor of two, and improve propulsive efficiency, all this by flow control of various sorts, reducing drag or increasing mixing, on the wings, fuselage and inside the engine.

Flow control is in its infancy. What is envisioned are, surfaces covered with micro-devices that can sense the state of the flow, and actuators that can influence the flow, introducing disturbances at just the right time to increase or reduce the mixing of high- and low-speed fluid, (making the flow follow the contour of a wing, for example, or increasing the rate at which combustion takes place in an engine) or reducing the drag. One of the most important aspects of this process is the interpretation of the sensor input, and the decisions regarding what disturbance to introduce, when and where (known as the control algorithm). This requires an acute understanding of the structure of the flow; such an understanding is obtained by the use of dynamical systems theory, which allows the construction of relatively simple (though still complicated) models of the flows.

We may mention here noise pollution and abatement or control of fluid mechanically-induced sources. There are two principal applications: the first is aircraft and aircraft engine noise. For example, noise abatement or control is a key to the feasibility of any future supersonic transport. Without special treatment, the engines of a supersonic transport are so noisy that current regulations prohibit its operation from US airports. To meet the regulations, the noise level must be very substantially reduced; to bring this about, we need some way to greatly increase the mixing of the heated jet from the engine with the surrounding air, to cause the jet to expand much faster, and slow down considerably. Exotic nozzle shapes have been tried without much success, and current efforts are considering active control of the flow, in the manner described above. The second application concerns ships and hydromachinery. Here, fluid-mechanical noise production is not only a major source of noise pollution, affecting passengers and workers, but a major source of damage as well. Much of the noise produced in liquids is associated with cavitation, the local vaporization of the liquid in regions of reduced pressure, and the subsequent collapse of the vapor bubble as it is carried into regions of higher pressure; the collapse of the bubble on a surface generates pressures high enough to damage steel. Marine propellors typically fail because of cavitation damage. Detection of submarines and torpedoes is usually by their acoustic signature; in this case, the vessels are usually designed to avoid cavitation, which is extremely noisy; however, the turbulent boundary layers excite structural vibrations which can radiate noise to great distances. The turbulent boundary layer also generates pressure fluctuations (known as self-noise) which confuse the vessel's own listening apparatus. A great deal of research goes on in an attempt to reduce these effects. We can also mention here naturally occurring sound in oceans and lakes, which is of interest partly because it obscures sonar detection, and partly because the sound produced by falling rain, for example, can provide a useful route to remote monitoring of weather.

Many natural and technological flows are vortex-dominated, and such flows are a subject of special study. A vortex is a tube of fluid which is strongly rotating; a tornado is a dramatic example. Other high-energy and large-scale vortices are hurricanes and the polar vortex (the ozone hole). In supporting the weight of an aircraft, the wing generates a vortex, which trails behind the aircraft from the wingtips. The intensity of these vortices is proportional to the weight of the aircraft. These vortices close behind very large aircraft are strong enough to flip a light plane over, and are the reason for the required separation between take-offs at airports. Additional vortices are shed from maneuvering aircraft. To understand this we have to consider how fluid moves over a surface. Since fluid adheres to any surface with which it is in contact, in order to move past the surface the fluid must roll forward. This rolling is called vorticity. A vortex is concentrated vorticity. When the aircraft maneuvers, the flow sometimes leaves the surface, and it carries with it the vorticity that was generated next to the surface, which is rolled up by the flow into a vortex. The generation, interaction and dispersal or mixing of vorticity plays a profound role in a wide class of applied, geophysical and fundamental fluid flows. A better ability to predict and control flows will arise from a deep understanding of the processes leading to the formation (cyclogenesis), evolution, and persistence of coherent vortex structures in flows in which distributed vorticity is present. Such an understanding would make possible data assimilation in prediction codes and signal feedback for control of aircraft, ship and chemical process performance. Imagine forecasting meteorological or oceanographic events in which local environmental measurements and remote (e.g. satellite) observations are fed back into local space-time regions of the computer simulation code. This has the potential for reducing errors and improving the reliability of predictions. Similarly for man-made flows, we may have sensors located within the flow which provide feedback signals to force the flow in a stable manner.

As we have suggested, in most devices, and especially land, sea, and air vehicles, drag and fluid resistance take place in a very thin layer of fluid near the moving solid object. This is known as the boundary layer. In addition to being the source of drag, the processes in this thin region are subject to dramatic alterations that cause phenomena like the sudden loss of lift — or stall — in airplanes, and a concurrent sudden increase in drag. This is usually due to a massive change in the airflow near the wings in which the flow no longer smoothly follows the contour of the object but is violently torn away from it in a process called boundary layer separation, a process we have already mentioned. Much progress has been made in understanding this state of affairs and how to prevent it. It is an issue of major concern not only for economic reasons, but also for reasons of aircraft safety near airports and in flight, especially while manuevering. Instability of the boundary layer is the proximate cause for the transition of flow from laminar to turbulent, with consequent alteration of behavior. Similar issues of separation and instability of boundary layers arise in a vast variety of other flows, including internal flows in internal combustion, jet, and rocket engines, in medical equipment such as heart-lung machines, in manufacturing processes involving materials in a liquid or molten state, and so on. In most cases, these phenomena have major consequences on the performance and safety of these devices, and the prediction of motions in the boundary layer is a critical issue to the success of the associated technology.

The bulk of international commerce, both in raw materials and manufactured goods, is transported by sea. Seagoing vessels of all kinds face harsh and dangerous conditions, especially because of the power of ocean waves. Improvements in design of such vessels, and also important fixed ocean structures like offshore oil platforms, require understanding and predicting the interaction between the structure and waves. Water waves also are a major source of drag on ships, and this is a major factor limiting the speed and setting the cost of ocean transportation. Understanding of some aspects of this wave resistance has led to important design improvements, such as the bulbous bow now universally used to reduce wave drag on cargo vessels. Much more needs to be done to produce better designs for ships and fixed station platforms, to understand the effects of waves when ships are maneuvering both in open water and in harbor areas, and to deal with extreme wave states that threaten the survival of the ships, platforms, and, of course, their passengers and crews.

Coastal areas are densely populated, and of economic importance because they provide access to sealanes and shipping, to fisheries and the other resources of the oceans, and recreation. Coastlines are moveable, slowly, by waves, currents and tides. The interaction of waves with coastal installations and harbors, and the movement of sediment in the turbulent, wave-buffeted surf zone that causes the coastline to change its shape, are among the concerns of coastal engineers. Waves and their effects are difficult to predict, especially when the waves are high, and the consequent effects most impressive. Important progress has been made in understanding the origin, growth, and propagation paths of waves in the deep ocean but many critical issues in this process remain unknown. This is even more so as waves enter the shallower water near the coasts, where they are strongly affected, and help to drive strong currents, and where they are subject to the turning effects of decreasing depth. While some effects of wave and current action are relatively slow, like the reshaping of the shoreline, others are sudden and catastrophic, like tidal waves (tsunamis). The destruction of property and life following tsunami impact often has been devastating. Now, understanding of how tidal waves are born and grow has reached a level that permits tracking of these waves and early warning of populations in their paths. Prediction of damage requires understanding of the waves at their largest amplitudes, and this remains a challenging open problem.

Accurate prediction of the weather is an everyday concern, with enormous ramifications for most human activity and economic impacts too massive to tally. This is the realm of meteorology, which has always posed some of the most fascinating and difficult of fluid mechanical problems. The oceans play a key role in this process; "el Niño" has become a household name. The fluid mechanics and concomitant heat transport in the ocean are the realm of the physical oceanographer, and so it is the coupled ocean-atmosphere fluid system that controls the weather, and its long term trend, the climate. The fluid mechanics of these processes share many common features, and these fields, and related fluid processes in the Earth sciences, are now often collectively termed "geophysical fluid dynamics." The related areas involve fluid mechanics in stars and the giant gaseous planets, in other astrophysical fluid dynamics questions, and in fluid mechanics of the Earth's interior, which shape the distribution and drift motion of the continents, volcanic activity, and the generation of the magnetic field of our planet by the dynamo motions of its molten iron interior. Processes such as the breaking of wind waves in the ocean and in lakes cause bubbles of gas from the atmosphere to be mixed into the surface layers, where the gas enters into solution in the water. These processes are vital; for example, the oxygen levels and therefore the biological productivity of the seas and lakes are determined by this exchange of gas between the atmosphere and the surface waters. Similarly, the levels of greenhouse gases in the atmosphere are strongly influenced by the transfer of these gases to the ocean, which has an enormous capacity to absorb them; in this way, gas transfer plays a significant role in the important debate on global warming.

Understanding of fluid processes is key to a wide spectrum of environmental questions. Here one is concerned with siting of power plants and other installations that are sources of toxic chemicals or require large flows of water for cooling and other purposes, the river or lake source of which may be degraded in the bargain. Other concerns include protection against and prediction of spills of liquid pollutants (such as the Exxon Valdez catastrophe) or heavier-than-air gases (such as the Bhopal catastrophe). Ecologically sensitive coastal areas and river estuaries are often heavily used, and the prediction of flows in these systems is critical to planners concerned with avoiding their contamination. Groundwater and its motion and quality are major public health matters. The surface impacts of volcanism raise extremely difficult issues that need to be understood. These problems, and many other problems of environmental fluid mechanics are novel, complex, often poorly understood and inadequately studied. They are central to planning a complex society, and to anticipating the consequences of, and preparing for, the natural and manmade disasters that continually visit us.

Combustion of fossil fuels powers most aircraft, ship, and automobile engines, and produces much of our electrical power and home and industrial heating. Improvements of these combustion processes reduce fuel consumption and pollutant generation. Some notable examples of fluid mechanical research have contributed to this end, with massive economic benefit. For example, it was found that imparting swirl to the air in jet engine combustors improved fuel economy substantially. This innovation has found its way into the design of new, high efficiency home oil burners, extending the economic benefit considerably. Many other examples of innovative engine design based on an understanding of the fluid mechanics underlying the engine can be cited. Combustion research involves experimental measurements in an environment that tries to melt the instruments, and requires expertise in chemically reacting, heat releasing, variable density particle-laden flows; the scientific and engineering challenges are formidable.

It can be a happy, or a disastrous, circumstance when small changes produce large effects. This is the case with fluid motions, which tend to be very sensitive and responsive, sometimes even to minute alterations of flow rates, boundary shapes, boundary temperatures; in fact, to virtually all conditions of the motion. This sensitivity is due to the tendency of fluid motion states to be unstable. From a practical standpoint, it affords an opportunity to fine-tune designs and industrial processes to achieve a desired result with small alterations. Thus, for example, processes which produce sheets of material (metals, plastics, or other materials) typically pull the sheets rapidly from a molten state, and the surface quality of the sheets and films so produced, or the rates at which they can be produced, is affected by instabilities in the liquid sheet before solidification occurs. Similarly, crystal manufacture, such as silicon used in computers and most modern electronic equipment, is achieved through crystallization from a crucible of moving liquid, and the fluid instabilities affect production rates and product quality. The general problem of transition from laminar to turbulent motion, with all of the ramifications associated with transition, is a long-standing problem of fluid instability. The instability of a fluid motion can have positive or negative effects, depending on whether the result of the instability produces or destroys a desired property of the flow. Thus, for example, one may wish to avoid or delay transition to turbulence to reduce vehicle drag, or one may wish to promote it to enhance mixing in combustion processes in engines. While the economic benefits of understanding and controlling fluid instabilities are well known in the industries mentioned, an awareness of their potential for improving production quality and rates is virtually nonexistent in others. The introduction of this area of fluid mechanical science to many industrial sectors where it is not known could have valuable consequences. In the following paragraph, which broaches another topic, several examples relevant to this paragraph will nevertheless be found.

Magnetohydrodynamics (MHD), which deals with the combined effects of fluid mechanical and electromagnetic forces, is an exciting but, at the moment, only moderately active area of research and development that has not been exploited to nearly its full potential. This relatively low level of present activity is regrettable considering that a large variety of flow phenomena can be modified in a dramatic way through the controlled application of electromagnetic forces. Well-known examples include the damping, modification and even suppression of turbulence in a variety of flows; also, the use of electromagnetic stirrers in a bath of molten metal, as in steel casting, which provides the only non-intrusive method (that is, not requiring the introduction to the bath of a device, which would likely melt) currently available for keeping the contents of the bath well-mixed. Currently, the most promising area of MHD application appears to be in the materials processing industry where, for instance, a magnetic field can be used to modify the flow patterns which occur naturally in the production of single crystals of semiconductors, thereby insuring that the composition of the product (that contains trace amounts of other elements to make it electrically active) is uniform.

The naturally occurring flow patterns referred to above arise because the flowing material is from place to place lighter (tending to rise) or heavier (tending to sink), because the temperature and composition are not uniform. A flow produced by these effects is called buoyant convection. Buoyant convection occurs in many environmental flows. Examples include: convection in room fires, in energy storage systems, and in atmospheric and oceanic systems. In view of their frequent occurrence, these flows deserve special attention. The forces which drive convective flows can also be used as controls to optimize the operation of various processes involving, for example, crystal growth or chemical vapor deposition, and, depending on the application, either to enhance or to suppress flow instabilities within the system.

The production of high performance structural materials and coatings (such as the carbon fiber reinforced plastic used in golf clubs, tennis rackets and bicycle frames) also involves complicated phenomena where the discrete molecular nature of the gas cannot be ignored, especially in the manufacture of microelectromechanical devices. These phenomena are complicated because conditions in the gas are so extreme, and changes so rapid, that the internal state of the gas never catches up to its surroundings. As a result of the importance of these phenomena in such production, there has been a resurgence of interest in the field of rarefied gas dynamics which was associated traditionally with the flight of aircraft and missiles at high altitudes. In fact, the design of tiny machines having dimensions of the order of microns requires the implementation and modification of rarefied gas dynamical computational techniques which were originally developed for a completely different application.

Many of the computational techniques referred to above aim to construct solutions to more or less exact equations describing the flow of gases under rarefied conditions. In recent years, however, important advances have been made using the method of molecular dynamics (MD), which applies to liquids as well as to gases. Here the behavior of a fluid under particular circumstances is determined by computing simultaneously the motion of all the individual interacting molecules. This, of course, is possible only on the largest computers. Important insights have thereby been obtained into situations in which flow dimensions are of the order of inter-molecular dimensions, for example the rupture of a thin liquid film, as occurs when a gas bubble approaches a liquid-vapor interface, or the dynamics of the moving edge of a liquid drop spreading across a solid substrate. Such calculations provide extremely useful information concerning the point at which we must abandon the usual image of a fluid as a seamless continuum, and must consider it instead as a collection of molecules. We have noted before that fluid usually sticks to any solid surface. This is an excellent approximation so long as the fluid flows over the surface like sand over a beach-ball—that is, so long as the inter-molecular dimensions are small relative to the dimensions of the surface. As the two become comparable, however, like sand flowing past a pin-head, the simple condition at the surface no longer works, and more sophisticated conditions must be applied; MD can help to determine what those are. Similarly, MD offers an opportunity for studying flows that involve two fluids that mix a little on a molecular level, so that they are not separated where they meet by a sharp interface, but are diffusing into each other while they are flowing. Situations like this occur in many industrial chemical processes and in the kitchen; imagine mixing molasses and cream. MD also allows us to investigate phenomena involving an interface between two fluids, one that is strongly influenced by surface impurities, another situation that occurs frequently in industrial processes. All of these flows are much too complicated to compute from equations, and this type of molecule-by-molecule simulation is the only way to find out what is happening.

Those flows involving two (or more) fluids that do not mix, or may mix a little, are called two-phase (or multi-phase) flows. Another two-phase material which plays a key role in a variety of natural and industrial processes is a suspension of solid particles in a liquid. Examples include coal slurries, biological suspensions, high-energy composite fuels for space propulsion and colloidal suspensions for making films as well as coatings for electronic applications, in addition to fluids containing suspended particles that can be influenced by imposed electric fields, so that the nature of the flow can be changed by applying an external electric field. In particulate flows one wishes to predict the bulk behavior of the suspensions from a knowledge of the fine structure or, conversely, to construct suspensions having prescribed flow behavior (called rheology). This requires that the many factors which contribute to the rheology of such systems, i.e. the influence of one particle on a neighbor through the motion of the fluid around it, the forces due to bombardment of the particles by the surrounding molecules, the surface forces on the particles, etc., be properly accounted for. Furthermore, particles tend to wander, from regions of large particle concentration to regions of low, but also from regions in which the layers of fluid are moving rapidly relative to one another to regions in which this is not true. This has been shown to play a key role in these flows, since the flow causes the particle concentration to change, and this changes the properties of the fluid, causing marked changes in the flow. This kind of interaction makes the flow exceedingly difficult to compute. Newly developed experimental techniques as well as more sophisticated numerical simulations have provided new insight on how particles in suspension rearrange themselves under flow conditions to produce the observed phenomena.

All of the discussion above, with the exception of multi-phase flows, concerns problems involving gases or liquids that contain small molecules, like water, where the bulk properties of the fluid (like density and viscosity) are independent of the flow conditions. Another large and important branch of fluid mechanics is concerned with liquids that are often referred to as "complex," in recognition of the fact that these materials exhibit much more complicated behavior. Examples of complex fluids can be found in any kitchen, bathroom, playroom or garage, and include egg white, cake batter, silly putty, proprietary oil additives, blood, mucous and many, many others. Most of these fluids either consist entirely of large molecules, or have large molecules floating in them, as well as particles or droplets. Most plastics in their liquid state fall in this category. This branch of fluid mechanics is often called non-Newtonian to distinguish it from the classical work on small-molecule (or Newtonian) fluids. Although this class of fluids is common in nature, in a variety of technologies, and as the liquid-state precursors of many important types of advanced materials, the status of our understanding of their behavior and our ability to predict their motions, is at a very early stage of development. In general terms, the difference between complex fluids, and the single component, Newtonian fluids, is that in the latter case, the mathematical formulation is known but the macroscopic physical processes are complex and often not well understood, especially for turbulent flow conditions; for complex fluid, even the appropriate governing equations and conditions at the boundaries (do these fluid stick to solids or is it more complicated?) are still not well understood. To compound the difficulty, the model equations that have been proposed are extremely difficult to solve; and standard methods of computational fluid dynamics generally do not work for this class of problems.

This paucity of understanding extracts a substantial economic penalty. The production and processing steps leading to a finished product employing advanced materials are most often carried out via deformation (stretching, squeezing) and flow in the liquid state. Although largely empirical procedures have historically been used in the design of new processes, future economic competition, as well as requirements for improved product quality, reproducibility and precision, all demand the development of a deductive basis for process design and control. For example, the inability to predict the behavior of polymer liquids in an extrusion molding process precludes prediction of the final shape of the solidified product — thus the design of each mold must be done by a trial and error process costing tens to hundreds of thousands of dollars, and much time, for each new part, and limiting our ability to produce precision parts. Since thousands of new injection molding processes are developed each year, the cost of our ignorance amounts to hundreds of millions of dollars for this one technological application.

Qualitative phenomena observed in non-Newtonian flow experiments are often dramatically different from expectations based on similar observations for small-molecule liquids (e.g., a finger dipped in many of these fluids will spin a thread when withdrawn, and the forces involved are quite different from those produced when the fluid is rubbed between the palms).

Development of new experimental techniques are needed to provide much more comprehensive characterization of the rheological behavior of complex fluids, and for characterizing the microstructural state of a non-Newtonian liquid undergoing a flow, since this determines the properties of any product which results from the flow process.

If we succeed in answering these fundamental questions, the potential pay-offs in the area of advanced materials processing are many-fold. (1) They will form a basis for computer-aided design of processing systems for manufactured parts eliminating time-consuming and costly trial-and-error development. (2) Major increases in production rates in manufacture of fibers may be possible. Instabilities of the bulk flow (leading to unsteadiness and fiber non-uniformity), and apparent breakdowns in boundary conditions, etc., currently constitute the major limitations of both production rates and product quality, but no one knows how to minimize or eliminate these instabilities, or even whether it is possible. (3) One critical feature of complex fluids is that their microstructural state, and thus their macroscopic properties, can be altered when they undergo a flow. Thus, there is the possibility of developing products from a complex fluid with properties that can be predetermined or optimized by modification of the processing flow conditions, e.g., polymers may yield very light weight and moldable electrical conductors, but only if we can understand how to process them into highly oriented and stretched configurations. Although the potential economic pay-off is enormous in terms of light weight materials of high strength, high conductivity, etc., the technology is today largely empirical and extremely limited in scope. (4) Finally, a key route to new materials with specified properties, which is generally much less expensive than chemical synthesis, is by mechanical blending of two (or more) fluid or fluid and solid components. Given a set of constituent materials, and their properties, there is clearly a major economic incentive to develop the ability to predict the outcome of a blending process, as well as to predict how to modify the process to achieve a desired morphology. Beyond the applications of complex fluids as precursors of new materials, or materials-based products, there are many additional technological applications for suspensions, emulsions and multiphase (gas-liquid) fluid flows. Among these one may mention multiphase flows in oil reservoirs, or in groundwater percolation processes; cavitation phenomena which lead to noise production, and to many well-known and expensive structural failures ranging from propellors on ships to dam spillways, due to cavitation damage, and in thin, viscous films to lubrication breakdown in hydromachinery; the use of multiphase fluids in heat transfer processes that are intimately connected to the cooling processes in nuclear power plants, and pipeline transport processes involving slurries. One common feature in some of these applications is still the overall macroscopic flow properties, but in other applications it is important to understand the details of motion at the microscale. For example, in an oil reservoir, it is important to be able to predict overall pumping costs of any secondary or tertiary recovery process, but control of the morphology of the boundaries separating oil and water is also critical to the production of oil rather than water. As another example, the details of the interfacial regions in multiphase boiling heat transfer determine success, or failure via the development of local hot spots due to "dryout" of the solid heated boundary.

Finally, the intersection of fluid mechanics and biology in the area of biofluid mechanics offers the opportunity for many important applications, both in better understanding of normal biological processes (for example, cell, tissue, cartilage or even bone growth in response to fluid stresses, transport processes, etc.), but also in the development of therapeutic medical procedures. Among a long list of biofluids research with clear medical implications, we may cite: (a) fluid mechanics' role in the growth of atherosclerotic tissues in the circulatory system, and an understanding of mechanisms for localization of atherosclerotic lesions, based upon the response of the biological system at the cellular level to fluid stresses; (b) heart and heart valve function and the design and performance evaluation of artificial replacements (prosthetic cardiac valves); (c) cardiovascular flow measurement methods: although much current development is directed toward research applications, there is clearly a major medical pay-off in improved diagnosis of vascular disease and in the design and evaluation of therapeutic interventions; (d) pulmonary flows — interesting fluid-structure interaction problems in understanding physiological phenomena such as "wheezing" — possibly leading to better treatment methods for asthmatic conditions, etc. Also the role of fluid films, surfactants and airflow in such pathologies as "Sudden Infant Death Syndrome" or "Crib Death." It is clear that this is a field at early stages of impact. The problems are (often) more microscopic, at the level of cells or micropores, than is characteristic of other areas of fluid research. The fluids, apart from water and air, are often more complex.

The editors hope that the general summary given above has at least suggested the vitality of the field of fluid mechanics and that the reader with some scientific background will be motivated to gain further insight by studying the chapters which follow.

APPLICATION OF DYNAMICAL SYSTEMS THEORY TO FLUID MECHANICS

Hassan Aref
Theoretical and Applied Mechanics
University of Illinois
Urbana, IL 61801-2935

INTRODUCTION

The scientific discipline known as dynamics has been with us in something resembling its modern form since the time of Newton. The idea that motion is due to forces, and the embodiment of this causal relationship in a system of differential equations, is basic to the subject. The refinements required to apply Newtonian mechanics to a continuum such as a fluid also again belong to the classical heritage of the subject.

Around the turn of the century it was realized that the equations of motion of even relatively simple dynamical systems, such as two pendula linked by a spring, could have extremely complicated solutions. The problem of three interacting point masses in astronomy, the famous three-body problem, became a focus of attention, and it was ultimately shown by the mathematical physicist H. Poincaré and others that a solution of this problem 'in closed form', i.e., in a format similar to the well known Kepler problem of two bodies, does not exist. In effect, the first examples of simple mechanical problems that display *chaos* had been found. The surprise and consternation that these seminal discoveries should have caused were at the time largely buried in the excitement over several new physical theories that were thrust upon the stage of science: relativity, quantum mechanics, and atomic physics.

Not until the early 1960s did the first applications of the notion of chaotic behavior to fluid mechanics appear. In 1963 the meteorologist E. N. Lorenz published his well known paper on 'deterministic, nonperiodic flow', in which a severely truncated model of the equations describing two-dimensional convection, which he studied as a simplified model of climatic variations, was shown to possess very complex behavior. The 'Lorenz system' as it is called today, is dissipative, and the nature of its chaotic behavior is fundamentally different from that seen in a non-dissipative system such as the three-body problem. The study of Lorenz gave rise to a large body of work, analytical, computational and experimental, that we can characterize in retrospect as the topic of *temporal chaos* in fluid systems. The new 'chaos theory' that emerged, backed by careful experimental observations and comprehensive computational studies, has changed our understanding of the nature of transition from laminar to turbulent flow, although it has not yet yielded a complete picture.

One of the key features of chaos is that the underlying equations leading to this state of motion need not themselves be terribly complicated. A simple quadratic map of a line segment or of the plane onto itself can produce chaotic solutions with very complicated time dependence.

An even more straightforward generalization of the ideas of Poincaré to fluid motions occurred a few years after Lorenzs work, in two brief papers by the mathematician V. I. Arnold and the astronomer M. Hénon on the possible complexity of flow kinematics in certain three-dimensional, steady flows. Intriguing precursors to this work can be found in a 1955 paper on ocean circulations by the oceanographer P. Welander, and at a more qualitative level in the results of various studies of fluid mixing in chemical engineering, mantle convection, and so on. Most of this work was generally ignored by the fluid mechanics community. Arnolds and Hénons studies contained the germ of an idea for *chaotic transport* in a fluid flow. They noticed that the kinematic equations describing the motion of individual fluid particles were, in general, complex enough to yield chaotic particle paths even in rather simple flows. The flow fields required for chaotic transport need not have the complicated time dependence produced in temporal chaos.

Roughly speaking this is where the field of chaos applied to fluid flows stands today. We have achieved important inroads into distinct and important aspects of fluid flows, with several potential application areas clearly in sight. On the other hand, there are equally clear limitations. There are flow regimes, such as fully developed turbulent flow, that combine aspects of temporal and spatial chaos in ways that we still do not understand.

It is important to stress another aspect of the dynamical systems approach to fluid motions that originated also in the early to mid-1960s. Whereas the surprise in chaos was the understanding that very simple systems can display immensely complicated behavior, the surprise in the discovery of *solitons*, by the plasma physicists M. D. Kruskal and N. J. Zabusky in 1965, was that seemingly complicated, fully nonlinear, systems with infinitely many, strongly coupled degrees of freedom could sometimes behave in entirely organized and regular ways. The notion of solitary waves and solitons has had a profound impact on our understanding of the formation of stable or slowly varying patterns in fluid motions. The emergence of large-scale *coherent structures* in turbulent flows, identified in experiments performed in the late 1960s to early 1970s, should probably be viewed in this light.

What is 'dynamical systems theory'?

We will not give a mathematically comprehensive definition of 'dynamical system.' A rather general class of dynamical systems are described by a finite number of time-dependent variables that evolve according to a set of ordinary differential equations (ODEs). A mechanical system with a finite number of degrees of freedom, such as a compound pendulum, is a dynamical system in this sense, but the concept is broader and includes systems from ecology and economics, among others, that would not ordinarily be associated with the word dynamics. A very important role in the theory is played by discrete mappings of some mathematical space onto itself, e.g., the mappings of sequences of two symbols. Sophisticated abstract mathematical definitions of the general structure that is called a dynamical system exist. The reason for broadening the scope of dynamical system in this way is that one finds certain common features, such as bifurcations and chaos to be discussed below, that are shared across this large family of constructs, related only by the mathematical format that we use to describe them.

Systems described by one or more partial differential equations (PDEs), as we commonly encounter in fluid mechanics, are intrinsically more complicated than those described by finite systems of ODEs. A PDE is conceptually the equivalent of an infinite set of ODEs. Thus, a comprehensive understanding of a fluid mechanical situation from the point of view of dynamical systems theory is typically limited to cases where, for one reason or another, the PDE behaves like a finite set of ODEs. Such situations are more common than one might *a priori* assume. In many cases, the behavior of a fluid, and the solutions of the governing PDEs, can be represented by a finite collection of modes. The time-dependent amplitudes of these modes are the dependent variables in the finite set of ODEs that make up the dynamical system used to describe the fluid motion. The particularly interesting part of this story is, of course, that the behavior of a finite system of ODEs, even a rather small one involving as few as three variables, can display chaos. The chaotic solutions define new flow regimes, intermediate between the well established regimes of laminar and turbulent flow. It is within these new regimes that one should look for scientific discoveries and technological inventions.

Instability, bifurcation and chaos

As a parameter such as the Reynolds number or Rayleigh number is varied, transitions occur through the loss of stability by one state and the emergence of an alternate, stable state. The description of individual instabilities right at the onset is described by a well-established and time-honored framework known as *linearized stability theory*. The initial amplification of an unstable perturbation is captured by this theory, but the eventual saturation of the instability into a new pattern of motion requires a generalization

known as *nonlinear stability theory*. The transition from an original state
of motion (now unstable) to a new state as a certain value of some system
parameter is exceeded is referred to as a *bifurcation*.

In the dynamical systems approach the overall network of interconnected
bifurcations, the *bifurcation diagram*, provides a road map to the fate of
the system as a parameter is increased. The bifurcation diagrams of several
important flows, such as Rayleigh-Bénard convection or Taylor-Couette flow,
have been measured in some detail and certain features have been compared
with analytical and numerical calculations. A good ODE model of a fluid
system should capture a sizeable portion of the bifurcation diagram of the
system considered, at least in a qualitative sense.

The first bifurcations in a dynamical system, starting from a simple state
such as the state of rest or some simple steady motion, are sometimes to
states of periodic motion. The next bifurcation will then often introduce
a second period. The resulting motion is then a quasi-periodic state if the
second period is incommensurate with the first. Sometimes a new period
will lock onto the old one. In principle, one could expect this process to
continue through an infinite sequence of bifurcations, each adding a new
frequency to the roster, along with all the sums and differences with existing
frequencies due to nonlinear interactions. In due course, system behavior
of great complexity would be established, with the multitude of frequencies
being difficult to distinguish from genuinely stochastic behavior, although,
in principle, the time dependence is just that of an almost-periodic function.
This, indeed, was the earliest dynamical systems picture of how turbulence
develops in a fluid-mechanical system. It is usually associated with the names
of the mathematician E. Hopf and the physicist L. D. Landau.

In 1971, a few years after Lorenz had discussed his model, but unaware
of that work, the mathematicians D. Ruelle and F. Takens suggested, on the
basis of general mathematical considerations, that after only a small number
of bifurcations a chaotic state could emerge. The mathematical suggestion
has been verified in computer experiments on systems of a few ODEs, and
more significantly in laboratory experiments on real fluid systems, such as
Rayleigh-Bénard convection and Taylor-Couette flow.

The appearance of chaos in a fluid system is often not a terribly dramatic
event, and occurs well before the onset of large-scale turbulent flows. As the
discussion above has indicated, the onset of chaos can be as 'mild' as the
frequency mismatch of a few wave-like disturbances in the system, and one
may have to measure very carefully in order to detect the appearance of this
state.

It should be stressed that the 'chaos approach' to complex behavior in
fluid motions is fundamentally different from the approach adopted by what
is known as the statistical theory of turbulence. In the former complex, noisy,

fluctuating behavior arises out of *deterministic* equations of motion, and a key objective is to elucidate the nature of the complexity. In the latter, the turbulent state is assumed *a priori* to be so complicated that a deterministic description is precluded, and a *statistical description* of the flow must be adopted from the start. This is similar to the approach of statistical thermodynamics, where one abandons any hope of dealing with the deterministic mechanics of interacting atoms or molecules due to the overwhelming number of equations that would have to be solved. Instead, certain statistical assumptions are made, which have turned out to have surprisingly powerful predictive capabilities. In the case of turbulence, similar statistical theories have been tried for many years, but they do not carry the same conviction, nor do they have the same degree of universal predictive success as statistical thermodynamics does. It is natural to inquire whether the basic assumptions of the statistical approach to turbulent flow are not in need of additional ideas from fundamental theory, and it would then seem imperative to go to a framework in which the nature of stochastic behavior is considered a theoretical 'output' rather than an 'input'.

Highlights of the use of dynamical systems theory in fluid mechanics

As a quantitative model of convection the Lorenz equations leave much to be desired. As a qualitative model, however, they opened up unprecedented and largely unanticipated opportunities in fluid mechanics. Not only did it become possible to achieve noisy, complex behavior suddenly, at a sharply defined value of the governing parameters, it appeared that this highly complicated behavior was governed by a small system of ODEs. The notion that deterministic mathematical systems can produce apparently stochastic output was, of course, known in principle from the algorithms that generate random numbers on computers, but it was a surprise that a continuously variable system in space and time, such as a fluid, would display this kind of behavior. Lorenzs (1963) paper, in fact, showed how a mapping with a single maximum was associated with the chaotic behavior seen in the system of ODEs. Mappings of this kind were well known to generate random numbers at appropriate settings of the parameters. Suddenly the esoteric mathematical topic of discrete mappings, which the mathematicians long had realized as intimately connected with the qualitative behavior of ODEs, became a working tool for the fluid mechanician.

The theory of one-hump maps of an interval onto itself, which had been studied in considerable detail by mathematicians, mathematical ecologists, and physicists turned out to have a spectacularly successful, quantitative theory of transition to chaos. The ecologist R. M. May (1976) gave a beautiful synopsis of what was known about such maps just before the physicist M.

Feigenbaum (1978, 1979) discovered the quantitative, metric universality that governs their repeated 'subharmonic' bifurcations. This behavior has been termed 'universal' because it appears within an extended family of similar maps. In many cases the low-dimensional behavior also dominates systems that *a priori* might have been thought to involve a larger number of degrees of freedom.

The theory of subharmonic bifurcations, and related theories of 'universal' transition scenarios, along with the experimental observation of such transitions in fluid systems by the physicists J. P. Gollub, A. Libchaber, H. L. Swinney and others, must be counted among the highlights of our subject. This work has forever changed our approach to the problem of transition to turbulence. More recent generalizations of this approach to 'open' flows of particular engineering interest, such as jets, wakes, and boundary layers provide clear research agendas for the future.

The notion of chaotic transport, with its applications in fluid mixing, materials processing, dynamo action in magnetic fluids, and its potential for contributing to topics such as sound radiation from vortex flows and coherent structure dynamics in turbulence, promises to become a mainstay of the subject as well. Flow kinematics has permanently 'lost its innocence' through this work. Some of the contributions that initiated 'modern' interest in these problems were made by H. Aref, E. A. Novikov, J. M. Ottino, M. Tabor, and many others. Studies of both temporal and spatial aspects of chaos in fluids are routinely reported at major national and international conferences.

The subject of *pattern formation*, in which one seeks to understand the spatial patterns arising from bifurcations, is also a central one for fluid mechanics. At the simplest level patterns arising out of *amplitude equations*, related to but usually not systematically derived from the governing Navier-Stokes equations, have received considerable attention. We have included a discussion of some results here, since this promises to be exciting and fertile ground for the immediate future, and a stepping stone for finally tackling the full problem of spatio-temporal chaos, a hybrid of complex temporal and spatial phenomena that is relevant to some types of turbulent motion.

Given the late appreciation of chaos and the relatively sophisticated instrumentation and numerical methods required to study it, one might assume chaos to be a rare phenomenon in fluid dynamics. However, such an assumption would be incorrect. As with dynamics in general, chaos is more the rule than the exception. A wide variety of fluid systems show chaotic behavior, and an even larger number contain chaotic behavior embedded within more complex phenomena such as turbulence.

TRANSITION TO CHAOS

Single-point measurements; time series

The recognition of chaotic dynamics in the mid-1970's required several important elements from the experimental point of view. First, it was necessary to learn how to control fluid systems precisely so that the possible effects of noisy boundary conditions could be eliminated, and so that closely spaced bifurcations could be detected. Second, it was necessary to develop precise low-noise probes of local velocity and temperature. Third, it was in practice necessary to have automated data acquisition so that extended time series of 10,000-100,000 points could be collected for later processing. Since instabilities frequently introduce discrete modes with definite frequencies, digital spectral analysis provides a powerful technique for detecting bifurcations. Furthermore, the initial appearance of broadband features in a spectrum is still often the simplest method of detecting the onset of chaos. Thus, the role of computing in experimental studies of chaotic dynamics has been just as important as it has been for numerical studies.

Examples of systems displaying chaotic behavior

Chaotic phenomena were initially discovered and studied in Rayleigh-Bènard convection and Taylor-Couette flow. These systems were selected because they could be carefully controlled, so that chaotic dynamics could be distinguished from noisy boundary conditions. Interest was focussed initially on the problem of demonstrating the existence of chaos, and finding the routes by which the transition to chaos occurred. Within a few years, the number of systems showing chaotic behavior had expanded markedly. For example, interfacial waves in simple laboratory geometries were demonstrated to be chaotic along with convection in mixtures, where the density gradients reflect compositional variations in addition to temperature gradients. One might add to this list rotating fluids sustaining baroclinic waves of the type relevant to geophysical fluid dynamics. Eventually, the problem of chaos in open systems such as wakes and jets was addressed. However, these systems required additional concepts for a proper understanding (see below).

Phase-space reconstruction, fractals, and Lyapunov exponents

The concept of chaotic dynamics relies on a phase space description of dynamics, in which a point represents the instantaneous state of the system, and the trajectory represents the time history of the system. Application of this notion to experiments was initially inhibited both by the difficulty of deciding how to define the relevant degrees of freedom, and by the difficulty of measuring those variables. However, in an important simplification first suggested by Takens, a single time series $f(t)$ can be used to obtain a number of variables by using delayed versions $f(t + n\tau)$ of the signal, where n is an integer. Roughly speaking, these delayed variables correspond to derivatives of $f(t)$, and they are sufficient (under some circumstances) to characterize the dynamics. By plotting the system's trajectory in a phase space formed from these variables as coordinates, a geometrical picture of the dynamics could be constructed. Since fluid systems are usually dissipative, the trajectories in phase space converge toward a limit set that is termed an attractor.

Diagnosing chaotic dynamics required testing for two important properties of chaotic or 'strange' attractors: (a) They are generally fractal objects, i.e., their dimensionality is fractional. Roughly speaking, this means that the number of boxes of size e required to cover the attractor in phase space varies as a fractional power of ε. (b) Two trajectories on a chaotic attractor that are initially close to each other will, on average, deviate exponentially from each other in the course of time. It is this property, known as *sensitive dependence on initial conditions*, that makes chaotic states in practice unpredictable: Any round-off error in a computation will grow quickly, leading to rapid loss of information about the actual state of the system.

Scenarios for the transition to turbulence - universality

There are a very large number of ways in which a system can become chaotic. However, several 'routes to chaos' are particularly common, and can be related to the behavior found in nonlinear mappings. The most famous of these is the period-doubling cascade, in which the dominant periodicity, T, in the time dependence of a system doubles successively as a parameter (such as the Reynolds number) is varied. These bifurcations accumulate in a geometric series, until chaos begins at a well defined value Rt. The remarkable character of this cascade is that *if* it occurs, the way it happens is essentially independent of the nature of the system. Several exponents that define the accumulation rate of the cascade and the amplitude ratios of various peaks in the power spectrum are *universal*. The detection of this process and its explanation proved to be powerful tools in demonstrating both the existence of chaos and the utility of the theory of mappings to real fluid problems.

Another route to chaos that turned out to be universal involved the interaction between disturbances with several incommensurate frequencies: these are termed quasi-periodic states. The transition to chaos via quasi-periodicity can be described quantitatively in terms of a class of two-parameter mappings known as circle maps.

One of the weaknesses of dynamical systems theory is that it provides essentially no knowledge of when the transition might occur (i.e., at what

Reynolds number). Furthermore, though several of the routes to chaos turned out to be universal, there are many others that are not universal. It is quite possible for a system to make a sudden and discontinuous transition to chaos without any warning as a parameter is varied. This type of transition does not have a system-independent description.

Open versus closed systems: absolute and convective instability

In closed systems such as Rayleigh-Bènard convection, instabilities are typically amplified until the resulting state fills the system completely. External noise then becomes insignificant in the sense that the statistical properties of the dynamics are not significantly affected by the addition of a small amount of external noise. Many open systems, especially jets, wakes, and mixing layers, behave quite differently. These systems are generally 'convectively unstable', which means that disturbances can be carried downstream more quickly than they are amplified. In this circumstance, a very small amount of noise near the 'source' can affect the flow for a substantial distance downstream, and the character of the fluctuations reflects *both* chaotic dynamics and external noise. The resulting patterns are often called noise-sustained structures to indicate the importance of external noise to their persistence. In these situations chaos is often operative, but its detection and investigation is more complex.

SYSTEMS WITH MANY MODES

Spatially extended systems

The discovery that some fluid systems behave as if they have only a few degrees of freedom is remarkable. When this happens, the fluctuations are correlated essentially over the entire system, and the dynamics are then essentially *tempora*. However, this circumstance is not the one of greatest interest from a practical point of view. Many systems are inherently extended, and are much larger than the typical scale of the initial instability. A typical example might be a thin flowing film, or convection in a layer that is much larger laterally than its depth. For these systems, a profitable approach has been first to study the formation and evolution of steady patterns near the onset of the initial instability, and then to investigate how these patterns become disordered or chaotic.

Mathematical models of pattern formation

Near the onset of the instability that gives rise to a steady pattern, nonlinear effects are weak. In that case, the phenomenon can typically be described mathematically in ways that are fairly independent of the details of the system. For patterns varying in *only one* dimension, for example, the pattern can usually be described by a complex amplitude function $A(x, t)$ that varies *slowly* in space and time. In Rayleigh-Bénard convection, it might represent the envelope of the spatially periodic temperature field, θ, in the midplane of the fluid layer, which can be obtained through the relation

$$\theta(x, t) = Re\{A(x, t)e^{ikx}\}.$$

Sometimes it is possible to construct simplified dynamical equations to describe the space and time variations of the amplitude function, by forming a systematic power series solution of the basic equations of motion (the Navier-Stokes equations for fluids), and then keeping the lowest order terms in the powers and gradients of the amplitude function. This is fortunate, since the full equations are often hopelessly difficult to solve. For patterns in one dimension, the resulting amplitude equation often turns out to be essentially the same as the Ginzburg-Landau model used for describing the degree of order near a phase transition. It looks like this:

$$\tau_0 \frac{\partial A}{\partial t} = \epsilon A - \gamma |A|^2 A + \zeta_0^2 \frac{\partial^2 A}{\partial x^2}.$$

The coefficient ϵ of the linear term in this equation changes sign at the onset of the instability, causing A to grow exponentially in time, until limited by the cubic nonlinear term to a steady state amplitude that varies as the square root of ε. The derivative term basically causes patterns to spread or contract diffusively. It is important for patterns which are *non-uniform* due to boundaries, defects, or initial conditions. This equation is of course much simpler to work with than the full equations of motion, and gives a concise summary of the ways in which some simple one-dimensional patterns evolve.

One consequence of this particular evolution equation is that patterns described by it evolve toward the minimum of a certain 'potential function' F, which may be expressed by a spatial integral over A and its gradient. This phenomenon is analogous to the minimization principles of equilibrium statistical mechanics, and would be a remarkably useful unifying principle if it were generally true. Although there seem to be many circumstances in which a minimization principle is at least approximately at work for pattern-forming systems, there seem to be an equal number where obvious violations occur and time evolution takes the system 'uphill' so to speak. Therefore, we are forced to look at the behavior of individual systems to understand their dynamics. Nonlinear systems cannot simply be reduced to a variational principle.

For patterns evolving in *two* spatial dimensions, a great deal of work has been done to devise appropriate amplitude equations. The motivation is twofold: (a) The Navier-Stokes equations are hard to integrate numerically over the very long time scales that are characteristic of evolving patterns; (b) a purely numerical solution gives little insight into the physical processes involved. Various two-dimensional amplitude equations have been proposed to describe experiments, with some success. It is fair to say that for several systems not too far from the threshold of the initial instability, we now have a pretty good understanding of the major features of pattern formation and evolution. In particular, it is possible to understand at least approximately the role of point and line defects, and how they evolve in time. However, the model equations are probably increasingly inaccurate as one goes further above the threshold.

For systems where the initial instability is toward traveling waves, as in the case of binary fluid mixtures, the relevant dynamical equation is a bit more elaborate. There are strong grounds for believing that the resulting patterns can be largely described by an equation similar to the Ginzburg-Landau equation, but with complex coefficients. This complex Ginzburg-Landau equation,

$$\frac{\partial A}{\partial t} = \epsilon A + (1 + ic_1)\frac{\partial^2 A}{\partial x^2} + (1 + ic_3)|A|^2 A - (1 - ic_5)|A|^4 A,$$

is exceedingly rich and interesting, and it is believed to contain many of the phenomena one sees experimentally in these systems, such as pulses and fronts and chaotic dynamics. It will be probably several years before all the phenomena it contains have been explored mathematically and numerically.

Spatio-temporal chaos

In spatially extended systems, an initially cellular pattern often breaks up into domains separated by defects or imperfections. As the nonlinearity is increased, a kind of chaotic motion called 'defect mediated turbulence' ensues, in which the defects fluctuate erratically. A further increase in the parameter may lead to *spatio-temporal chaos*, in which the long range order of the pattern is destroyed. This phenomenon has been studied in many fluid systems, and in partial differential equations viewed as simplified models of fluid systems. These include the complex Ginzburg-Landau equation mentioned above, the Kuramoto-Sivashinsky equation, and others. These models provide some qualitative insight, and they allow one to avoid brute force solutions of the Navier-Stokes equations, which are more demanding numerically than the model equations.

Many experimental and numerical examples of spatio-temporal chaos have been discovered. At present, though many of the examples have common features, we do not have either a unified empirical summary of the transition to spatio-temporal chaos, or an adequate theoretical framework.

It is clear that the phenomenon involves many degrees of freedom, and that dynamical systems theory in its basic form is an inadequate framework for a full understanding.

CHAOTIC TRANSPORT

Eulerian and Lagrangian representation of fluid flow

We are accustomed to the idea of tracking individual particles in the theories of Newtonian or classical mechanics. In the mechanics of fluids this kind of particle-based description is known as the *Lagrangian representation*. It is possible to write the equations of motion for a fluid using this approach, but it turns out to be quite cumbersome, and an alternate approach, the progenitor of the now ubiquitous field theories of physics, is usually adopted. This approach is called the *Eulerian representation* of fluid motion. It works with a number of fields, in particular the velocity field $\mathbf{V}(\mathbf{x}, t)$ of the fluid, the value of which at any space-time point $(\mathbf{x}, t)$ gives the velocity of the fluid at that spatial point x at instant t. Since different fluid particles pass through x at different times, the velocity field does not in general contain information on individual particle velocities (without further processing).

Kinematics and 'chaotic advection'

Let us write down the simple kinematic statement that the fluid particle located at $\mathbf{x} = (x, y, z)$ has velocity $\mathbf{V}(\mathbf{x}, t)$. If the component of $\mathbf{V}$ are called (u, v, w), we obtain a set of three ordinary differential equations

$$\begin{aligned} \dot{x} &= u(x, y, z, t), \\ \dot{y} &= v(x, y, z, t), \\ \dot{z} &= w(x, y, z, t). \end{aligned}$$

These *advection equations* are the simplest expression of the link between the Lagrangian and the Eulerian representation of fluid flow. Already for steady flows in three dimensions, i.e., for the case when u, v, and w do not depend on time, these equations are rich enough to produce chaotic particle motion within the fluid. Similarly, for two-dimensional *unsteady* motion, i.e., for the case when $w = 0$ and u and v depend only on x, y and t, chaotic particle motions can be produced.

Chaotic motion of fluid particles, or of passively transported (advected) foreign particles in a flow is known as *chaotic advection*. It is important to stress that it is entirely possible to have chaotic advection in a laminar flow, even at very low Reynolds number. This has led to an interesting range of applications in chemical engineering and in low-speed applications in mechanical engineering and materials processing. The field has grown rapidly, and new areas of application are being added continuously.

Experimental verification; applications to fluid stirring

The immediate and intuitive corollary of being able to set up chaotic particle motions in a fluid is that enhanced mixing and transport can take place. Typically, a group of particles in chaotic motion will tend to deviate exponentially from one another, and this leads to substantial mixing with the surrounding fluid in a much shorter time than would be expected for regular motion. Tremendous stretching of finite fluid regions takes place which paves the way for an enhanced activity by diffusion and conduction due to the greater surface area available for these processes.

Several experimental verifications of the mechanism of chaotic advection are now available. These include direct visualization of patterns of mixing in very viscous liquids (Stokes flow), where the time-dependent variation of flow required in the two-dimensional case can be effectively established by moving the flow boundaries in prescribed ways. At higher Reynolds numbers one has less direct control of the flow pattern, which is typically a secondary flow such as convection rolls or the secondary vortex motion produced by centrifugal forces in a curved pipe. Experiments on such configurations again show enhanced transport due to chaotic motions. For example, in one set of experiments it was shown that the efficiency of a common heat exchanger tube could be increased by 10%, with essentially no penalty in pressure drop, simply by altering the geometry in such a way that more chaotic transport occurred. In several cases the notion of enhanced chaos in fluid mixing provides a rationalization of empirical practice that was incompletely understood before. In other cases, the enhanced understanding points to interesting new design strategies that are only beginning to be explored.

Chaotic vortex motion

The Eulerian representation of fluid motion uses as its basic ingredients the fields of velocity and pressure (for an incompressible fluid). However, the *vorticity field*, formally given as the curl of the velocity, $\omega = \nabla \times \mathbf{V}$, and physically to be interpreted as proportional to the local angular velocity of fluid particles, is an essentially Lagrangian entity. Its equation of motion, the vorticity equation, states that the vorticity of a fluid particle is rotated and stretched due to the local rate-of-strain within the fluid, and is diffused by viscous forces. In fact, the kinematic viscosity of the fluid can be interpreted as the diffusion coefficient of vorticity.

For a certain idealized model of vortex motion, which arises if the vorticity can be assumed to be concentrated in a set of parallel vortex tubes, the vorticity equation reduces to a system of coupled ODEs, and chaotic motion of the vortices can be readily demonstrated. It is tempting to extend these theoretical insights to actual concentrated vortex structures that appear as 'coherent structures' in many turbulent flows. Both the mixing and transport of advected scalars, such as temperature and concentration, and the mutual interaction of the vortices themselves can then be addressed within the framework of chaotic dynamics, promising a new approach to many basic questions in intermediate and high Reynolds number flows. Research along these lines, aimed at a more deterministic understanding of transition and turbulence, is just beginning.

CONCLUSIONS AND OUTLOOK

Successes of the dynamical systems approach

It is fair to say that the dynamical systems approach to fluid mechanics has had a profound effect on our understanding and description of the entire flow regime that encompasses the transition to turbulence. The older dichotomy of 'laminar' and 'turbulent' has yielded to a richer and more complex classification of flow states. The notion of repeated subharmonic bifurcations with universal properties as a route to chaos, and the concept of a finite number of bifurcations before the onset of a chaotic flow state are definite advances related to the dynamical systems point of view.

The appreciation of chaotic kinematics in a laminar flow, and the application of this idea to fluid mixing, vortex interactions, flow-structure interactions, dynamo action in a conducting fluid, and sound radiation from flows will lead to important new insights and new possibilities for engineering design.

The main conceptual advantage of the dynamical systems approach is that the transition to a stochastic, turbulent flow state is not postulated and input in an *ad hoc* fashion in the theoretical description, but is assumed to be latent in the description of fluid motion by the Navier-Stokes equation. This point of view is altogether more fundamental, and should in the long run lead to deeper understanding and greater predictive capability.

Limitations of the dynamical systems approach

The main limitations of the dynamical systems approach as we see it applied to fluid mechanics today are: (1) the inability to deal in detail with systems involving more than a few degrees of freedom, and (2) the lag in development of effective calculational tools relative to statistical field theories, where such tools have been under development for decades. While one can anticipate a steady evolution of calculational techniques, the limitation of current understanding to low-dimensional systems appears more serious, and seems to demand radically new ideas. The understanding of multi-degree of freedom systems from a dynamical systems point of view, promises to be as large an intellectual challenge as the fusion towards the turn of the century of kinetic theory, statistical mechanics and classical thermodynamics.

Research Needs

We have reached a milestone in which the phenomenon of chaotic dynamics in its simplest forms has been well elucidated. Researchers are beginning to apply these ideas successfully to more complex flows of great interest to both fundamental science and applications. We may cite the following research needs:

- Experiments on and models of spatio-temporal chaos in extended systems that are weakly turbulent. This work could lead to a better understanding of the chaotic behavior of nonlinear partial differential equations in general.

- Studies aimed at applying dynamical systems theory to the coherent structures observed in turbulent flows.

- Efforts to better understand the relationship between chaotic advection and turbulent transport.

- Attempts to explain and understand postulated mechanisms in statistical theories of turbulence, such as the notion of a turbulent cascade, self-similarity, intermittency, fractal structure, etc., in terms of dynamical systems concepts.

- Continued effort to turn the insights gained from the dynamical systems approach into new strategies for flow management and control, and to use these insights as the basis for new inventions and processes across a broad spectrum of applications.

SUGGESTIONS FOR FURTHER READING

Aref, H. (1984) Stirring by chaotic advection. *J. Fluid Mech.* **143**, 1-21.

Aref, H., Jones, S. W. & Thomas, O. M. (1988) Computing particle motions in fluid flows. *Computers in Physics* **2**(6), 22-27.

Baker, G. L. & Gollub, J. P. (1990) *Chaotic Dynamics: An Introduction.* Cambridge Univ. Press.

Cross, M. C. & Hohenberg, P. D. (1992) Pattern formation outside of equilibrium. *Rev. Mod. Phys.* (to appear).

Feigenbaum, M. J. (1978) Quantitative universality for a class of nonlinear transformations. *J. Stat. Phys.* **19**, 25-52.

Feigenbaum, M. J. (1979) The universal metric properties of nonlinear transformations. *J. Stat. Phys.* **21**, 669-706.

Gollub, J. P. & Libchaber, A. (1983) Laboratory experiments on the transition to chaos. In *Chaotic Behavior of Deterministic Systems*. North Holland.

Guckenheimer, J. (1986) Strange attractors in fluids: Another view. *Ann. Rev. Fluid Mech.* **18**, 15-31.

Lanford III, O. E. (1982) The strange attractor theory of turbulence. *Ann. Rev. Fluid Mech.* **14**, 347-364.

Lorenz, E. N. (1963) Deterministic nonperiodic flow. *J. Atmos. Sci.* **20**, 130-141.

May, R. M. (1976) Simple mathematical models with very complicated dynamics. *Nature* **261**, 459-467.

Ottino, J. M. (1989a) The mixing of fluids. *Sci. Amer.* **260**(1), 56.

Ottino, J. M. (1989) *The Kinematics of Mixing: Stretching, Chaos and Transport*. Cambridge Univ. Press.

Swinney, H. L. & Gollub, J. P. (eds.) (1985) *Hydrodynamic Instabilities and the Transition to Turbulence*. 2nd ed. Topics in Applied Physics, vol. **45**, Springer-Verlag.

TURBULENT BOUNDARY LAYERS

Peter Bradshaw
Mechanical Engineering Department
Stanford University
Stanford, CA 94395

INTRODUCTION

Turbulent boundary layers are found on the outside of aircraft, ships and road vehicles, on the inside of turbomachines, and in many other situations in engineering and the earth sciences. Turbulence is what mixes the cream in a coffee cup; roughly half the drag of a typical airliner is due to the turbulent boundary layer; and the wind near the Earth's surface, which controls local pollutant dispersion, behaves like a boundary layer one or two km thick. The prediction or control of turbulent boundary layers and other turbulent flows, on a wide range of scales, is thus a major activity in fluid dynamics, both at research level and in engineering or environmental practice. Turbulent flow is so complicated that, although it obeys the same equations of motion as other fluid flows, numerical solution of those equations is still too expensive for general application. Simplified "model" equations, usable in everyday predictions of turbulent flow, contain empirical information and must be calibrated using data from experiments – or from direct numerical solutions ("simulations") regarded as research tools.

In steady viscous "laminar" flow over a solid body, the effects of viscosity are usually confined to a thin layer next to the surface, growing in thickness with distance along the body. If the body is "streamlined" then, by definition, the boundary layer remains thin until the rear end of the body, when it leaves the surface to form a narrow wake. In the case of a "bluff body" then, again by definition, the boundary layer separates before reaching the rear end of the body, forming a broad wake with recirculation of the flow. In either case, the subsequent rate of growth of the wake is small, as in the boundary layer itself. Laminar jets also spread slowly with downstream distance, and although the growth rate of a laminar shear layer depends on Reynolds number and on the type of shear layer, the shear-layer thickness δ, however defined, grows at a rate $d\delta/dx$ much less than 1. The reason is that – once more by definition – if the Reynolds number is high, the ratio of typical inertia forces to typical viscous forces is large, and the rate of reduction of momentum of the flow by viscous stresses is small. The smallness of $d\delta/dx$ allows some simplifications

31

to be made in the equations of motion, and these are collectively known as the "boundary-layer approximation" (a slightly misleading term since the same approximations apply to jets, wakes and other thin shear layers).

The basic property of unsteady, eddying turbulent motion is that it mixes mass and momentum very much more quickly than a purely laminar flow. However, rates of growth of turbulent boundary layers, wakes, etc. are still small enough compared to unity for the boundary-layer approximations to be applied, perhaps with caution in some cases. Turbulence in thin shear layers is just as complicated as turbulence in more general flows, but the restricted range of geometry means that it is possible to derive empirical formulas and data correlations which will apply, say, to the whole set of attached boundary layers with adequate engineering accuracy – though the empirical coefficients may have to be changed before analogous formulas can be applied to the wake or jet.

There are many turbulent flows which obey the boundary-layer approximation only in restricted regions. The flow over a backward-facing step is a simple example of the many real-life flows in which a boundary layer first separates and then reattaches. The flow between separation at the step corner and reattachment roughly six step heights downstream consists of a mixing layer between the outer, downstream-going flow and the low-speed recirculating flow next to the surface. The larger eddies in a turbulent boundary layer have "lifetimes", measured as distance traveled downstream, of ten times the boundary-layer thickness or more. Therefore the mixing layer in the step flow continues to bear traces of the initial boundary layer: just because the flow has the boundary conditions of a mixing layer does not instantaneously convert its statistical properties to those of a fully-developed mixing layer (even if we ignore the other differences between the curved mixing layer behind a step and the canonical mixing layer between two uniform parallel streams). Likewise, reattachment does not immediately convert the mixing layer back into a canonical boundary layer: indeed, the rapid changes in mean velocity (specifically, rapid changes in mean-velocity gradient) near the reattachment point apply further perturbations to the flow. The flow does not return to the behavior expected of a turbulent boundary layer in zero pressure gradient (the "flat plate" boundary layer) until at least 50 step heights downstream of the step, although a general prediction method for turbulent boundary layers would become applicable much sooner after reattachment.

We see that although boundary layers and other thin shear layers are a very important class of turbulent flows, not all flows can be synthesized from thin shear layers alone, and in particular it is found that quite moderate distortions of thin shear layers, too small to invalidate the boundary layer approximation, are sufficient to cause changes in turbulence properties which cannot be predicted by simple calculation methods.

BASICS OF TURBULENT BOUNDARY LAYERS

A detailed review has been given by Ligrani (1989).

The simplest of all turbulent boundary layers is that over a "flat plate" mounted along the stream so that the external-stream velocity is constant. The only explicit length scale in the problem is the distance from the leading edge of the plate, but the location of transition from laminar to turbulent flow in the boundary layer depends on the external conditions, and so the only meaningful length scale for the turbulent boundary layer is its local thickness. As in laminar flow, the velocity asymptotes to the external-stream velocity U_e, and although the probability of a turbulent eddy extending out to infinity is infinitesimal, the definition of the boundary-layer edge is arbitrary, usually being taken as the distance from the surface, y, at which the velocity has risen to 0.99 or $0.995U_e$. Two common non-arbitrary definitions of length scales are the displacement and momentum thicknesses: they are roughly $1/10$ to $1/5$ of the "0.99" thickness. The former is simply the distance by which the external streamlines are displaced by the slowing down of the fluid in the boundary layer as it grows, while the momentum-(deficit) thickness is the thickness of a layer of external flow whose momentum flux equals the deficit of momentum in the boundary layer. In the case of a boundary layer on a flat plate, the rate of growth of momentum thickness is equal to $c_f/2$, where c_f is the skin-friction coefficient, a suitable dimensionless form of the surface shear stress. Surface shear stress, leading to drag of a wing or pressure drop in a pipe, is the most important single quantity in turbulent flow, so the Reynolds number based on momentum thickness is the one most commonly used for correlating the properties of boundary layers, whether on flat plates or otherwise. Nevertheless, the total thickness of the boundary layer, being a rough measure of the size of the largest turbulent eddies, is more useful in qualitative discussion.

For distances from the surface less than about 0.1 to 0.2δ, the size and behavior of the turbulent eddies are influenced more by the presence of the wall than by the turbulence in the outer layer. In this region, the shear stress is closely equal to the wall value, and although the wall shear stress is eventually due to the presence of the external stream, it is a meaningful scaling factor for the flow in the inner layer. The external-stream velocity certainly does not affect the flow near the surface directly, although any acceleration of the external flow is accompanied by a streamwise pressure gradient which does affect the flow close to the wall. However in small or moderate pressure gradients, it is found empirically that the flow close to the solid surface is indeed uniquely determined statistically once the distance from the surface, the surface shear stress, and the fluid density and viscosity are specified. By dimensional analysis and simple physical arguments we obtain a supposedly universal "law of the wall", which, except in the viscosity-dependent region very close to the surface, is a logarithmic relation between velocity and distance from the surface. Boundary-layer prediction methods which do not explicitly invoke the logarithmic law for the inner layer are almost invariably arranged to be compatible with it: since the velocity at the outer edge of the logarithmic region may already be as high as $0.7U_e$, this almost guarantees adequate predictions of two-dimensional boundary layers not too close to separation. Obviously, the basic assumption that the flow near the surface is totally independent of upstream history and outer-layer behavior, except in so far as they determine the surface shear stress, cannot be exactly correct; and the accuracy and/or limits of validity of the "law of the wall" results are currently under debate.

The outer layer of a turbulent boundary layer is strongly dependent on upstream history and on the external boundary conditions, as are most other turbulent flows. In principle it is not directly influenced by viscosity, but in fact significant changes appear at low Reynolds number (less than about 5000 based on momentum thickness). This is because the viscous diffusion layer that forms the irregular instantaneous outer edge of the turbulent flow becomes thicker and, being wrinkled, occupies a large fraction of the outer layer volume. There are special cases where the history is in some sense uniform so that local scaling rules can be used ("self-similar" flows), but they are rare in practice.

Fortunately, the relations between different statistical properties, collectively known as the "structure" of the turbulence (Robinson 1991), are more nearly universal, and turbulence modeling relies on this. It is clear from experiment that different kinds of shear layer have different structure, which casts doubt on models that assume universality.

Research Needs

(a) Better understanding of inner-layer structure and the limits of law-of-the-wall scaling

(b) Better understanding of low-Reynolds number effects on turbulent/nonturbulent interfaces.

FREE TURBULENT FLOWS

In free turbulent flows (i.e. flows far from solid surfaces; jets, wakes, mixing layers, etc.) there are no simple scaling regions analogous to the law of the wall, so free turbulent flows are in principle a much more stringent test of a prediction method. In practice, the variety of free turbulent flows of engineering interest is considerably smaller that the variety of boundary layers; and in most cases the flows are either sufficiently close to the canonical self-similar ones for simple correlation formulas to apply, or they are so strongly distorted – by impingement on a solid surface or otherwise – that even the most advanced prediction methods are not very reliable guides to behavior. The most important free shear layer in aerospace and mechanical engineering is the mixing layer, specifically the mixing layer between two nominally non-turbulent streams of chemical reactants such as vaporized fuel and air. Here again the statistics of the wrinkled diffusion-dominated interfaces between turbulent and nonturbulent fluid (Sreenivasan 1991) are important because reaction depends on mixing at molecular level.

Research Needs

(a) Scalar interface statistics including fractal properties
(b) Turbulence structure in impinging jets

ORDERLY STRUCTURE

The early view of turbulence was that it was a completely random motion (albeit with continuous velocity distributions) with spatial scales sufficiently small that analogies with molecular motion (leading to gradient-diffusion formulas) were a reasonable approximation. It was pointed out about 40 years ago that the covariances between the velocity components at 2 points ("correlations") were necessarily determined, for large distances between the points, by the shapes of the largest eddies in the flow. It was found by experiment that the large eddies usually filled the flow: their shapes varied from flow to flow, but were always such as to contribute to the Reynolds shear stress (so that the large eddies maintained themselves by acquiring part of the production of turbulent energy).

The contribution of the large eddies to the shear stress in most flows is significant but not overwhelming, but mixing layers are dominated by the large eddies, whose shape is related to the eigenmodes of the inviscid infection-point instability, namely spanwise vortex rolls: by turbulence standards these large-eddy rolls are remarkably orderly and persistent. In recent years much effort has been devoted to the study of orderly or "coherent" structures in different kinds of turbulent flows, and there has been a tendency to call almost any identifiable flow pattern in turbulence an "orderly structure". Originally, the name implied a structure whose lifetime was significantly longer than that expected for a typical turbulent eddy. Clearly the turbulent eddies which contribute most of the shear stress and therefore most of the turbulent energy production are expected to, and indeed must, have a lifetime of the order of (turbulent energy)/(energy production rate). There are of course other definitions of "lifetime" and it is possible for an eddy to remain observable, by flow visualization or otherwise, after several complete transfusions of turbulent energy, but the lifetime based on energy exchange is the one that must appear in prediction methods. It is rather remarkable that virtually none of the research that has been done on orderly structure has had an impact on turbulence modeling: in particular the idea that turbulence could be represented as a combination of orderly structures with fine-scale effectively-random turbulence has proved very difficult to implement in quantitative terms. Nevertheless the search for statistical order in turbulence is the approach most likely to lead to reliable (statistical-average) models.

THREE-DIMENSIONAL TURBULENT FLOWS

The largest fraction of basic research, both in experimental work and in turbulence simulation (solution of the full time-dependent Navier-Stokes equations for the complete turbulent motion) is still devoted to two-dimensional flows, but turbulence model development, and experimental/simulation work intended to support it, is being concentrated more and more on three dimensional flows. The simplest of the three dimensional flows (Bradshaw 1987) are boundary layers, in which the thickness δ changes slowly in both the x (roughly streamwise) and z (roughly spanwise) directions, so that the boundary-layer approximation can be applied to the x- and z-component momentum equations. The shear stress now has two components, in the x and z directions, and their behavior is not well predicted by turbulence models developed for two-dimensional flows. If an initially two-dimensional turbulent boundary layer acquires a z-wise component of mean velocity, the x-wise component of shear stress tends to decrease while the z-wise component increases more slowly than would be predicted by extensions of two-dimensional models. The failure of two-dimensional models is not spectacular, but it is of concern in high-tech industries.

A more severe test of turbulence models is the group of three-dimensional "slender" flows, in which gradients in the z direction are of the same order as those in the y direction: examples include vortices embedded in turbulent boundary layers, and flows in non-circular ducts. The flow in a long straight non-circular duct will acquire contra-rotating vortex pairs in the vicinity of each corner, actually generated by Reynolds-stress gradients. This phenomenon is unknown in laminar flow, and cannot be predicted by turbulence models that assume an isotropic effective "eddy" viscosity. These "stress induced" secondary flows are quite weak, easily overwhelmed by secondary flows caused by bends, fins or other disturbances in the duct, and indeed have attracted more attention in the laboratory than in real life.

Vortices form an important class of turbulent flows, and provide a rich variety of ill-understood phenomena such as vortex breakdown. The strength (circulation) of the wing tip vortices trailing behind an aircraft is set by the lift of the wing, but the details depend on the development of the boundary layer and wake of the wing, because the vortex is composed of wake fluid. A boundary layer approaching a tall solid obstacle on the surface rolls up into a "horseshoe" or "necklace" vortex, wrapped around the body with its legs trailing downstream. There are many other occasions in which longitudinal vortices interact with boundary layers: usually the interaction is a nuisance, leading to higher drag, but the small plate vortex generators protruding from the wings or engine nacelles of some aircraft are intended to introduce longitudinal vorticity into the boundary layer to increase mixing and discourage flow separation, at least in some flight conditions.

Research Needs

- (a) Detailed measurements of turbulent structure in carefully-chosen three-dimensional flows to explore the structural differences that lead to discrepancies in the prediction methods, and corresponding simulations.

- (b) Less-detailed measurements in more practical configurations, to act as test cases for turbulence models.

- (c) More studies of vortex flows, especially the mechanisms of turbulence damping by rotation in the vortex core, and of vortex breakdown.

"COMPLEX" TURBULENT FLOWS

Complex flows (Purtell 1992) are shear layers perturbed by externally-imposed distortion or by interaction with another turbulence field. The distortions may be so severe as to violate the boundary-layer approximation, but shear layers are remarkably sensitive to distortion even within the limits of the boundary-layer approximation. A notorious example is the effect of streamline curvature (as on a highly-cambered turbomachine blade). Streamline curvature introduces extra terms into the boundary-layer equations, which must now be written in curvilinear coordinates, but the effects of streamline curvature are larger, by an order of magnitude, than these explicit terms would indicate. In a boundary layer on a concave surface, the angular momentum decreases outward from the center of curvature, and as a result the flow is unstable, leading to the formation of quasi-steady longitudinal vortices superimposed on, and interacting with, the turbulent motion. Conversely, on a convex surface the boundary layer is stabilized and turbulence intensities decrease, and in extreme cases the outer layer becomes essentially non-turbulent. Beyond these simple facts, and despite a considerable amount of experimental work, the detailed mechanisms are not well understood and are not well represented by current calculation methods: explicit extra correction factors, depending on a dimensionless form of the streamline radius of curvature, are needed in most cases. Lateral (spanwise) divergence or convergence also has a significant effect on turbulent shear layers, respectively increasing and decreasing the level of turbulent activity. The same applies to the effects of bulk compression and dilatation in compressible flow.

In many practical cases, the stream external to a shear layer is not the ideal uniform non-turbulent "free stream" of simple experiment and simple theory, but may itself be sheared and/or turbulent. Little work has been done on the growth of a shear layer below an external stream which is itself sheared. An important case in which the external flow is both sheared and turbulent is "internal" layer which forms at a change of surface roughness. This is a very common situation in meteorology, where the wind passes from the (smooth) ocean to the (rough) land, or from open countryside to a city.

Free-stream turbulence in nominally-unsheared flow is an important problem in many forms of turbomachinery. The most extreme case is the flow out of the combustion chamber of a gas turbine, in which the turbulence intensity is designedly high in order to promote complete mixing and combustion. This flow then passes over the turbine blades, and the free-stream turbulence produces an unwelcome increase in heat transfer to the blades. Recent research work has shown that the effects of high (15-30 percent) turbulence on skin friction and heat transfer is larger than predicted by extrapolation of formulas intended for lower-intensity free-stream turbulence, but a confusing factor is that the high turbulence intensities have been generated experimentally by means which tend to produce inhomogeneities or shear in the flow, so that it is not clear if the effects observed are entirely attributable to free-stream turbulence.

Research Needs

- (a) Experiments and simulations on flows with extra rates of strain and/or free-stream turbulence to elucidate physics and to provide data for turbulence models.

- (b) Particular attention to high-intensity free-stream turbulence typical of turbomachines, with or without accompanying mean distortion.

COMPRESSIBLE FLOW

Because the velocity fluctuations in a boundary layer are generally one or more orders of magnitude smaller than the mean velocity, the Mach number based upon these fluctuations is generally small until the mean Mach number has reached roughly five. Prediction of compressible flows requires the addition of equations to describe the temperature field, thus giving the density for insertion into the momentum and continuity equations, but it is found that low-speed turbulence models are adequate up to $M = 5$. The law-of-the-wall analysis can be plausibly generalized to compressible flow, and seems to give good results, although some turbulence models fail to reproduce the compressible form of the log law because of the way in which they treat density gradients: however, the effect on predictions seems to be fairly small.

Mixing layers between two parallel streams, or between one stream and still air, are of considerable importance in compressible flow. This has been highlighted by recent work on supersonic combustion connected with the National Aerospace Plane project, in which the mixing layer is that between the fuel and the oxidant in a supersonic ramjet. The mixing layer is a much more highly turbulent flow than a boundary layer, and direct compressibility effects due to large Mach-number fluctuations appear at a mean-flow Mach number of about one, above which the spreading rate decreases. Only recently have plausible attempts been made to allow for this effect in calculation methods, but necessarily the allowances involve empirical coefficients. The optimum correction for mixing layers seems to produce an unwanted reduction in skin friction and spreading rate in the boundary layer, and a compromise is needed. It could be argued that the boundary layer and mixing layer are such intrinsically different flows that one could not reasonably expect the same prediction method to apply accurately to both, even in incompressible flow. There is currently some uncertainty in mixing-layer spreading rate: recent carefully-conducted experiments show a significantly larger decrease in spreading rate with increasing Mach number than a consensus of previous data.

Bulk compression of a turbulent flow by passage through a shock wave generally increases turbulent intensity, as usual by a factor an order of magnitude larger than that predicted by the extra terms in the equations of motion. Even distributed compressions produce significant defects, while expansions produce a reduction in turbulent intensity and shear stress. The mechanism is, broadly, that the vorticity transport equations for compressible flow express the conservation of ω/ρ so that compression increases both mean and fluctuating vorticity.

There is remarkably little reliable information on turbulence structure in compressible flow. The presence of large temperature fluctuations make hot-wire measurements more complicated, to say nothing of mechanical difficulties, while laser-doppler velocimeter measurements are complicated by refraction of the light beams and the general difficulty of access to supersonic wind tunnels. So far, simulations of compressible flow have been confined either to ideal homogeneous cases, including flows with shock waves or combustion, or to mixing layers; but simulations of compressible boundary layers (without shock waves) are within the reach of current technology.

Research Needs

- (a) Turbulence structure measurements in mixing layers at Mach numbers up to, say, five.

- (b) Turbulence structure measurements in supersonic boundary layers, and more reliable measurements, even of mean flows only, in boundary layers in the hypersonic range $M > 5$.

COMBUSTION AND CHEMICAL REACTION

Chemical reactions obviously require mixing of the reactions down to molecular scale, and this is to be distinguished from larger-scale "stirring" although the difference is hidden by ordinary time (Reynolds)-averaged statistics. Because of the very large increase in mean temperature of the gases during combustion, kinematic viscosity can increase to the point at which the local flow Reynolds number becomes quite low, and the effect of this is to damp out the fine-scale motion of the turbulence which is responsible for the final stages of mixing to molecular level. This is a particular confusion in laboratory-scale experiments in which even the initial Reynolds number is not very high.

There are very few comprehensive data sets for turbulent combustion, and it appears likely that, in the near future at least, basic understanding is likely to come more from simulations than from experiments. However, even comparatively simple chemical reactions such as those between oxygen and the lower hydrocarbons go through many stages, and each of the intermediate species requires its own storage in a simulation. Currently, simulation of realistic chemistry in a true three-dimensional simulation is impractical and two-dimensional simulations, in which the w-component velocity is assumed to be zero at all times, are currently in use. Unfortunately, the essential vortex-stretching mechanism which cascades energy from the large-scale eddies to the small-scale motion (which mixes the reactants) is missing in instantaneous two-dimensional flow, so these simulations cannot be expected to throw light on the detailed statistics of the fuel-oxidant interface. Reviews, with some emphasis on modeling, are given by Pope (1987) and Bilger (1989).

Research Needs

- (a) Detailed measurements of the joint velocity/concentration statistics in a combusting flow, or even in a non-combusting scalar-mixing flow, would be very desirable but, as in the case of compressible flow, there are considerable practical difficulties. There are comparatively few detailed measurements, including fine-scale structure, even in low-speed flows with passive scalar contaminants.

HEAT TRANSFER AND ENVIRONMENTAL FLOWS

Because molecular diffusivities are unimportant in turbulent flows at large local Reynolds numbers, results for heat transfer can be applied to pollutant transfer, and conversely, providing the density differences are small. Meteorology is of course dominated by turbulence processes, and the pioneering work in turbulence simulation was done by meteorologists. The principal additional problem to those discussed above is the behavior and condensation of water vapor in the atmosphere. A good deal of work is currently in progress on particle-laden flows, but most basic experiments deal with non-evaporating particles, especially those which are large enough that they fail to follow the flow and thus can significantly affect the turbulence at high concentration. The effects of buoyancy on the planetary boundary layer are qualitatively analogous to those of streamline curvature: if the density increases upwards, as when the earth's surface is warmed on a sunny day, longitudinal-vortex "roll cells" can form. If the density decreases upwards, as a result of heat transfer from the surface by radiation during the night, turbulent activity is suppressed and layers of mist may form. As in the case of streamline curvature, understanding of buoyancy effects is incomplete, and extra correction factors are needed for turbulence models.

Research Needs

(a) Since turbulence is the main agency for mixing pollutants in the atmosphere and ocean, its understanding and prediction on all scale from local to global is essential. Measurements in the atmosphere and ocean are of course essential, but more can be done to read across the results of laboratory experiments on scalar mixing, whether directly or in the form of turbulence models based on laboratory data.

CONCLUSIONS

Turbulence has been described as the "pacing item" in the use of the national aerodynamic simulation facility and has long been regarded as the chief outstanding difficulty of fluid dynamics. Particularly in aerospace engineering, boundary layers and other thin shear layers are the most important kinds of turbulent flow, but they are frequently subjected to external influences like distortion or free-stream turbulence which are difficult to treat to good engineering accuracy in prediction methods. This being so, experimental work on turbulence appears to have an assured place in future research, supported by computer simulations.

REFERENCES

Bilger, R.W. (1989) Turbulent diffusion flames. *Ann. Rev. Fluid Mech.* 21, 101.

Bradshaw, P. (1987) Turbulent secondary flows. *Ann. Rev. Fluid Mech.* 19, 53.

Ligrani, P.M. (1989) Structure of turbulent boundary layers. *Encyclopedia of Fluid Mech.*, ch. 5 (N.P. Cheremisinoff, ed.), Gulf Publishing.

Pope, S.B. (1987) Turbulent premixed flames. *Ann. Rev. Fluid Mech.* 19, 237.

Purtell, L.P. (1992) Turbulence in complex flows - a selected review. *AIAA* 92-0435.

Robinson, S.K. (1991) Coherent motions in the turbulent boundary layer. *Ann. Rev. Fluid Mech.* 23, 601.

Sreenivasan, K.R. (1991) Fractals and multifractals in fluid turbulence. *Ann. Rev. Fluid Mech.* 23, 539.

LAMINAR BOUNDARY LAYERS AND SEPARATION

Susan N. Brown
Department of Mathematics
University College London
Gower Street
London WCIE 6BT
England

INTRODUCTION

The theory of laminar boundary layers represents an extremely exciting and rewarding field of study. It is a branch of fluid mechanics that embraces rational, even at times rigorous, analytical work, minor and major numerical investigations and production runs, and painstaking and dramatically revealing experiments. Thus workers in any one of these branches have the anticipation and satisfaction of comparing their results with those of the others, and very profitable many of these comparisons and subsequent collaborations have turned out to be.

To a modern analytical fluid dynamicist looking for a challenging problem on which to exercise his skills in asymptotics the title of this section may be permitted to encompass any study of the Navier-Stokes equations involving a limiting process that results in a region of the flow in which changes are rapid. In such a region the equation may simplify so as to permit an exact solution, or at least a solution whose salient features, in particular breakdown or singularities, may be identified and described. Such licence allows the investigator to trespass on the domain of almost all the articles in this collection except those in which the flow is turbulent. However, to contain the field, attention will be limited to those topics which have developed, directly or otherwise, from Prandtl's (1908) concept of a viscous boundary layer adjacent to a body in a flow at high values of the Reynolds number R, the majority of the flow field being essentially inviscid.

The most obvious and dramatic developments of Prandtl's theory have been, and are still being, applied to aerodynamics. It is here that, quite rightly, much funding has been directed, and the concepts of faster and safer conventional flight, and of space flight, have a glamour and challenge that have appealed to many able fluid dynamicists. Their results, from universities, private companies and government funded research establishments, have been made available through the literature and by collaboration for analysis

43

and comparison. Such collaboration is extremely valuable, one obvious result being that the benefits of expensive equipment can be spread and capitalised on. The phenomenal development in flight technology this century illustrates not only that large sums have been invested, but that they have been invested with results that must have been beyond expectation. The subject has attracted many of the ablest theoreticians, computational fluid dynamist and experimentalists, particularly in the US and the UK, and in Russia where on many occasions investigations have preceded, or independently coincided with, those in the west. Apart from the obvious one to flight, the theory of laminar boundary layers has application to many problems of environmental and industrial importance, in oceanography and meteorology, rotating and stratified fluids, medicine and confined flows, turbo-machinery and acoustical problems, to name but a few. Its interdisciplinary tendons are everywhere. Of particular interest are the recent advances in the study of the stability of steady and unsteady flows, and the genesis of an explanation of the early processes of transition.

The boundary layers to be considered here can be divided roughly into five types and in each case the fluid may be compressible or incompressible. Firstly, there is the classical non-interacting boundary layer which responds to a prescribed gradient. This theory, still very much alive as unsolved problems remain, is a direct development of the seminal ideas of Prandtl in the early part of the century. Secondly, there is the far reaching Messiter-Stewartson-Neiland triple-deck theory, initiated in the late 1960's, in which the pressure gradient is related to the displacement thickness and is to be determined as part of the calculation. Early ideas of Lighthill led to the eventual consistent formalism of the multi-structured layer. Both these theories have the advantage that they apply to many practical examples that are amenable to analytic treatment. However, as more powerful high-speed computers have become available, the possibility has arisen from horizons to be widened to include large-scale computation of both unsteady and three-dimensional flows. This has provided further opportunity for both the computational fluid dynamicist and the numerical analyst to confer with the experimentalist who always has to contend with these effects. Thus three more types of attack of boundary-layer character may be identified. These are, respectively, the study of interacting flow in which the equations are considered to hold right across the boundary layer, the solution of parabolised Navier-Stokes equations to permit integration in a forward (streamwise) direction, and computation of the solution of the Navier-Stokes equations themselves at high but finite values of the Reynolds number. The difficulties of such computations, even at moderate values of R, should not be underestimated.

The results are extremely valuable as they give opportunity for comparison not only with analytical 'convention' boundary-layer solutions, but also with experiments.

Discussion of the above five types of flow will be divided into two fairly broad summaries, namely steady flows and unsteady flows. Each summary will include, as appropriate, two or more dimensions, compressibility, and analytic, numerical and experimental approaches. Emphasis will be placed on topics that are felt to be at the forefront of world science, and those where extra funding is desirable to stimulate progress, or help to complete investigations already initiated. Much of the discussion will concern the phenomenon of separation, a meaningful precise definition of which is probably not common to all workers in the field. To the theoretical or computational fluid dynamicist it may mean that his equations have become singular and are no longer adequate to describe the flow, to the experimentalist that there has developed a small or massive region of reversed flow, and to the pilot that his lifting surface or helicopter blade has entered the stall regime. Its prediction and control is of vital importance to all these investigators.

STEADY LAMINAR FLOWS AT HIGH REYNOLDS NUMBERS

Although the concept of the classical laminar boundary layer is now in its ninth decade, it is by no means sterile or dead. The Oxford University Press publication 'Laminar Boundary Layers' of the 1960s now looks very dated, so extensive and rapid has been the development of the subject. A classical boundary layer is defined to be one in which the pressure gradient is prescribed and the velocity components are to be determined to match with a pre-determined outer inviscid flow. The wall may be impermeable or have a prescribed blowing velocity. The equations, in two dimensions at least, are parabolic, and conditions must be prescribed at an initial station. Regions of reversed flow cause considerable difficulty as information is now carried upstream.

The classical two-dimensional boundary layer in an adverse pressure gradient has essentially no way of responding to downstream conditions so reacts by developing a singularity at the separation point, the well-known Goldstein singularity. For incompressible flows at least, its form is well understood, and it is surprising how small an adverse pressure gradient is necessary to initiate separation. For a stationary wall it is usually sufficient to identify separation with the point of zero skin friction and the departure of the streamline from the surface. Thus for the computational fluid dynamicist of such a two-dimensional layer his task is over when the separation point occurs, as his equations are inadequate to describe the flow further downstream. However, for a study of a three-dimensional boundary layer the situation is more complicated and there are still many questions to be addressed and fully an-swered. In this case not only is information from the wall and the edge of the boundary layer felt at all points along the normal, but at each station the hyperbolic nature of the equations is the directions parallel to the body lead to the existence of a wedge of influence bounded by the limiting directions of the velocity vector at all positions from the wall to the outer parts of the layer. This is known as Raetz's principle and is built into many modern approaches for computing three-dimensional boundary-layer solutions.

The procedure can perhaps best be illustrated by consideration of the flow past a prolate spheroid whose incidence is gradually increased. At zero incidence an integration from the forward attachment point leads to a sep-aration line that is symmetrically situated, and is generally understood to be an envelope of limiting streamlines and described by an analysis similar to that of a Goldstein singularity. As this line becomes tangential to the in-viscid streamline through the attachment point the separation line develops an arrow-like vertex and a wedge-shaped region of inaccessibility is formed. There does not seem to be agreement as to the form of the breakdown occur-ring on the leeward side of the wedge, and further three-dimensional integra-tions are highly desirable. Here is an area in which numerical and analytical work ideally complement each other. What, for example, is the structure of the singularity in the immediate neighbourhood of the vertex of the arrow? Is this some type of marginal separation? It is possible that these ques-tions could be answered by an integration for a case that differs little from a two-dimensional problem. Limiting streamline patterns can be found using integral methods which give insight into the possible type and distribution of the geometrical singularities; although these methods often give valuable qualitative information, they cannot be used to address the question of the precise form of the breakdown.

Of considerable practical importance is the avoidance of massive regions of reversed flow leading to vortex-dominated regions and to dynamic stall. Boundary-layer control by suction or blowing is a possibility. Compressibility adds further complexity to the problem although of degree rather than kind for subsonic flow. For supersonic flows the computational fluid dynamicist has the extra difficulty of the transonic regime with its equations of mixed type and the necessity of developing successful techniques for shock-capturing and for shock-wave/boundary layer interaction. However, numerical simula-tions of flow past a complete aircraft is no longer an impossibility.

Even quite simple boundary-layer flows, for example the incompressible Blasius flow past a finite flat plate, or the supersonic flow ahead of a ramp, contain regions in which it is not sufficient to consider the pressure gradi-ent to be prescribed, i.e. the boundary layer to be classical. Such a region occurs at the trailing edge of the flat plate and in the neighbourhood of the separation point ahead of the ramp, and the resulting interactive equations, the derivation of which in the 1960s was a great breakthrough, provide the means by which the parabolic inner-layer equations obtain information about the geometrical conditions downstream. Triple-deck theory, a rational analysis of these high Reynolds number flows, continues to prove remarkably fruitful in the elucidation of their properties and has shown, in many cases, excellent agreement with experiments carried out at quite moderate values of the Reynolds number. For the supersonic flow ahead of the ramp, smooth separation is predicted as is the plateau pressure and the final rise to the downstream pressure level. Various techniques have been developed for the successful handling of the region of reversed flow for which it is necessary to incorporate as asymptotic downstream boundary condition. For the flat plate at incidence stall angles can be predicted.

For subsonic flow the Goldstein singularity cannot be removed by an interactive triple-deck approach except in the situation of marginal separation, i.e. that in which the skin fraction becomes zero at a point and then immediately recovers. There are still open problems where marginal separation theory might be of application but has not been fully exploited. These are isolated examples described of theoretical application to three-dimensional flows but very little comparison with computation or experiment, and there are still outstanding problems that might be looked at in two dimensions, even in steady flow. An alternative structure of separation with the inviscid flow given by Kirchoff free-streamline theory has been proposed and attempts made to embed this in a complete picture of flow past a cylinder at high Reynolds number. Although it appears that, instead, the Sadowski vortex describes the high Reynolds number limit in the case of the cylinder, other geometries deserve study.

The question of the breakdown of steady interactive boundary layers is itself of interest. Possible contexts are flow over humps or in channels with rigid or compliant walls. There is always the possibility of the singularity being removed in a further interactive structure.

The current upsurge of interest in high-speed flight has led to a revival of studies in hypersonic flow. Although formulated before the triple-deck era, the problem of hypersonic flow past a sharp-nosed body with an attached shock for example, is an interactive problem. The boundary layer behind the shock is divided into an inviscid and a viscous shock layer, the unknown surface between them being surrounded by an adjustment layer. Accurate numerical solutions of even the simplest problem, with the assumption of a perfect gas and Chapman's or Sutherland's law for the viscosity-temperature relation, are difficult to obtain. The upstream influence present means that it is necessary to apply a downstream boundary condition, firstly to ensure departure from the strong-interaction solution that holds in the neighbourhood of the leading edge, and then to prevent the solution from locking on to a branch that terminates in a singularity at a finite distance. Highly cooled walls and heat transfer are an added complication. Further computations would be desirable as would a study of real gas effects and of the region where the fluid can be no longer regarded as a continuum. It may be that a parabolised Navier-Stokes approach would have some advantages in this situation.

A three-dimensional effect that has proved to a certain extent amenable to analysis is that of the collision of boundary layers such as occurs at the equator of a rotating sphere, on the leeward side of a cone at incidence, or in a curved pipe, to name a few examples. Sometimes the phenomenon can be interpreted as a lift-off of the main boundary layer as the secondary boundary layers collide underneath it. Sometimes the boundary layer must be turned inviscidly as in the plume at the upper vertex of a heated sphere, or at a wing-fuselage junction where the terminal conditions of one boundary layer must supply the initial conditions for another. Although in some cases the form of the singularity is known, the possibility of interaction in this situation does not seem to have been fully exploited, nor does the interpretation of this sort of collision as a termination line of a general three-dimensional boundary layer.

Finally, in this sub-section on steady boundary layers, brief mention will be made of those that occur in rotating and stratified fluids. In the former, horizontal Ekman and vertical Stewartson layers follow as a consequence of the Taylor-Proudman Theorem, and oceanographers and meteorologists emphasise different goals from those of the aerodynamicist. The theory of rotating flows has important industrial applications as does that of convective flows in cavities with sidewalls. Interesting breakdown properties can occur in the horizontal and vertical boundary layers in the latter, connected as they are by regions of multi-directional flow. Considerable difficulties are likely to be associated with the computation of such flow.

UNSTEADY LAMINAR FLOWS AT HIGH REYNOLDS NUMBERS

Unsteady laminar boundary layers are fully expected to represent a fundamental stage on the path to transition to turbulence. This, in addition to the analytical, computational and experimental challenges presented by the study, goes a long way towards explaining the tremendous progress in this field in the last twenty years. Before this time very little was known apart from the few exact solutions for impulsively-started flows or oscillating plates. Indeed it was not generally believed that the unsteady boundary-layer equations would develop a singularity in a finite time.

It is difficult to draw a dividing line between the topics loosely covered by the term unsteady boundary layers in which the time dependence is due to forcing, and stability theory in which the oscillations are generally free. In modern work similar techniques are applied to either, and unsteady boundary layers are themselves examined for their stability, possibly to oscillations of a very different frequency from those of the underlying flow. Non-linearity is a vital consideration. An important realisation is that singularities now are most likely to be in the interior of the fluid rather than on the wall; an example of this, although steady in the appropriate Newtonian frame, is the Sychev singularity in the flow over a moving wall, the analogue of the Goldstein singularity for a fixed wall.

When the flow is unsteady it becomes feasible, indeed in many situations advantageous, to work in Lagrangian coordinates and follow the fluid particles, rather than in Eulerian coordinates and sit and watch the fluid go by. A dramatic illustration of this is the van Dommelen-Shen calculation and description of the breakdown of the boundary layer on an impulsively-started circular cylinder, one of the earliest demonstrations of a finite-time singularity in a boundary layer. This also confirmed the existence of the Moore-Rott-Sears (MRS) singularity, originally proposed for Eulerian coordinates, occurring at a point in the flow at which the shear vanishes and which moves with the speed of the fluid there. More recent work has extended this to three dimensions and to include the effects of compressibility. Thus the generic characteristics of three-dimensional unsteady separation have been identified, but the complexity, and indeed paucity, of results even in steady three-dimensional separation indicates that this is an area still requiring large-scale investigation.

The merits of two systems of viable coordinates extend from the theoretician to the computational fluid dynamicist. Opportunities now exist for the cross-checking of theories or of numerical results. It may be advantageous to change approaches as the time-step increases. The intrinsic beauty of the Lagrangian formulation is the fact that, in these coordinates, the momentum boundary-layer equation exhibits no singularity at separation. The singularity is confined to the continuity equation which predicts an unbounded normal coordinate as a fluid particle is 'squeezed out' of the boundary layer.

After the question of whether a classical unsteady boundary layer can break down in a finite time has been answered in the affirmative, it is natural to ask the same question about an interactive layer. The answer again is yes in the sense that the equations have an analytically acceptable structure in which a singularity is generated at a finite time. As in the classical situation it moves with a representative fluid speed, and in a limiting case in which the pressure is equal to the displacement thickness, these quantities satisfy an integro-differential equation with break-up properties. In the inviscid limit it becomes Burger's equation with the consequential shocks. When a more general case was studied it became clear that as the singularity was approached the skin friction became large and negative with the pressure gradient (or even the pressure) exhibiting a discontinuity on the scale considered. Numerical results, for separation induced by a potential vortex near a wall in an impulsively started flow, have showed the viscous dominated regions pushed out of the boundary layer. The analysis predicts vortex formation with closed streamlines and provides a possible path to a description of transition. Applications are expected to be to flow over pitching airfoils, to bluff-body shedding and in turbomachinery. It may even be that the eruptions can be interpreted as the genesis of the hairpin vortices familiar in turbulent boundary layers. This is an exciting development where the disciplines of analysis, computation and experiments complement each other and further work could bring worthwhile results. An extension to three dimensions requires integration of a non-linear partial differential equation of mixed type, with boundary conditions partly to be determined as a requirement of consistency of the solution. This is a mathematical and computational challenge.

Associated with finite-time breakdown as described above is the question of whether further interaction can remove the singularity. Examples are known in which further interaction actually increases the strength of the singularity.

As the Mach number increases the complexity grows. Shocks form and are present in both external and internal flows. Self-sustained oscillations occur and there immediately arises the question of stability. Instability may be temporal or spatial depending whether the unforced oscillations are assumed to be harmonic in space or time. A positive temporal or spatial growth rate implies instability, or which spatial instability may be absolute or convective, a distinction which superficially seems simple enough. In appropriate moving coordinates these are the same, but the distinction in initial value problems requires great care in tracing the roots of the dispersion relation in the complex plane. Except in the simplest of cases this, in itself, turns out to be an interesting mathematical problem.

The study of receptivity in a laminar boundary layer could serve as an important introduction to the same problem in a turbulent one. Both experiments and theory have been carried out to determine how the very long wavelengths of imposed free-stream disturbances are reduced to the very short Tollmien-Schlichting wavelengths of boundary-layer disturbances. The development of these quasi-steady perturbations can be attributed to non-uniform or non-parallel mean-flow effects, caused either by the slow growth of the viscous boundary layer or by some geometrical anomaly. Far downstream, the flow is described by a combination of normal modes with the arbitrary multipliers, the coupling coefficients, determined. The relation of these with an alternative set of eigensolutions, proposed for this far-downstream behaviour some years ago, is not yet fully understood.

The stability of both steady and unsteady boundary layers is of fundamental importance in that it seems plausible that, through such investigation, an asymptotically rational description of the onset of transition may be achieved. Such, no doubt, is the dream of the worker in this field. The last decade has seen an enormous amount of effort concentrated on stability theory. This has been assisted by the development of high-speed computing facilities, and the realisation of the power of the Navier-Stokes equations to describe linear, weakly nonlinear and fully nonlinear instability. Analysts have been prepared to derive complex expansions and multi-structures to describe perturbations that are fundamentally of Rayleigh or of Tollmien-Schlichting type. Viscous, time-dependent and nonlinear critical layers have been studied for neutral and near-neutral instability. Nonlinear amplitude equations have developed from that of the early Stuart-Watson theory to integro-differential equations of a kind that also occur in other branches of theoretical physics. Some of these are history-dependent, some have built-in upstream influence, and many exhibit finite-time or finite-position breakdown.

A very interesting recent development in stability theory is that of vortex/wave interaction. To understand this, one may imagine a steady two-dimensional boundary layer, classical or interactive, having a velocity profile that is developing an inflection point as it proceeds downstream. At some station the Rayleigh equation associated with this profile admits a non-zero neutral eigensolution. From this point on, the flow develops in a three-dimensional fashion with, for example, periodicity in the span-wise direction. If high frequency temporal oscillations are prescribed, then a short wavelength streamwise disturbance develops, whose quadratic interaction in the critical layer determines the order of magnitude of the wave-pressure perturbation. The amplitude of this pressure perturbation is related to the transverse vortex (steady) velocity component through a jump condition on the stress across the critical layer. The vortex components satisfy equations of boundary-layer type, and the wave pressure continues to satisfy Rayleigh's equation with the streamwise velocity component of the vortex as forcing. Thus the perturbation, which remains neutral, drives and is driven by the mean flow as both develop downstream. This is an exciting theory of which solutions so far are restricted to a successful description of the initiation of the process. Numerical solutions of the further development are required as well as an analysis of this flow far downstream. Of particular interest is the possibility of breakdown at a finite station. This breakdown may be characterised by a critical layer that develops from being linear and fairly passive to unsteady and/or non-linear. Cats-eyes are a feature of such critical layers, familiar in meteorology and rotating and stratified flows.

When the boundary layer is developing on a curved surface a possible form of spatial instability is due to the formation of Gortler vortices. These are of practical importance in aerodynamics and the flow overwings and turbine blades. As they are steady they perhaps belong more properly in section 2, but the techniques to describe them are similar to those of temporal stability theory. Spanwise periodicity is assumed with a streamwise-dependent wave number that for free disturbances must be determined in terms of the Gortler number. The governing perturbation equations are parabolic, and hence there is no unique neutral curve as the solution depends on the initial conditions; this is in contrast to Tollmien-Schlichting waves where the governing equations are elliptic. Non-linear studies demonstrate the control of the mean flow by the vortex structure which is confined, at large distances downstream, by two shear layers. Temporal instability studies predict that the outer of these is the first to break down. Steady receptivity investigations are also of interest and have been initiated, as has a small amount of research into Gortler instabilities in a three-dimensional boundary layer.

A final comment on high-speed flow ends this subsection. It is known that the large Mach number flow past a flat plate supports acoustic modes, confined between the wall and the sonic line, and a vorticity mode that is concentrated in the neighbourhood of the adjustment layer. Present theories do not describe sustained oscillations of the leading-edge shock, a problem of practical importance as well as theoretical interest. In addition, perturbations of wake boundaries and of separation positions merit further study in both high and lower speed free-streamline flows.

SUMMARY

Although boundary layers, in the sense of regions of rapid change, are prevalent in almost all branches of theoretical physics, it is in fluid mechanics that the concept is most familiar. The rewards of careful analysis, whether it be theoretical, numerical or experimental, of these regions are many, including often that of an understanding of the flow in a parameter range wider than that strictly implied by the limiting process under consideration.

A few decades elapsed between Prandtl's famous paper proposing the existence of a viscous layer in the neighbourhood of a body in a fluid of small viscosity, and the subsequent rapid development, formal or otherwise, of the method of matched asymptotic expansions. Now it is a powerful technique with which most applied mathematicians and theoretical engineers are familiar. Laminar boundary layers, steady and unsteady, are studied in aerodynamics, where the ideas originated and have probably led to the most spectacular results, in combustion theory, meteorology, oceanography and, perhaps more recently, in medical applications such as blood flow. Such flow fields may separate, break down at a finite time, or in other ways become unstable, and again boundary-layer techniques are profitably employed to describe the instability. An important goal is a complete rational description of transition to a fully turbulent state.

As computing facilities improve, more complex problems come within the reach of the fluid dynamicist. Funds are required for full Navier-Stokes calculations, boundary-layer calculations and parabolised Navier-Stokes calculations, the last-named being a promising compromise between the former two. Methods for reversed flow calculations, with the consequent difficulty of downstream conditions and upstream influence, require further development. In addition, fluid mechanics is at base a practical subject, and well-equipped laboratories and carefully-thought-out experiments are of vital importance.

Collaboration between the various workers in this field must be encouraged. Comparison between analytical, numerical and experiment results is vital for a healthy and profitable development of this important subject. The study of laminar boundary layers is certain to prove extremely rewarding both in the immediate future and for some time to come.

SUGGESTED READINGS

Adamson, T.C. & Messiter, A.F. (1980) Analysis of two-dimensional interactions between shock waves and boundary layers. *Ann. Rev. Fluid Mech.* 12, 103-138.

Anderson, J. D. (1984) A survey of modern research in hypersonic aerodynamics. *AIAA* paper no. 84-1578.

Cheng, H.K. (1993) Perspectives on hypersonic viscous flow research. *Ann. Rev. Fluid Mech.* 25, 455-484.

Cebeci, T., Khattab, A.K. & Stewartson, K. (1983) Three-dimensional laminar boundary layers and the ok of accessibility. *J. Fluid Mech.* 107, 57-87.

Drazin, P.G. & Reid, W.H. (1981) Hydrodynamic Stability. Cambridge University Press.

Elliott, J.W., Smith, F.T. & Cowley, S.J. (1983) Breakdown of boundary layers (i) on moving surfaces (ii) in semi-similar unsteady flow (iii) in fully unsteady flow. *Geophys. Astrophys. Fluid Dyn.* 25, 77-138.

Fornberg, B. (1988) Steady viscous flow past a sphere at high Reynolds numbers. *J. Fluid Mech.* 190, 471-489.

Goldstein, M.E., Leib, S.J. & Cowley, S.J. (1987) Generation of Tollmien-Schlichting waves on interactive marginally separated flows. *J. Fluid Mech.* 181, 485-517.

Goldstein, M.E. & Hultgren, L.S. (1989) Boundary-layer receptivity to long-wave free-stream disturbances. *Ann. Rev. Fluid Mech.* 21, 137-166.

Hall, P. (1990) Gortler vortices in growing boundary layers: the leading edge receptivity problem, linear growth and nonlinear breakdown stage. *Mathematika* 37, 151-189.

Hall, P. & Smith, F.T. (1991) On strongly nonlinear vortex/wave interactions in boundary-layer transition. *J. Fluid Mech.* 227, 641-666.

Kachanov, Y.S., Ryzhov, O.S. & Smith, F.T. (1993) Formation of solitons in transitional boundary layers: theory and experiments. *J. Fluid Mech.* 251, 273-297.

Kleiser, L. & Zang, T.A. (1991) Numerical simulation of transition in wall-bounded shear flows. *Ann. Rev. Fluid Mech.* 23, 495-538.

Messiter, A.F. (1983) Boundary-layer interaction theory. *J. Appl. Math.* 50, 1104-1113.

Moretti, G. (1987) Computation of flows with shocks. *Ann. Rev. Fluid Mech.* 19, 313-337.

Rubin, S.G. & Tannehill, J.C. (1992) Parabolized/Reduced Navier-Stokes computational techniques. *Ann. Rev. Fluid Mech.* 24, 117-144.

Shen, S.F. & van Dommelen, L.L. (1982) Unsteady separation. AIAA paper no. 82-0347.

Smith, F.T. (1982) On the high Reynolds number theory of laminar flows. IMA *J. Appl. Math.* 28, 207-281.

Smith, F.T. (1982) Concerning dynamic stall. *Aero. Quart.* 33, 331-352.

Smith, F.T. & Brown, S.N. (eds) (1987) Boundary-layer separation. Springer.

Stewartson, K. (1974) Multi-structured boundary layers on flat plates and related bodies. *Adv. Appl. Mech.* 14, 145-239.

Sychev, V.V. (1980) On certain singularities in solutions of equations of a boundary layer on a moving surface. P.M.M. USSR 44, 587.

COMPUTATIONAL AERODYNAMICS

David A. Caughey
Cornell University
Mechanical and Aerospace Department
Upson Hall
Ithaca, NY 14853-7501

INTRODUCTION

Computational aerodynamics is the inter-disciplinary field directed at the development and application of numerical (computational) techniques for the solution of problems in aerodynamics. The field is concerned with the development of mathematical models for the description of fluid phenomena, the design and analysis of discrete numerical approximations for the solutions to the problems thus posed, and their implementation and use on digital computers. Because of the technological importance of aerodynamics, efforts in this field have, in many respects, led the development of the more general field of computational fluid dynamics (CFD).

The past two decades have seen a great deal of activity in the field of computational aerodynamics. Much of this activity has focused on the development of methods to compute steady flows past aerospace vehicles at low to moderate speeds (say, from incompressible to low supersonic Mach numbers). The principle fluid mechanical model has consisted of the Reynolds-Averaged Navier-Stokes equations for a perfect gas (or, in some cases, their inviscid approximation, the Euler equations) with a turbulence model where appropriate. Currently available computer codes are capable of predicting the overall features and integrated forces for relatively complex geometries, so long as there are not too large regions of separated flows, although these computations are still quite expensive to perform. For flows with large regions of separation, the methods perform less well; this is thought by most researchers to be due to the limitations imposed by the turbulence models used. Also, the prediction of laminar-to-turbulent boundary layer transition remains a stumbling block to the prediction of flows for which the vehicle performance is sensitive to transition location. At higher Mach numbers, the chemistry of non-equilibrium, high temperature air must be taken into account and, for the analysis of propulsion systems the chemistry, in some cases non-equilibrium, of the combustion process must also be modelled. At higher altitudes (implying lower Knudsen numbers) the Navier-Stokes equations themselves must be extended or abandoned.

The problems associated with geometrical complexity have driven discretizations from structured grids (boundary-conforming grids having an underlying Cartesian structure, or i,j,k ordering) to multi-block implementations of structured-grid algorithms, and to completely unstructured grids consisting of tetrahedral cells of essentially arbitrary connectivity (equivalent to the meshes for linear basis functions used by the finite-element community). As computational power continues to increase, unsteady problems will receive more attention, as will the coupling of computational models for various phenomena (such as structural mechanics, aero-acoustics, and electromagnetic signature prediction, for example).

Research in the field of computational aerodynamics can be subdivided into the areas of:

1. Grid generation;

2. Algorithmic development;

3. Model development;

4. Applications.

As with any categorization, the divisions are somewhat arbitrary, and there is considerable overlap and interaction between these areas. For example, algorithms are intimately connected with the grids on which the solution is to be described (and if the grid is to be dynamically adapted to features of the evolving solution, the two are inextricably linked), the success of applications to flows of ever increasing geometrical and fluid mechanical complexity drives the requirements for grid generation and feeds back into model development, and so on. These remarks notwithstanding, I believe this categorization will be useful for the discussion of research directions in the field.

GRID GENERATION

In spite of intense development in the past decade or more, the generation of grids for complex three-dimensional flow problems remains a major item pacing the use of computational techniques and forms the most labor-intensive aspect of the application of these techniques to engineering problems. Continued efforts are needed to make this process more efficient, including the development of efficient and general techniques for the definition of complex three-dimensional surfaces. Graphical interaction with the grid generation tools is likely to become more and more important, in concert with the development of interactive tools for the visualization and

interpretation of computed three-dimensional fields. The application of artificial intelligence (smart systems) to the grid generation problem may prove
to be a promising approach.

Major advances are needed in techniques to allow the grid to adapt to
features of the solution, including the development of criteria to drive the
adaptivity and to allow assessment of the quality of the resulting grid. It
may, of course, prove impossible to judge the quality of the grid independently of the quality (accuracy) of the solution. There is also the question of
whether it is better to require robustness of the flow solver capable of dealing with arbitrarily stretched and/or discontinuous grids, or to build into
the grid sufficient smoothness and/or regularity based upon properties of the
flow solver. Optimally adapted grids for fluid mechanical problems are unlikely to be determined on the basis of geometry alone, so, e.g., the use of
wave modelling characteristics to adapt the grid may result in triangulations
which are not necessarily optimal from a purely geometric point of view (e.g.,
Delaunay triangulations).

ALGORITHM DEVELOPMENT

An important goal of algorithmic research will be the development of
truly multi-dimensional extensions of the Riemann-based schemes developed
so successfully for the Euler equations. These will be especially important
for unstructured grid applications, since a triangular grid guarantees that
at least one cell face will not be well-aligned for a (locally-one-dimensional)
Riemann problem. If solution adaptivity produces large gradients in cell size,
it will be necessary for the flow solver to handle these without degradation
in accuracy, and continued improvement of algorithms capable of treating
highly-stretched grids, such as those needed to resolve high-Reynolds number
boundary layers, will be needed. The development of higher-order accurate
schemes, including schemes which have small phase, as well as amplitude,
errors will be important for such applications as aero-acoustics and for the
prediction of transition. There will be an ever greater emphasis on robustness – including insensitivity to grid quality – as the aerodynamic analysis
becomes embedded in a larger system which is coupled to structural analysis,
aero-acoustic or electromagnetic signature codes, and/or optimization codes
for design.

As scalable parallel architectures overtake large vector-processing machines in total throughput, development of algorithms suitable for these
(primarily) distributed memory machines will become increasingly important. Some implementations of explicit time-marching algorithms may be
relatively straightforward; implicit schemes will require more ingenuity if
they are to remain competitive. The interplay between the data structures

implied by the grid generation, the memory assignment among the processors, and its access by the flow solver will require a truly interdisciplinary approach to produce effective algorithms.

The extension of algorithms to treat non-equilibrium hypersonic flows (and chemically reacting flows with finite-rate kinetics) will also be an important emphasis.

MODEL DEVELOPMENT

The Navier-Stokes equations are generally accepted to be an accurate model for aerodynamic phenomena, except at very low fluid densities. The principal issues in modelling thus center on the effects of turbulence and of non-equilibrium thermodynamics. Direct Numerical Simulations (DNS) and Large Eddy Simulations (LES) of turbulent flows will continue to provide important information for the building and testing of turbulence models, but are not likely to be useful for practical engineering computations in the foreseeable future.

Turbulence models continue to be the weak link in many computational aerodynamic analyses, especially for flows with significant regions of separation, flow separations caused by shock-wave/boundary-layer interaction, and for unsteady flows. Improved models for engineering use are needed for these situations, and attention must be paid to their implementation on unstructured grids (where a natural boundary-layer-like coordinate is absent).

Models for the prediction of boundary-layer transition location will be critical to the success of Laminar Flow Control (LFC) technology as well as for the prediction of performance of the National Aero-Space Plane (NASP) in some parts of its flight envelope.

The development of models for non-equilibrium thermodynamic and chemical phenomena at high temperatures will be important, including homogeneous multi-temperature models for vibrational and rotational non-equilibrium as well as surface interaction models to predict ablation and spallation. Experiments to validate these models will be particularly important.

APPLICATIONS

Application to the design of aerospace vehicles is the ultimate motivation for the development of computational aerodynamic techniques. Work in the past has focused on inverse formulations and constrained optimization techniques, and much remains to be done here, especially for multi-point design problems. The extension to multi-discipline design, including structural considerations (such as flutter), aero-acoustic considerations, and constraints on electromagnetic signature will require extremely robust aerodynamic algorithms, and may require the development of linearized models for various phenomena. The increasing use of computational techniques for aero-acoustic applications and their continued development for unsteady flows will be major areas of activity. The use of expert systems to formulate and implement design constraints remains a largely un-explored area.

Hypersonic flows and other reacting flows with finite-rate kinetic effects will be an important area with applications to conventional energy conversion processes as well as to both the National Aero-Space Plane and to elements of the Strategic Defense Initiative (SDI). For these high-energy flows for which ground facilities are inadequate for testing in many of the parameter ranges of interest, the validation of computational techniques becomes especially important; it will be important independently to develop sensitivity analysis to assess uncertainties due both to numerical effects and to modelling assumptions and/or approximations.

The use of visualization must continue to be developed, with the emphasis shifting from the use of three-dimensional color graphics to illustrate expected features to more quantitative presentations and to the development of techniques, perhaps based on artificial intelligence and pattern recognition, to detect unexpected features in the results of computations.

SELECTED REFERENCES

Caughey, David A. & Hafez, Mohamed H. Eds., (1994) Frontiers of Computational Fluid Dynamics – 1994, John Wiley & Sons, Chichester.

Glowinski, Roland & Pironneau, Olivier (1992) Finite Element Methods for Navier-Stokes Equations, Ann. Rev Fluid Mech., Vol. 24, pp. 167-204.

Hardin, Jay C. & Hussaini, M. Y. Eds., (1993) Computational Aeroacoustics, Springer-Verlag, New York.

Jameson, Antony (1989) Computational Aerodynamics for Aircraft Design, Science, Vol. 245, 28 July, pp. 361-371.

Jameson, Antony The Numerical Wind Tunnel – Vision or Reality? AIAA Paper 93-3021, 24th Fluid Dynamics Conference, Orlando, Florida, July 6-9, 1993.

Mehta, Unmeel B. (1991) Some Aspects of Uncertainty in Computational Fluid Dynamics Results, Trans. ASME, J. Fluid Engineering, Vol. 113, No. 4.

Roe, P. L. (1986) Characteristic-Based Schemes for the Euler Equations, Ann. Rev. Fluid Mech., Vol. 18, pp. 337-365.
Weatherill, N. P., Gaither, K. P. & Gaither, J. A. (1995) Building Unstructured Grids for Computational Fluid Dynamics, CFD Journal, Vol. 1, No. 4, pp. 1-28.

PARTICULATE FLOWS AND SEDIMENTATION

Robert H. Davis
Department of Chemical Engineering
University of Colorado at Boulder
Engineering Center, ECCH 1-43
Campus Box 424
Boulder, CO 80309-0424

INTRODUCTION

Particulate flows and sedimentation occur in a wide variety of natural and industrial processes. Familiar examples include sediment transport by rivers, pulp and paper processing, blood flow, pumping and clarification of waste water, and paint application. Further examples from critical technology areas include advanced materials such as composites and coatings, energy applications such as coal slurries and fluidized catalyst particles, and biology applications such as the processing of cell cultures and the selective separation of malignant cells from healthy ones.

A chart showing typical sizes for several types of particles is given in Figure 1. In general, suspended particles are smaller than approximately 100 microns (0.1 millimeter) in size, since larger particles rapidly settle out of suspension due to gravity. The Reynolds number for flow around suspended particles is usually small compared to unity, and so inertial effects may be negligible relative to viscous forces. Colloidal particles are those smaller than about one micron in size. Because of their large surface area to volume ratio, colloidal particles are subject to Brownian motion and attractive and repulsive interparticle forces.

The motion of small particles in viscous flows represents one of the earliest classes of problems studied in fluid mechanics, dating at least from the classical analyses of G. G. Stokes in 1851 on the viscous drag on pendulums and of A. Einstein in 1905 on the effective viscosity of dilute suspensions. Much of the early fundamental research in this area is reviewed in the text by Happel and Brenner (1965). A renewed and growing interest in the study of particulates and colloids is evidenced by several recent texts which are included in the list of references. Much of the recent research has focused on predicting macroscopic properties of suspensions from the microscopic behavior of individual and interacting particles in the suspending fluid. This

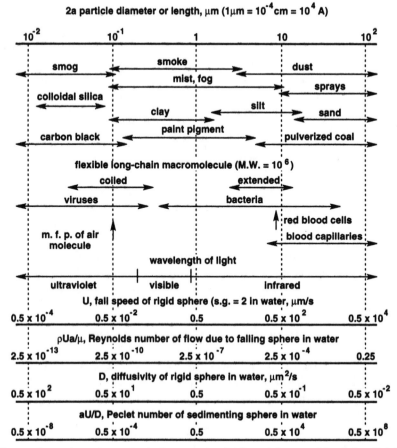

Figure 1: Orders of magnitude for typical colloids and fine particles (after Batchelor, 1976).

research field is now called microhydrodynamics, a term coined by G. K. Batchelor.

The current and anticipated future research activity in microhydrodynamics derives from synergistic advances involving both new applications of particulate and colloidal suspensions, and new theoretical and experimental tools for the fundamental study of such systems. This report briefly reviews several of these advances, together with their implications for possible future research directions.

FUNDAMENTAL RESEARCH DIRECTIONS

Rheology

During the 1960's and 1970's, there was considerable research activity on the rheology of dilute suspensions (Brenner, 1970, 1974). Much of this research focused on the interplay between the orienting effects of flow and the disorienting effects of Brownian motion, with considerable progress made on predicting the bulk rheological properties of dilute suspensions of spheroidal particles from the resulting distributions of particle orientations. More recent research has concerned semidilute and concentrated suspensions in which particle-particle interactions affect the rheological behavior of suspensions of spheres and suspensions of rod-like particles (Shaqfeh and Koch, 1990; Milliken et al., 1989).

Research Needs

A major challenge involves the elucidation of the coupled relationship between the microstructure of the particle distribution and the bulk flow behavior. Further research is also needed on polydisperse suspensions and on mixtures of high and low aspect particles, as found in the processing of fiber-reinforced composites. Other considerations include the use of non-Newtonian suspending fluids, with the possibility of evaporation or cooling, and particles with internal degrees of freedom (such as polymer molecules or flexible fibers or cells).

Particle Migration and Hydrodynamic Diffusion

Individual particles in a suspension undergo fluctuations in their motion due to interactions with other particles. These cause a net particle migration from regions of high concentration to regions of low concentration, and from regions of high shear to regions of low shear (Leighton and Acrivos, 1987a). This is referred to as hydrodynamic diffusion, because it results from random hydrodynamic interactions between particles, and has been observed in both sheared and sedimenting suspensions (Phillips et al., 1992; Davis and Hassen, 1988). Hydrodynamic diffusion plays important roles in viscous resuspension and transport of sediments and slurries, and it influences rheological properties by affecting the distribution of particles in space. Shear-induced particle migration is a likely cause of inhomogeneities in composite materials. Minimizing these inhomogeneities in structural materials and solid-rocket fuels is of critical importance for safety and performance reasons.

Research Needs

Particle migration and hydrodynamic diffusion are not well-understood, with only limited data available for monodisperse suspensions of spheres (Leighton and Acrivos, 1987b; Ham and Homsy, 1988). Considerable future research is expected to involve a variety of particle and flow types. An understanding of the role that small repulsive forces and surface roughness elements play in causing irreversibilities in the particle interactions is needed. Of additional importance is hydrodynamic diffusion among particles of different sizes and shapes. Particle migration and hydrodynamic diffusion in particulate flows and sedimentation represent a fruitful area for theoretical as well as experimental research.

DYNAMIC SIMULATION

During the past decade, numerical techniques have been developed for simulating particulate flows and sedimentation. These techniques are similar to molecular dynamics simulations of gases and liquids and have been given the name Stokesian dynamics (Brady and Bossis, 1988). Particles are allowed to interact through hydrodynamic forces, Brownian forces, interparticle attractive and repulsive forces, and external forces such as gravity. For dynamic simulations, the evolution of the suspension microstructure is predicted by solving a matrix which describes the motion of the interacting particles at a given time, and then using a time-stepping procedure to determine the particle locations and orientations at the next time step. Stokesian dynamics has considerable promise for providing quantitative information on both microscopic and macroscopic properties of particulate suspensions subject to various flows. Because it is computationally intensive, however, early work has been limited to simulations of small numbers of spherical particles, often constrained to a monolayer. These limitations are being relaxed as high-speed, parallel computers become available.

Research Needs

Considerable future research is expected to involve fully three-dimensional simulations of large numbers of particles, polydisperse suspensions, and the addition of external fields, such as are important in electrorheological fluids. Advances in boundary-collocation and other numerical techniques have also made possible the study of interactions of particles with complicated shapes. It is hoped that these advances will allow dynamic simulation techniques to provide a better, predictive understanding of the rheological and diffusional properties of concentrated, complex suspensions such as are important in composite materials and coatings for electronics applications.

SEDIMENTATION

Sedimentation is commonly used in the petrochemical and environmental industries to separate particles from fluid or to separate particles with different settling speeds from each other. It also forms the basis of some laboratory techniques for measuring the size distribution in a particulate dispersion. Fundamental research on sedimentation in recent years has included hindered settling due to particle interactions, sedimentation of polydisperse suspensions, and the use of inclined channels to greatly enhance the rate of separation by sedimentation (Davis and Acrivos, 1985).

Research Needs

Further research is needed on the sedimentation of nonspherical particles, the development of suspension microstructure during sedimentation, and hydrodynamic diffusion in sedimenting systems. Another important problem that is not well-understood involves gravity flow of a concentrated sediment layer down an inclined surface, as the behavior of such flows is very sensitive to the interplay between rheology, hindered settling, and hydrodynamic diffusion within the concentrated layer (Nir and Acrivos, 1990). Of further interest is the coupling between sedimentation and aggregation.

AGGREGATION

Particle aggregation in suspensions is desirable for enhancing particle separation from fluid and for the production of granules, but is undesirable if a stable, uniform dispersion of small particles is required. In order for particles in suspension to aggregate, they must be brought close together by Brownian motion, differential sedimentation, or bulk flow, and they then must experience an attractive force which is sufficiently strong to overcome any repulsive forces and the lubrication resistance to relative motion. Considerable recent research on Brownian-induced, shear-induced and gravity-induced aggregation has extended the early collision models developed by M. von Smoluchowski in 1917 to include the effects of hydrodynamic interactions and colloidal forces (Schowalter, 1984). Another area of intensive research has involved the prediction and measurement of fractal structures of colloidal aggregates (Meakin, 1988).

Research Needs

Further fundamental research is needed on aggregation of nonspherical particles and on the nature of adhesive forces and their dependence on particle surface structures. Also not well-understood is hydrodynamic and colloid processes leading to the formation of large agglomerates from smaller aggregates. A closely-related need is a better understanding of the adhesive strength of aggregates and agglomerates, and of how they can be broken apart by shear stresses. Of practical importance is the potential use of controlled aggregation of cells for selective separations in biological systems.

COLLOIDS

Since many suspensions of practical importance contain micron-sized or smaller particles, the study of colloidal phenomena in recent years has been extensive. In addition to active research on the rheology and stability of colloidal suspensions, efforts have focused on Brownian motion, electrophoresis, and other mechanisms of colloid transport (Saville, 1977; Russel, 1981; O'Brien and Ward, 1988; Anderson, 1989).

Research Needs

Because of their importance in a variety of electromechanical parts, the study of electro-rheological fluids containing colloidal particles which may be oriented by electric fields is expected to receive greater attention in the future (Gast and Zukowski, 1989). Further research is needed on the behavior of colloid particles with moderate double-layer thicknesses, and on the effects of flow on concentrated colloidal suspensions which undergo sol-gel or order-disorder transitions.

EXPERIMENTAL METHODS

Optical techniques, such as holography, velocimetry, interferometry, and light scattering, have been widely used to measure particle size, orientation, and velocity distributions in fluids (Adrian, 1991). Often, the goal is to measure fluid flow patterns, and the particles being tracked serve as passive tracers. In studies of microhydrodynamics, however, the particle motions and microstructure are of direct interest. For transparent suspensions which are either dilute or have matched refractive indices, flow visualization experiments to obtain this information may be made using video-recording and image-processing equipment. Laser Doppler anemometry techniques may also be used for measuring velocity and concentration profiles in concentrated suspensions, provided that the refractive indices of the particles and fluid are closely matched. For colloidal suspensions exhibiting structural anisotropies, optical techniques based on birefringence, dichroism and small angle light scattering may be used to infer information on particle orientation and suspension rheology (Fuller, 1990). For opaque suspensions, x-ray radiography and nuclear magnetic resonance (NMR) imaging techniques are being developed to measure particle motion and local concentration (Mondy *et al.*, 1986; Graham *et al.*, 1991).

Research Needs

Video and image processing equipment is rapidly becoming both sophisticated and inexpensive, and they are expected to be widely used to study particulate flows and sedimentation in the future. The use of compact video discs is also expected to gain popularity, as these eliminate the need for a separate step of digitizing frames from video tape. Further research is needed on improving the resolution of x-ray, NMR, and acoustic techniques for probing the microstructure of opaque suspensions.

CONCLUSIONS

The field of microhydrodynamics, involving particulate flows and sedimentation, is very active, with numerous fundamental advances made in the last few years. These advances have occurred because of the rapid development of new applications of particulate and colloidal suspensions, together with the concomitant development of new computational and experimental tools for their study. These trends are expected to continue.

REFERENCES

Adrian, R. J. (1991) Particle-imaging techniques for experimental fluid mechanics. *Ann. Rev. Fluid Mech.* **23**:261-304.

Anderson, J. L. (1989) Colloid transport by interfacial forces. *Ann. Rev. Fluid Mech.* **21**:61-91.

Batchelor, G. K. (1977) Developments in microhydrodynamics, in Koiter, W. T., ed., *Theoretical and Applied Mechanics*, North Holland. 33-55

Brady, J. F. and G. Bossis (1988) Stokesian dynamics. *Ann. Rev. Fluid Mech.* **20**:111-57.

Brenner, H. (1970) Rheology of two-phase systems. *Ann. Rev. Fluid Mech.* **2**:137-76.

Brenner, H. (1974) Rheology of a dilute suspension of axisymmetric Brownian particles. *Int. J. Multiphase Flow* **1**:195-341.

Davis, R. H. and A. Acrivos (1985) Sedimentation of noncolloidal particles at low Reynolds numbers. *Ann. Rev. Fluid Mech.* **17**:91.

Davis, R. H. and M. A. Hassen (1988) Spreading of the interface at the top of a slightly polydisperse sedimenting suspension. *J. Fluid Mech.* **196**:107-34.

Fuller, G. G. (1990) Optical rheometry. *Ann. Rev. Fluid Mech.* **22**:387-417.

Gast, A. P. and C. F. Zukowski (1989) Electrorheological fluids as colloidal suspensions. *Adv. Colloid Interf. Sc.* **30**:153-202.

Graham, A. L. *et al.* (1991) NMR imaging of shear-induced diffusion and structure in concentrated suspensions undergoing Couette flow. *J. Rheology* **35**:191-201.

Ham, J. M. and G. M. Homsy (1988) Hindred settling and hydrodynamic dispersion in quiescent sedimenting suspensions. *Int. J. Multiphase Flow* **14**:533-46.

Happel, J. and H. Brenner (1965) *Low Reynolds Number Hydrodynamics,* Prentice-Hall.

Hunter, R. J. (1989) *Foundations of Colloid Science,* Vols. I & II, Oxford Science.

Jeffrey, D. J. and A. Acrivos, A. (1976) Rheological properties of suspensions of rigid particles. *AIChE J.* **22**:417-32.

Kim, S. and S. J. Karrila (1991) *Microhydrodynamics: Principles and Selected Applications,* Butterworths.

Leighton, D. T. and A. Acrivos (1987a) Shear-induced migration of particles in concentrated suspensions. *J. Fluid Mech.* **181**:415-39.

Leighton, D. T. and A. Acrivos (1987b) Measurement of shear-induced self-diffusion in concentrated suspensions of spheres. *J. Fluid Mech.* **177**:109-131.

Meakin, P. (1988) Fractal aggregates. *Adv. Colloid Interf. Sci.* **28**:249-331.

Milliken, W. F. *et al.* (1989) Effect of the diameter of falling balls on the apparent viscosity of suspensions of spheres and rods. *PCH* **11**:341-355.

Mondy, L. A. *et al.* (1986) Techniques of measuring particle motions in concentrated suspensions. *Int. J. Multiphase Flow* **12**:497-582.

Nir, A. and A. Acrivos (1990) Sedimentation and sediment flow on inclined surfaces. *J. Fluid Mech.* **212**:139-53.

O'Brien, R. W. and D. N. Ward (1988) Electrophoresis of a spheroid with a thin double layer. *J. Colloid Interf. Sci.* **121**:402-13.

Parsi, F. and F. Gadala-Maria (1987) Fore-and-aft asymmetry in a concentrated suspension of solid spheres. *J. Rheology* **31**:725-32.

Phillips, R. J. *et al.* (1992) Constitutive equation for concentrated suspensions that accounts for shear-induced particle migration. *Phys. Fluids A* **4**:30.

Probstein, R. F. (1989) *Physicochemical Hydrodynamics*, Butterworths.

Russel, W. B. (1981) Brownian motion of small particles suspended in liquids. *Ann. Rev. Fluid Mech.* **13**:425-55.

Russel, W. B., D. A. Saville, and W. R. Schowalter (1989) *Colloidal Dispersions*, Cambridge University Press.

Saville, D. A. (1977) Electrokinetic effects with small particles. *Ann. Rev. Fluid Mech.* **9**:321-337.

Schowalter, W. R. (1984) Stability and coagulation of colloids in shear fields. *Ann. Rev. Fluid Mech.* **16**:245-61.

Shaqfeh, E. S .G. and D. L. Koch (1990) Orientational dispersion of fibers in extensional flows. *Phys. Fluids A* **2**:1077-93.

Van de Ven, T. G. M. (1989) *Colloidal Hydrodynamics*, Academic Press.

von Smoluchowski, M. (1918) Versuch einer mathematischen theorie der koagulationskinetik Kolloider Lösungen. *Z. Phys. Chem.* **92**:129-68.

ISSUES IN NON-NEWTONIAN FLUID MECHANICS AND RHEOLOGY

Morton M. Denn
Department of Chemical Engineering
University of California
Berkeley, CA 94720-9989

INTRODUCTION

The production, processing, and utilization of advanced materials often requires deformation and flow in the liquid state, followed by a solidification process. Typical examples are the extrusion and molding of high-performance polymers and polymer composites and the molding of concentrated slurries for the manufacture of ceramics. The common characteristic of these processes is that the fluid has a microstructure which ranges in scale from tens of nanometers to microns or even millimeters. The microstructure is itself affected by the flow process, and the properties of the final solid object are dependent on the microstructural state at the cessation of flow. A molten polymer, for example, is made up of macromolecules which are highly entangled. (A bowl of spaghetti is a common analogy.) The stress state of any material element during flow is dependent on the local average orientation of polymer chains, while the orientation depends in turn on the history of deformation of that material element (e.g., the flow pattern at each point in a mold). Because the structure must deform to allow flow, the response to an imposed stress will exhibit features which are characteristic both of viscous liquids and of elastic solids, hence is termed viscoelastic. The study of the relation between stress and deformation, which is one focus of this section, is known as rheology. For a given molecular system the mechanical properties following solidification are largely determined by the chain orientation distribution. Low molecular-weight, single-phase liquids are typically Newtonian, where the local stress state is linear in the local rate of deformation. Microstructural liquids, in contrast, are highly non-Newtonian; the stress is usually non-linear in the rate of deformation and often depends on the complete history of deformation. Flow and stress fields for non-Newtonian fluids may therefore differ qualitatively from those for Newtonian fluids under comparable imposed external conditions. Flow instabilities which can have a major influence on manufacturing processes occur in non-Newtonian fluids

69

where stable flow would exist for a Newtonian fluid. Accurate simulation of stress and structure development in processing flows requires the solution of mathematical and computational problems which are poorly understood because of the interaction of microstructural and stress states. In this section we will address research needs in the fluid mechanics and rheology of non-Newtonian liquids, including constitutive equations (rheological equations of state), experimental needs, and numerical issues.

CONSTITUTIVE EQUATIONS FOR FLUIDS WITH MICROSTRUCTURE

The recent focus in the development of constitutive equations has been in the use of microstructural models for homogeneous polymeric liquids. These are treated in several texts (e.g., Bird *et al*, 1987; Doi and Edwards, 1986). The conceptual framework of most models is that polymer chains move more easily in the direction of the backbone than in transverse directions (often called *reptation*); the stress is sometimes related to the deformation through a tensor which reflects the local state. These models and the equations derived from them capture many of the flow features of polymer melts and concentrated solutions, but are inadequate to describe some important rheological phenomena. Many non-Newtonian fluids of commercial interest are multi-phase. Polymer composites are typically processed as suspensions of rigid or semi-rigid particles (calcium carbonate flakes, glass or carbon fibers) in a polymer matrix, for example, and many unfilled polymers are incompatible blends. Continuum theories based on microstructural models are well-developed for dilute suspensions of rigid anisotropic particles in Newtonian fluids (Tucker, 1991), but the dilute range in which particles do not interact (which requires a volume fraction of less than 1% for a particle aspect ratio as small as 10) is rarely encountered in practice. The theories are much less satisfactory for the semi-dilute range, where particle interactions must be considered but coordinated motions are unimportant (Rosenberg *et al*, 1990), and there is no adequate theory for the concentrated suspensions of primary interest. Continuum constitutive equations have not been developed even for dilute suspensions in which the continuous phase is non-Newtonian. Nematic liquid crystals, which are locally anisotropic because of a molecular structure which consists of rigid segments of high aspect ratio, have become increasingly important in electronics applications and as high-performance polymers. The rheology of low-molecular-weight nematics appears to be well-described by a continuum theory of Leslie and Ericksen (deGennes, 1974), which relates the stress to a director field. The director can "tumble" in flow under certain conditions, establishing orientation boundary layers, and isolated defects resulting from singularities in the director field are possible. The rheology of polymer nematics is far more complex. The microstructure at rest consists of micron-scale regions of high local orientation, with transitions between these regions caused by local defect structures. The rheology at low deformation rates seems to be dominated by defects. At intermediate rates there is a transition from director "tumbling" to "non-tumbling," which causes a qualitative change in the rheological properties. Only preliminary steps have been taken thus far to deal with the role of the micron-scale structure in continuum rheology (Larson and Doi, 1991).

Research Needs

Fluid mechanics research can provide important insight into constitutive equation development through study of the following topics:

a. The flow of concentrated suspensions, with homogenization to provide continuum formulations for the stress.

b. Mechanics of interacting particles in non-Newtonian fluids.

c. The migration and interaction of defects in the flow of nematic liquids.

FLOW OF NON-NEWTONIAN FLUIDS

There are few solutions for the flow of polymeric liquids described by elementary continuum equations under conditions characteristic of processing. The major exceptions are free-surface flows such as those in fiber formation and jet printing (e.g., Bousfield *el al.*, 1986). The existence of solutions has not been established in most cases, and the little that has been done (Elkareh and Leal, 1962) indicates that existence in confined flows may in fact be a problem. Similarly, some constitutive equations lead to a change of mathematical type in the flow field (e.g., Joseph, 1990), the consequences of which are largely unexplored. Attempts to obtain approximate solutions for single- and multi-phase flows of non-Newtonian fluids have sometimes been based on a "variational principle" for which a proof is lacking (Binding, 1988). The results have been useful in analyzing some complex flows, and proper attention to the theoretical basis of the approach is needed. The strength of the singularity in the stress for flow past a corner is unknown for most viscoelastic constitutive equations (Davies, 1988; Keiller and Hinch, 1991). There has been speculation that the predicted stresses will exceed the strength of a covalent bond over finite length scales, and that the singularity may sometimes be non-integrable. This unsolved problem plays an important role in computational issues described below. The proper boundary conditions for many non-Newtonian fluids are unresolved. There is evidence that the no-slip boundary condition fails for polymer melts at stresses of order 0.2 MPa, and that this failure is related to flow instabilities that limit polymer extrusion operations (Denn, 1992). Excluded-volume considerations and stress-induced diffusion almost certainly cause composition gradients near a wall in otherwise homogeneous microstructural fluids (Doi and Onuki, 1992), and these gradients may manifest themselves experimentally as apparent slip. The formulation of boundary conditions for viscoelastic fluids over inflow surfaces is also unresolved (Renardy, 1990). The non-linearity and history-dependence of the stress leads to flow instabilities which differ qualitatively from those in Newtonian fluids. These manifest themselves both in low-Reynolds Number rheometrical flows for characterization and in processing situations such as flow through a contraction. Interfacial instabilities are of particular interest in co-extrusion processes, where many layers may be extruded from a single die. Larson (1992) has recently published a comprehensive review of flow instabilities for non-Newtonian fluids. An important area which has received little attention is flow-induced phase change, which can occur because of order induced in the microstructure by the stress field. One example is the apparent onset of crystallization at high temperatures during the spinning of polyester textile fibers from the melt at very high speeds (Ziabicki and Kawai, 1985). Most simulations of polymer processing operations ignore the effect of the fluid mechanics on the solidification mechanism and rate.

Research Needs

a. Studies of existence, boundary-condition formulation, and the effect of change-of-type for viscoelastic constitutive equations in geometries with changes in cross-section.

b. Determination of the stress singularity at corners for non-Newtonian fluids.

c. Exploration of the causes and consequences of real and apparent wall slip for processing flows.

d. Identification of mechanisms for flow instabilities at low Reynolds number in rotational, converging, and interfacial flows.

e. Better understanding of the role of flow-induced stress on the thermodynamics and kinetics of phase changes.

f. Analysis of variational principles for viscoelastic fluids.

EXPERIMENTAL ISSUES

There are two major outstanding problems in rheological measurement. One is the measurement of extensional stresses of mobile fluids such as polymer solutions, which has been the subject of several recent international workshops (Sridhar, 1990). This measurement is essential for determining the predictive power of constitutive equations and the flow is closely related to many important processing situations. The other is the measurement of the full extra-stress tensor in shearing flows at high deformation rates. Lodge has exploited a difference in normal stresses measured by flush-mounted and depressed transducers (Lodge *el al.*, 1991) to measure both shear and primary-normal stresses at high rates for some non-Newtonian fluids, but a robust technique which will operate at temperatures and pressures characteristic of polymer melt flow is not yet available. The measurement of velocity profiles in optically transparent non-Newtonian fluids is routinely carried out using laser-doppler and laser-speckle velocimetry, and optical birefringence, dichroism, and scattering have been exploited to study microstructure development (Fuller, 1990). These techniques are not available for opaque systems. Nuclear magnetic resonance spectroscopy has been applied to some opaque systems (Nakatani *el al.*, 1990; Sinton and Chow, 1991; Barrall *el al.*, 1992), but the technique is in its infancy. Considerable progress is needed in this area.

Research Needs

Reliable experimental techniques for the following measurements:

a. extensional stresses in mobile liquids.

b. normal stresses in shear at high rates.

c. velocity and structure profiles in opaque liquids.

NUMERICAL ISSUES

Stokesian dynamics has provided insight into the flow of concentrated suspensions of spherical particles in Newtonian liquids (Brady and Bossis, 1988), and Kim and Karrila (1991) have explored the use of boundary integral methods for the direct solution of flows of suspensions containing non-spherical particles. The computational problems for flows of practical interest using these direct simulation methods are formidable, however. Most computational effort for non-Newtonian fluids has been based on continuum descriptions, using finite-element, finite-difference, and boundary-element methods. There are major limitations in the simulation of flow of viscoelastic liquids in confined geometries, where large stress gradients, often associated with corner singularities, prevent convergence under conditions of practical interest (Keunings, 1989). Solution of the corner singularity could circumvent this problem by enabling the construction of singular elements. Better understanding of the effect of change of type on numerical techniques is also essential for progress. These computational issues are intimately tied to flow problems discussed above.

Research Needs

a. Development of practical schemes for the direct simulation of structure development in concentrated suspensions.

b. Robust methods for treating large stress gradients in viscoelastic liquids.

ACKNOWLEDGMENT

We have benefited from three prior reports and a recent review article on this topic:

- The mechanics of fluids with microstructure, G. Leal and R. A. Brown, for the report The Future of Fluid Mechanics Research, January, 1987.

- Final report of the NSF workshop on the applications of fluid mechanics to materials processing, March, 1988.

- The mathematical sciences applied to materials sciences, National Research Council, 1991.

- Denn, M. M. (1990) Issues in viscoelastic fluid mechanics, Annu. Rev. Fluid Mech. 22, 13 - 34.

REFERENCES

Barrall, G. A., Frydman, L., & Chingas, G. C. (1992) NMR diffraction and spatial statistics of stationary systems, *Science* **255**, 714 - 717.

Binding, D. M. (1988) An approximate analysis for contraction and converging flows, *J. Non-Newtonian Fluid Mech.* **27**, 173 - 189.

Bird, R. B., Curtiss, C. F., Armstrong, R. C., & Hassager, O. (1987) *Dynamics of polymeric liquids*, 2nd ed., Vol. 2, New York: Wiley.

Bousfield, D. W., Keunings, R., Denn, M. M., & Marrucci, G. (1986) Nonlinear analysis of the surface-tension driven breakup of viscoelastic filaments, *J. Non-Newtonian Fluid Mech.* **21**, 79 - 97.

Brady, J. F., & Bossis, G. (1988) Stokesian dynamics, *Annu. Rev. Fluid Mech.* **20**, 111 - 157.

Davies, A. R. (1988) Reentrant corner singularities in non-Newtonian flow. Part I. Theory. *J. Non-Newtonian Fluid Mech.* **29**, 269 - 293.

deGennes, P.-G. (1974) *The physics of liquid crystals*, Oxford: Oxford Univ. Press.

Denn, M. M. (1992) Surface-induced effects in polymer melt flow, in Moldenaers, P., and Keunings, R., eds., *Theoretical and Applied Rheology* (Proc. XIth Int. Cong. Rheology), Amsterdam: Elsevier, pp. 45 - 49.

Doi, M., & Edwards, S. F. (1986) *The theory of polymer dynamics*, Oxford: Oxford Univ. Press.

Doi, M., & Onuki, A. (1992) Dynamic coupling between stress and composition in polymer solutions and blends, *J. Physique II*, **8**, 1631 - 1656.

Elkareh, A. W. & Leal, L. G. (1989) Existence of solutions for all Deborah numbers for a non-Newtonian model modified to include diffusion, *J. Non-Newtonian Fluid Mech.* **33**, 257 - 287.

Fuller, G. G. (1990) Optical rheometry, *Annu. Rev. Fluid Mech.* **22**, 387 - 417.

Joseph, D. D. (1990) *Fluid dynamics of viscoelastic liquids*, New York: Springer-Verlag.

Keiller, R. A., & Hinch, E. J. (1991) Corner flow of a suspension of rigid rods, *J. Non-Newtonian Fluid Mech.* **40**, 323 - 335.

Keunings, R. (1989) Simulation of viscoelastic fluid flow, in Tucker, C. L. III, ed., *Computer modeling for chemical processing*, Munich: Carl Hanser Verlag, pp. 403 - 469.

Kim, S., & Karrila, S. J. (1991) *Microhydrodynamics*, Stoneham, MA: Butterworth-Heinemann.

Larson, R. G. Instabilities in viscoelastic flows, *Rheol. Acta.* **31**, 213 - 263.

Larson, R. G., & Doi, M. (1991) Mesoscopic domain theory for textured liquid crystal polymers, *J. Rheol.* **35**, 539 - 563.

Lodge, A. S., Pritchard, W. G., & Scott, L. R. (1991) The hole-pressure problem, *IMA J. Appl. Math.* **46**, 39 - 66.

Nakatani, A. I., Poliks, M. D., & Samulski, E. T. (1990) NMR investigations of chain deformation in sheared polymer fluids, *Macromolecules* **23**, 2686 - 2692.

Renardy, M. (1990) An alternative approach to inflow boundary conditions for maxwell fluids in three space dimenions, *J. Non-Newtonian Fluid Mech.* **36**, 419 - 425.

Rosenberg, J., Keunings, R., & Denn, M. M. (1990) Simulation of non-recirculating flows of dilute fiber suspensions, *J. Non-Newtonian Fluid Mech.* **37**, 317 - 345.

Sinton, S. W., & Chow, A. W. (1991) NMR imaging of fluids and solid suspensions in poiseuille flow, *J. Rheol.* **35**, 735 - 772.

Sridhar, T., ed. (1990) Proc. of an int. congress on extensional flow, *J. Non-Newtonian Fluid Mech.*, **35**, Nos. 2 and 3.

Tucker, C. L. III (1991) Flow regimes for fiber suspensions in narrow gaps, *J. Non-Newtonian Fluid Mech.* **39**, 239-268.

Ziabicki, A., & Kawai, H., eds. (1985) *High-speed fiber spinning*, New York: Wiley.

THERMAL AND MASS
DIFFUSION DRIVEN FLOWS

Benjamin Gebhart
School of Engineering and Applied Science
University of Pennsylvania
231 Town Building
Philadelphia, PA 19104-6315

INTRODUCTION

Rapidly increasing concern in enhanced environmental and in techno-logical processes are beginning to call forth first-order improvements and sophistication in both our understanding of new fundamental processes, as well as in our ability to generate better modeling and prediction capabilities. Many important new initiatives have been begun. Some are already matur-ing. Others already show great promise. Many of the most recent advances amount to fresh beginnings, to seek further fundamental understanding of both physical processes, and of the underlying properties of transport. Some are very new departures, stemming from most recent insights.

Several very direct opportunities are mentioned here. A very promis-ing mechanism concerns active feed-back control, perhaps from the outside, of internal flow fields. Thereby, laminar and/or turbulent regimes may be produced or suppressed, in both forced and buoyancy driven flows.

Another opening field concerns applications wherein forced and buoyancy driven flow effects are of entirely comparable magnitude. Many new process-ing technologies operate in regimes where-in strongly multi-dimensional and spatially complicated flows are both an integrated feature of these processes.

A very important, yet largely neglected area of transport applications, concerns buoyancy driven flows in partial enclosures, with one or several openings. A common mechanism is penetrative convection. Again, multi-dimensional effects commonly dominate. However, general and realistic mod-eling often requires new kinds of formulations, along with very extensive computational capability, to realistically represent the wide range of the im-portant applications encountered.

Many rapidly advancing surface processing technologies increasingly re-quire the generation of improved capability, and of quality control, in flow deposition fabrication schemes. Flow dimensionality interacts with buoyancy, viscous and inertial effects. The specific effects which control such processes must be investigated in great detail, to obtain the required product quality. Solutal buoyancy, driving fabrication, must also be tightly controlled, in many processes.

The mechanisms of buoyancy driven flows in porous media have been extensively investigated, using simplified classical flow formulations, mostly in terms of permeability. Porous media processes surround us, in waste disposal, in fossil fuel deposits, in aquifers and in reactors of all kinds. The crucial need, at this time, is to provide strong and general bases concerning the specific properties and characteristics of such materials and processes. New and more sophisticated methodologies must be brought to bear, in this critical field of environmental management in processing, and in waste management.

The above brief summary, of the impact of these proposed extensions of current technology, indicate a number of the most promising avenues of increased research and development. These are among the most critical areas of immediate concern, in the service of technology, to our society and to the environment. The following five sections set forth these matters in specific detail.

ACTIVE CONTROL OF CONVECTION

In many situations, modification of the normally occurring flow structures is desirable. For example, in crystal growth processes, it may be desirable to operate at Rayleigh numbers higher than the one at which convection occurs and yet have no convection. In other processes, it may be desirable to suppress (laminarize) chaotic or turbulent motions and maintain a steady, time-independent flow in order to minimize flow unpredictability, remove temperature oscillations which may reduce process uniformity or exceed safe operational conditions, reduce contaminant and impurity transport, and/or reduce drag. In still other processes, it may be advantageous to induce chaos, under conditions in which it would not normally occur, so as to enhance mixing, heat transport, combustion, or chemical reactions.

Although feedback control strategies which exploit or suppress naturally occurring instabilities in flows hold considerable promise as practical, effective and flexible means of modifying flows so as to achieve a desired behavior, little attention has been paid in the scientific community to such strategies. However, in a recent theoretical and experimental work, we have demonstrated (Singer et. al., 1991, Bau, 1992, Wang et. al., 1992, and Tang and Bau, 1993) that in simple convective systems, it is possible to modify the bifurcation structure of the convective flow so as to stabilize (destabilize) otherwise non-stable (stable) flow structures through the use of feedback control strategies. For illustration purposes, we depict in Figures 1(a) and (b), respectively, the time-dependent temperature in a thermal convection loop with and without a controller. The strategy we used involved sensing

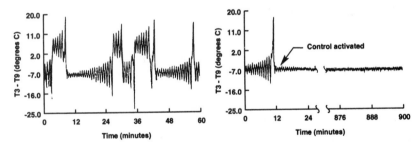

Figure 1: The experimentally observed temperature difference between positions 3 and 9 o'clock in a toroidal, thermal convection loop is depicted as a function of time. (a) Uncontrolled motion. The figure exhibits chaotic behavior. The figure has been reproduced from Singer J. et. al., 1991, Physical Review Letters, 66, 1123-1125. (b) Controlled motion. The controller was activated at time $t \approx 12$. Witness that for $t > 12$, the controller has rendered laminar the otherwise chaotic flow. The figure has been reproduced from Singer J. et. al., 1991, Physical Review Letters, 66, 1123-1125.

the deviation of a measurable quantity such as temperature or speed from its desired value and then modifying the boundary conditions, typically in a rather minor way, to compensate for this deviation. The figures clearly illustrate that one can render a chaotic (turbulent) flow laminar with the use of relatively simple control strategies.

Research Needs

i. Obtain a better understanding of the dynamics of uncontrolled and controlled convective flows driven by thermal, concentration, and/or surface tension gradients.

ii. Devise procedures which enable one to construct low-dimensional mathematical models capable of providing at least a qualitatively correct description of convective flow phenomena so as to facilitate the design and study of various control strategies.

iii. Devise control strategies for complex convective systems of industrial interest such as those used in material processing (i.e., the Czochralski and float zone processes).

iv. Develop algorithms capable of optimizing the control strategy for realistic systems.

v. Develop micro-sensors and micro-actuators for use in the control systems.

FORCED FLOWS WITH SIGNIFICANT BUOYANCY EFFECTS

Complex phenomena arise in forced flows of fluids with significant buoyancy forces. This is true even under relatively benign conditions; for example, moderate flow rates, i.e., moderate Reynolds numbers, moderate heating, i.e., moderate Rayleigh or Grashof numbers, straight channels, etc. The interaction between buoyancy and inertial forces often results in complex unsteady three dimensional fields. These effects are of importance in a number of areas including the manufacture of thin films, optical fibers, cooling of nuclear reactors, energy storage systems, etc.

In horizontal internal flows with gravity normal to the direction of the main flow, instabilities may appear (even with steady boundary conditions) which are a combination of transverse, traveling waves in the direction of the flow and longitudinal rolls associated with spanwise recirculation. These phenomena have adverse effects on, for example, the fabrication of microelectronic components; e.g., Jensen, Einset and Fotiadis (1991), Evans and Greif (1993). In particular, the goals of increased rates of deposition and increased uniformity of thin films can be difficult to achieve due to unstable and recirculating fluid flow which in many cases is due to the effects of buoyancy. In respect to the manufacture of optical fibers the effects of buoyancy and rotation may result in marked secondary flows; e.g., Lin, Choi and Greif (1992) which may have deleterious effects on these processes.

Most studies that have been carried out, both experimental and theoretical, have considered the condition when the buoyancy forces are small compared to the inertial forces. However, there is also great interest in the phenomena that occur when large temperature differences and strong buoyancy contributions are present. The velocity, temperature and concentration fields can be markedly non-uniform and can also vary strongly with time; indeed, it is often difficult to generalize or even characterize the flow or thermal fields. It is also noted that in symmetric systems, e.g, a cold wall vapor phase epitaxy reactor with symmetric boundary conditions, that the stable flow field is generally strongly asymmetric. In vertical flows (gravity parallel to the direction of the main flow) complex phenomena arise when gravity aids; e.g., a downward flow in a cooled channel, or when gravity opposes the flow. This includes flow reversals, i.e., flows which are locally in the opposite direction to the forced flow direction; asymmetries when the geometry and boundary conditions are symmetric; time dependence in cases for which the boundary conditions are steady and "early" transition to turbulence, i.e., transition at a lower Reynolds number than for the forced flow in the absence of buoyancy forces. For completeness, it is noted that the effects of rotation, mass transfer, chemical reactions, etc., are also of great importance and interest. Much work must be done if these effects are to be understood.

As the heating is increased the effects of buoyancy will lead to transition and turbulence and these mechanisms must also be studied.

Research Needs

a. Flow, thermal and concentration fields in the presence of large temperature differences.

b. Dependence of flow and thermal mechanisms on boundary conditions, configurations and chemical reactions.

PENETRATIVE CONVECTION, SEPARATION AND BUOYANCY DRIVEN FLOW IN ENCLOSURES WITH OPENINGS

There are several important areas in buoyancy-driven flows that have received very little attention in the literature, mainly because of the complexities resulting from open boundaries, reverse flow and opposing transport mechanisms. Three such topics are penetrative convection, separation and flows in partially open enclosures. These flow circumstances arise in many diverse problems of practical interest ranging from flows in the environment to those in rooms and buildings. They are also driven by a wide variety of sources for buoyancy input such as electronic components, fires, moisture transport and chemical pollutant discharge. Some of the basic considerations involved in these flows and the corresponding research needs are presented here.

PENETRATIVE CONVECTION

Penetrative convection refers to buoyancy-induced flows generated in a region where they are unstable and penetrating into a stable region. An example of this flow is a thermal plume generated by a heat source in an isothermal region and rising and penetrating into a stably stratified upper layer. The flow penetrates to a finite height and then drops to a stable position, while continuing to entrain fluid from the region adjacent to the penetrating flow. The characteristics of these flows and the resulting transport are of increasing interest and importance. Such flows arise in many practical circumstances, such as room fires, energy storage systems, electronic packages and environmental research. Typically a fire on the floor of a room generates a thermal plume, which is usually turbulent and which rises due to the input of thermal buoyancy. The room itself is often stably stratified due to the temperature and concentration of the combustion products that collect near the ceiling. A two-layer environment is frequently obtained, with the upper region at a much higher temperature, and containing the species resulting from combustion, and the lower layer providing a relatively cooler and cleaner environment. The stratification may also be the result of fire activity in adjacent rooms. The penetrative convection that arises is important in the mathematical modeling of the fire growth in the room in order to determine the spread of smoke and toxic gases. It is crucial to determine the height of penetration, the entrainment into the flow and the effect on the thermal environment in the room.

Similar problems arise in the environment due to buoyancy driven flows arising in a region of relatively uniform temperature and species concentration and penetrating into a stable region. Stratified air and water media frequently give rise to such circumstances, for instance, in the growth of a turbulent boundary layer in the atmosphere during early morning heating and the deepening of the upper-ocean mixed layer into the stably stratified pycnoline. Again, the penetrative distance and the flow field are of interest. Only a few experimental and numerical studies have been carried out on this problem, see Turner (1979). As such, there is a strong need for a detailed investigation to determine the dependence of the penetration characteristics on the governing parameters in the problem, such as buoyancy input and density difference between the two layers. It is important to study not only the penetration distance but also the nature of the flow near the stagnation region, the flow reversal and the final stable position attained. Both experimental and numerical efforts are needed, with an emphasis on turbulent flow, to understand the basic nature of this flow.

SEPARATION IN FLOWS WITH SIGNIFICANT BUOYANCY EFFECTS

The basic processes that govern separation of buoyancy driven flows from surfaces are very different from those in forced flow. If there is buoyancy input at the surface, a flow results adjacent to the surface and a stagnant or reverse flow region does not generally arise. In forced flow, the external pressure field drives the flow and an adverse pressure can lead to separation with the associated recirculation and reverse flow. The mechanisms driving the flow in natural convection are very different since the pressure is the consequence of the flow which is driven by buoyancy. Thus, the resulting flow is also quite different. However, the surfaces are of finite extent, as in a sphere, cylinder or flat plate, and the flow must ultimately leave the surface and rise as a buoyant plume. This is largely a realignment process as the flows from all sides come together and rise due to buoyancy. Very little work has been done on such separating buoyancy driven flows over finite-sized heated bodies such as cylinders and spheres. The absence of stagnation and reverse flow has been pointed out (Pera and Gebhart, 1972; Jaluria and Gebhart,

1975). But, clearly, detailed investigations are needed to understand the basic processes involved.

Opposing buoyancy effects in forced flow may also lead to separation. Such negatively buoyant flows have also received very little attention because of the complicated flow behavior that results from the interaction of the main flow stream with the reverse flow generated by the opposing buoyancy. Such flows arise, for instance, in heat rejection systems, in the environment, in mixed convection heat transfer from bodies, in continuous thermal processing systems like optical fiber drawing and hot rolling, and in room air conditioning systems. The buoyancy in these cases opposes the externally driven flow, resulting in the flow separating from the surface and reversing direction, see Turner (1966), Goldman and Jaluria (1986) and Jaluria (1992). The transport processes are strongly affected in the separated flow region and are not very well understood. Detailed experimental and analytical studies of negatively buoyant flows are needed to provide information on the basic characteristics of these flows and the resulting effect on heat transfer.

FLOWS IN ENCLOSURES WITH OPENINGS

Though a very large amount of work has been done on buoyancy-driven flows in enclosures, very little effort has been directed at partially open enclosures, i.e., enclosures with vents or openings. The flow and the associated thermal or mass transport are strongly influenced by the flow conditions at the opening. Such flows arise frequently in rooms and compartments due to the heat input due to fires and electronic components, see Quintiere (1977), Gebhart et al. (1988) and Jaluria and Cooper (1989). Similar flows also arise in furnaces and ovens for materials processing and solar energy collection systems. The interaction between the flow in the enclosure and the ambient conditions through the opening is a very important consideration since it affects the resulting flow in the enclosure. For instance, the flow generated in the enclosure is very different for a stably stratified ambient as compared to that for an isothermal one. It is, thus, important to characterize the basic mechanisms that govern the resulting flow.

Among the important questions to be answered in this flow circumstance are the nature and magnitude of the inflow and the outflow at the opening and the effect these have on the buoyancy-driven flow in the enclosure. Several approximate methods, based on the density difference across the opening, have been developed to estimate the inflow of the ambient fluid and the outflow of the buoyancy-induced flow from inside, across the opening. However, detailed experimental and analytical or numerical studies are needed to obtain the dependence of the flow inside the enclosure on the ambient conditions and on the opening dimensions and geometry. The flows that are usually of interest in practical circumstances are turbulent and, therefore, the appropriate turbulence model is also of interest. In addition, it is important to study the numerical imposition of the conditions at the opening to simulate typical physical situations.

BUOYANCY INDUCED FLOWS IN MATERIALS PROCESSING

Transport processes play a dominant role in materials processing. Significant strides in several important materials technologies, such as mass production of semiconductor crystals, thin films for integrated circuits, optical fiber drawing, and injection molding of aerospace and automotive components have been made possible partly due to advances in thermal engineering. In a majority of the processing domains of interest, flow effects are very important and impact directly on the capability of the process to produce these materials reproducibly and controllably, with the desired properties.

The flow encountered in most of the materials processing is typically forced convection, modified or perhaps, more appropriately, complicated by, buoyancy induced convection. Multiple complex flow structures commonly arise. Such flows, and their impact on quality of the materials grown, must be understood fully to exploit the existing and new materials of vital importance to the nations' technological infrastructure and competitiveness. The electronic and optical materials best represent this category of emerging materials. Two major technologies that play a significant role in the processing of these materials are the thin film deposition and bulk crystal growth. The discussion in the following sections is generally based on considerations of these technologies. The research issues addressed are, however, generic and apply to other materials and processes.

MULTIPLE STEADY AND TIME DEPENDENT FLOWS

The early 2-dimensional, and recently 3-dimensional models, have helped provide insight into the flow phenomena related to materials processing. Most of these models, however, use idealized/simplified processing conditions, geometric conditions and/or boundary conditions. In real systems, multiple complex flow structures are encountered. For example, in chemical vapor deposition (CVD) processing, non-linear coupling between buoyant, viscous and inertia terms may lead to multiple steady and time dependent flows. This phenomenon of bifurcation is very sensitive to reactor configuration and operating parameters; see for example, Jensen et al. (1991), Patnaik et al. (1989) and Fotiadis et al (1987).

As an illustration, consider Figure 2 (Patnaik et al.) which shows the calculated Nusselt number, streamlines and temperature plots for flow in a vertical organometallic vapor phase epitaxy reactor. In a narrow range of the susceptor temperature, T_s, $312 \leq Ts \leq 317°K$, the hysteresis phenomenon is manifested, and it is possible to have two steady flow patterns corresponding to the same susceptor temperature and inlet flow conditions. For low values of $\Delta T = T_s - T_o$, where T_o is the ambient temperature, forced convection dominates and Nu remains constant. For Ts corresponding to point "b" on the curve, a buoyancy driven clockwise cell appears which converts heat away from the susceptor thereby causing a drop in Nu. However, as T_s is decreased through the hysteresis loop, point "c", the buoyancy driven cell grows and Nu decreases. But if T_s is increased again past the second turning point, the buoyancy-induced cell dominates, and Nu increases with increase in T_s. Similar calculations for fixed T_s ($T_s = 973°K$) and different susceptor rotation rates Ω showed that for $320 \leq \Omega \leq 330rpm$, non-linear interactions between buoyancy and rotational convection resulted in two hysteresis loops and five possible steady state flows. As noted by Jensen et al. (1991) and Fotiadis et al. (1987), it is most likely that under actual reactor operating conditions, these multiple steady flows may transition to time periodic and fully three-dimensional flows. A comprehensive understanding of these phenomena is needed to delineate the optimum parameter ranges.

SOLUTAL BUOYANCY

In many new and emerging materials fabrication techniques, such as the growth of heterojunctions, composite layers of different density materials are grown. Generally, the requirement is that the interfaces be abrupt. The solutal buoyancy induced convective cells may arise when switching gases from one density to another, Palmateer et al. (1987). During deposition, these solutal flows, through interaction with thermal convection, may produce time periodic flows, over a range of thermal Grashof numbers. These time periodic solutal effects and the bifurcation effects discussed above can thus be simultaneously present in actual reactor systems and pose a formidable challenge in analysis. Fundamental investigations of these flows, pertinent to material processing techniques, need to be undertaken.

The solutal buoyancy effects in bulk crystal growth, e.g., in Czochralski growth of silicon, play an important role in the quality of the crystals grown. The solutal flow, combined with that due to thermal buoyancy, thermo-capillary and rotation of both the crystal and the crucible, result in complicated flow patterns. Different modes of instabilities arise that lead to interfacial oscillations. These contribute to fluctuations in the growth rate and in turn to undesirable dopant concentration striations in the crystal. The interactions between these various flow effects are not well understood and need to be investigated.

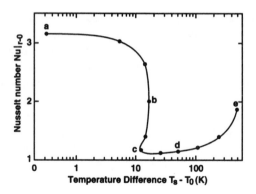

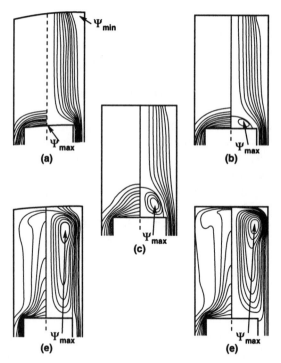

Figure 2: Calculated Nusselt number, streamlines and temperature contours for flow in a vertical organometallic vapor phase epitaxy reactor.

Research Needs

i. Fundamental investigation of 3-dimensional, time dependent flows pertinent to real materials processing equipment.

ii. Obtain a better understanding of the dynamics of the interacting flows involving the solutal buoyancy, thermal buoyancy, forced convection and surface tension induced flows.

iii. In modeling of transport in materials processing systems with solid/liquid phase change in complex geometries, the nature and modes of oscillations at the interface need to be investigated.

iv. Related to (i), (ii) and (iii) above, reliable data for the thermophysical properties for many of the processed materials is not available. Comprehensive studies must be undertaken to determine the material properties at the processing temperatures involved to develop reasonable solutions to the flow problems. For accurate prediction of the In many of the materials processing operation, fabrication process, processing.

BUOYANCY DRIVEN FLOW IN POROUS MEDIA

Single-phase flow in fully-saturated or unsaturated porous media is of considerable importance to a wide range of environmental applications, chemical engineering processes, and biological systems. Porous media typically consists of a rigid matrix with interconnected voids. Buoyancy driven transport in porous media is of interest in such technological problems as geothermal energy resources, long-term storage of thermal energy in aquifers, packed-bed chemical reactors, nuclear reactor safety, solidification of metal alloys, nuclear waste disposal, heat transfer in soils and numerous others. Because of practical, as well as fundamental significance, it is anticipated that these mechanisms will require a rapid expansion in research activity during the next few years.

Buoyancy driven convective transport in porous media is not a mature field. Before attempting to identify future research needs and opportunities it is desirable to delineate the present state of knowledge, in terms of the following four elements: 1) characterization of solid matrix structure, 2) thermophysical properties of solid matrix, 3) flow field regime, and 4) the transport of heat and the mass transfer regime. These elements are four corners of a polygon. The focus of contemporary research, on transport phenomena in porous media, is somewhere inside and occasionally along its periphery (Giorgiadis 1991).

Whether the flow is forced or buoyancy driven, three generic but important issues must be addressed. These include: i) an understanding of statistical properties of small scale flows, ii) the development of macroscopic conservation equations, and iii) refinement of existing and development of new diagnostic techniques. The issues identified are in concert with long standing recommendations. Future research needs should concern improved thermophysical data, better and more complete description of the structure of porous media, improved formulations of the governing equations to better relate microscopic and macroscopic behavior, and many experiments to verify theoretical models, establish engineering correlations, and provide more detailed insight into such convection phenomena (Torrence et al. 1982).

Even though the four elements relating to flow and transport in porous media do not pertain directly to buoyancy driven flows, they are an integral part of the formulation and solution of transport problems in liquid-saturated and unsaturated porous media. Significant progress has been made in understanding the transport processes which are fundamental to natural convection in porous media. However, a number of critical issues have been identified which required immediate research attention. The research needs are separated into four categories for clarity of presentation.

CHARACTERIZATION OF POROUS MEDIA

The description of realistic porous media have been accomplished by a number of models, but the model parameters are still difficult to relate to complex microstructures. A goal in microstructure characterization of porous media is to develop non-invasive imaging techniques which will produce statistical information, from given samples. Techniques such as x-ray scattering, nuclear magnetic resonance (NMR), thermal-wave spectroscopy and ultrasound scanning, coupled with sophisticated image processing tools, are to be used to extract successively more detailed microstructural information (Giorgiadis 1991). Characterization of anisotropic porous media is necessary. This is particularly critical for the study of the boundary effect (near solid and fluid interfaces). It is also necessary to explore the potential applicability of fractal geometry concepts, since they may offer an efficient way to model media with statistical heterogeneity and also may account for the fact that transport properties can be scale-dependent.

THERMOPHYSICAL PROPERTIES

Considerable uncertainty exists in the theoretical description of thermophysical properties of porous media. These are the basis for transport calculations. Presently, properties are being estimated using analytical and empirical models which relate values for the separate constituents to values for the fluid-saturated porous medium as a whole. Theoretical models which have been fully validated are needed for the permeability, effective thermal conductivity, and effective diffusion coefficient. These characteristics directly affect transport measurements and predictions. Validation of simple models to predict volume-averaged thermophysical characteristics is also needed.

FLOW FIELD AND REGIME

Significant progress has been made in the description of slow incompressible flow of Newtonian fluids through solid matrices, using ordered and disordered representations of homogeneous media. However, research on a self-consistent derivation of local volume-averaged conservation equations, including boundary effects, is required. The regimes of unsteady and turbulent flow through porous media is virtually unexplored. Models of anisotropic and heterogeneous porous media, and the fluid mechanics of boundary layers formed in the vicinity of solid walls require re-examination. Specifically, the uncertainties concerning proper hydrodynamic boundary conditions need to be resolved. Applications related to the manufacturing of composite materials, and to chemical engineering applications, point to the need to study non-Newtonian fluids and superfine porous media.

BUOYANCY-DRIVEN TRANSPORT

Natural convection in porous media is not nearly as well understood and characterized as its counterpart in pure fluids. Little is known about detailed flow patterns in porous media. The effects of inhomogeneity and anisotropy of porous media on transport phenomena have not been investigated. Combined thermal and solutal buoyancy driven convection in porous media needs to be studied. This latter problem is particularly relevant to the mushy region of solidifying metal alloy ingots and castings. Buoyancy driven transport in porous media (such as soils) for which the solid phase undergoes melting or sublimation, resulting in time dependent porosity and permeability, has received relatively little research attention. Natural convection during the phase change of liquid saturated porous media, and natural convection in unsaturated porous media in the presence of sublimation, condensation, evaporation and drying are crucial areas of future research.

Research Needs

a. Characterization of isotropic and anisotropic porous media.

b. Macroscopic and microscopic description of the structure of porous media including permeability, tortuosity, etc.

c. Theoretical, fully validated models for thermophysical properties such as effective thermal conductivity and effective diffusion coefficient.

d. Self-consistent derivation of local volume-averaged conservation equations, including boundary effects.

e. Flow structure and transport in porous media and the effects of inhomogeneity and anisotropy under natural convection conditions.

f. Buoyancy driven transport in porous media undergoing solid phase change (melting and sublimation).

g. Natural convection driven transport during phase change of liquid saturated and unsaturated porous media in the presence of sublimation, condensation, evaporation and drying.

REFERENCES

Bau, H.H., 1992, Controlling Chaotic Convection, (an invited paper) to appear in *Proceedings of the XVIIIth International Congress of Theoretical and Applied Mechanics*, Haifa, Israel, August 22–28, (J. Singer, A. Solan, S. Bonder, and Z. Hashin, editors).

Evans, G. and Greif, R. (1993) Thermally Unstable Convection with Applications to Chemical Vapor Deposition Channel Reactors, *Int. J. Heat Mass Transfer*, in press.

Fotiadis, D.I., Kremer, A.M., McKenna, D.R., and Jensen, K.F. (1987) Complex flow phenomena in vertical MOCVD reactors: effects on deposition uniformity and interface abruptness. *J. Cryst. Growth* 85: 154–165.

Gebhart, B., Jaluria, Y., Mahajan, R.L., and Sammakia, B. (1988) *Buoyancy-induced flows and transport*, Hemisphere Pub. Corp., Washington, DC.

Giorgiadis, J.G. (1991) Future research needs in convective heat and mass transfer in porous media. In: *Convective Heat and Mass Transfer*, S. Kakac, ed., Kluwer Academic Publishers, Dordrecht, The Netherlands, Vol. E196, pp. 1073–1088.

Goldman, D. and Jaluria, Y. (1986) Effect of opposing buoyancy on the flow in free and wall jets, *J. Fluid Mech.*, Vol. 166, pp. 41–56.

Jaluria, Y. (1992) Transport from continuously moving materials undergoing thermal processing, *Ann. Rev. Heat Transfer*, Vol. 4, pp. 187–245.

Jaluria, Y. and Cooper, L.Y. (1989) Negatively buoyant wall flows generated in enclosure fires, *Prog. Energy Combust. Sci.*, Vol. 15, pp. 159–182.

Jaluria, Y. and Gebhart, B. (1975) On the buoyancy-induced flow arising from a heated hemisphere, *Int. J. Heat Mass Transfer*, Vol. 18, pp. 415–431.

Jensen, K.F., Einset, E.O. and Fotiadis, D.L. (1991) Flow Phenomena in Chemical Vapor Deposition of Thin Films, *Annual Reviews of Fluid Mechanics*, 23, 197–232.

Lin, Y. T., Choi, M., and Greif, R., (1992) A Three-Dimensional Analysis of Particle Deposition for the Modified Chemical Vapor Deposition Process. *ASME J. Heat Transfer* 114, 735–742.

Palmateer, S.C., Groves, S.H., and Wang, C.A., Weyburne, D.W., and Brown, R.A. (1987) Use of flow visualization and tracer gas studies for designing an InP/InGaAsP OMVPE reactor. J. Cryst. Growth 83: 202–210.

Patnaik, S., Brown, R.A., and Wang, C.A. (1989) Hydrodynamics dispersion in rotating-disk OMVPE reactors: numerical simulation and experimental measurements. J. Cryst. Growth 96: 153–174.

Pera, L. and Gebhart, B. (1972) Experimental observations of the wake formation over cylindrical surfaces in natural convection flows, *Int. J. Heat Mass Transfer*, Vol. 15, pp. 177–179.

Quintiere, J. (1977) Growth of fire in building compartments, Fire Stds. Safety, ASTM STP 614, pp. 131–167.

Singer, J., Wang, Y. and Bau, H.H. (1991) Controlling a Chaotic System, *Physical Review Letters*, 66, 1123–1125.

Tang, J. and Bau, H.H. (1993) Stabilization of the No-motion State in Rayleigh-Benard Convection Through the use of Feedback Control, *Physical Review Letters*, 70, 1795–1798.

Torrance, K.E., Schoenhals, R.J., Tien, C.L., and Viskanta, R. (1982) Phase change heat transfer in porous media. In: *Proceedings of Workshop on Natural Convection*, K.T. Yang and J.R. Lloyd, eds., University of Notre Dame, Notre Dame, IN, pp. 37–45.

Turner, J.S. (1966) Jets and plumes with negative or reversing buoyancy, J. Fluid Mech., Vol. 26, pp. 779–792.

Turner, J.S. (1979) Buoyancy effects in fluids, Cambridge Univ. Press, Cambridge, U.K.

Wang, Y., Singer, J., and Bau H.H. (1992) Controlling Chaos in a Thermal Convection Loop, *J. Fluid Mechanics*, 237, 479–498.

RAPID FLOWS OF GRANULAR MATERIALS

James T. Jenkins
Department of Theoretical and Applied Mechanics
Cornell University
Kimball Hall
Ithaca, NY 14853

INTRODUCTION

Rapid flows of granular materials commonly occur in natural phenomena such as rock slides, debris flows, granular snow avalanches and underwater sediment slumps, and in industrial processes involving the rapid transport of bulk materials, for example, coal, ore, cereals, polymer pellets, and pharmaceuticals. The interest in rapid granular flows arises from an appreciation of their destructive potential, particularly as the expansion of urban areas into mountain foothills places increasing numbers of lives and amounts of property in the path of rock slides, debris flows, and snow avalanches (McPhee, 1988), and from the recognition of the enormous economic benefits to be gained by increases in efficiency in industries that depend on the handling of large quantities of particulate solids (Merrow, 1986). Also, there are indications that the mechanics of more complicated particulate systems involving particle interactions that are not collisional, but influenced or dominated by the interstitial fluid, may be understood by analogy with rapid granular flows. In a rapid granular flow the collisions alone are responsible for the transfer of momentum and the transfer and dissipation of energy. Momentum and energy are supplied to the flows by gravity and by forces applied at their boundaries. Understanding the mechanics of such flows would make it possible to predict the relationship between the forces applied to them and their speed and extent and could lead to the rational design of devices to initiate, sustain, and control the flow. The challenge to theory is to identify the variables that are relevant to the description of a rapid granular flow and to formulate balance laws, constitutive relations, and boundary conditions that govern their evolution. This activity is presently guided by numerical simulations of the Newtonian dynamics of interacting particles in simple flows as discussed by Goddard, Haff & Walton (1992) in a previous volume of this series. The simulations provide access to the numerical values of any instantaneous or averaged variable in the interior of the flow. A corresponding

92

advance in physical experiments is now taking place based on the development of instruments to measure velocities and concentrations within highly energetic systems of both transparent and opaque particles.

BACKGROUND

In 1954, Bagnold initiated the study of rapid granular flow with careful experiments on the rapid shear of neutrally buoyant spheres between concentric rotating cylinders. A flexible inner cylinder allowed him to measure both shear and normal stresses as functions of apparent shear rate over a range of particle concentrations. At high shear rates, he observed that both the shear stress and normal stress increased with the square of the shear rate. The ratio of the shear to the normal stress was constant and equal to about one-third. Bagnold's explanation of the quadratic dependence was based on collisions being the dominant mechanism of momentum transfer. In a collision between spheres moving with the mean velocity, the changes in momentum parallel and perpendicular to the flow are proportional to the mean shear rate; because the frequency of collisions is also proportional to the mean shear rate, the rates of change of both components of momentum are quadratic in the shear rate. For twenty years, these rudimentary constitutive relations served those interested in rapid flows of sands and sediments. However, because the constitutive relations were based on simple shear, they did not have the capacity to deal with inhomogeneous flows in which the concentration varies with position and in which the shear rate might vanish at some point in the flow. The recent renewal of interest in rapid granular flows developed from the convergence of three separate activities: renewed experiments on rapid flows of dry granular materials in shear cells (Savage & Sayed, 1982) and inclined chutes (Sayed & Savage, 1983), numerical simulations of the detailed dynamics of shearing flows of inelastic disks and spheres (Campbell & Brennen, 1985; Walton & Braun, 1986), and the exploitation of the analogy between the agitated grains and the colliding molecules of a dense gas (Ogawa, Umemura & Oshima, 1980).

THEORY AND SIMULATION

Progress beyond Bagnold's stress relations for simple shear began when the energy of the particle velocity fluctuations was recognized as being the source of the normal stress. When a balance law was introduced equating the local rate of change of fluctuation energy to its production by the working of the mean stress, its dissipation in collisions, and its transport by gradients, even rough statistical arguments permitted the calculation of the stress, rate of dissipation, and the energy flux in terms of the particle properties, the parameters governing a collision, and the mean fields of concentration, velocity, and fluctuation energy (e.g. Jenkins & Savage, 1983). Much of the development of theory for identical, frictionless, nearly elastic spheres was carried out in the context of the kinetic theory for dense gases (e.g. Chapman & Cowling, 1970). This led to improved averaging, excellent agreement of the predictions of the resulting theory with the results of numerical simulations and an appreciation for the limits of a theory based on the fluctuation energy. When the collisions are very dissipative, the theory for identical frictionless spheres has a more complicated structure (e.g. Richman, 1989). In this case, the deviatoric part of the second moment of the velocity fluctuations is as large as the isotropic part and it is necessary to include it as a field variable. A theory for simple shear that incorporates the full second moment in the velocity distribution function and uses a balance law to determine it, predicts stresses that agree well with numerical simulations, up to a point. As the size of the periodic cell in a simulation increases, the steady, homogeneous shearing flow is observed to break up into unsteady clusters (e.g. Hopkins & Louge, 1991). These clusters resemble those that are seen in a variety of particulate systems ranging from circulating fluidized beds to ocean sediments. They may result from a long wavelength instability in the simple shearing flow. An initial instability is predicted by the theory for nearly elastic spheres (e.g. Savage, 1992), so the clustering might be qualitatively understood in its context (e.g. Schmidt & Kytomaa, 1994). However, given the ubiquitous nature of particle clusters, it is important to have a theory that describes the gross dynamics of very inelastic grains. Consequently, a complete theory for identical, highly inelastic, frictionless spheres should be developed. Such a theory may not be possible to obtain using methods of the kinetic theory alone. In this case, the development of the appropriate phenomenology should be guided by numerical simulations. Friction in collisions provides an additional mechanism of dissipation. It also forces the consideration of the internal variables associated with the spin of spherical particles. Existing kinetic theories for frictional, nearly elastic collisions employ a collision model that is too crude. A more appropriate model, phrased in terms of the coefficient of friction exists (e.g. Foerester, Louge, Chang & Allia, 1994) and a kinetic theory based on it has recently been formulated for small friction (Jenkins & Zhang, 1995). Numerical simulations should be carried out to guide the incorporation of the rotational degrees into theories for large friction and asymmetrical particles in the simplest way possible. Shearing of mixtures of particles with different diameters inevitably results in their segregation by size. Such segregation is an important concern in many industrial processes and a common feature in geological deposits. Existing kinetic theory for binary mixtures of smooth, nearly elastic spheres (e.g. Jenkins & Mancini, 1989) should be employed to predict segregation in order to identify those mechanisms that are independent of the rotation of the particles. Again, with the guidance of numerical simulations, such theory should be extended to include polydispersity in particle diameters, greater inelasticity, and friction. Physical experiments and numerical simulations both indicate that the nature of the boundaries is crucial in determining the character of a flow. Indeed, boundary interactions are the source (or sink) of the fluctuation kinetic energy that determines the stresses. Using methods similar to those of the kinetic theory, boundary conditions have been derived for the slip velocity and the flux of fluctuation energy at boundaries that are bumpy and frictionless (e.g. Richman, 1988) or flat and frictional (Jenkins, 1992). This activity should be continued, informed by the results of numerical simulation (e.g. Louge, 1994). Of particular interest is the structure induced by the wall on the arrangement of the particles near it and the influence of this structure on the detailed dynamics of these particles as they interact with the wall. Whenever possible, boundary value problems should be solved and the solutions compared with the quantities that can be measured in the physical experiments. The solutions should also be used to design experiments that could indicate the limits of the theory or show it to be wrong.

EXPERIMENTS

Shear cell experiments, similar in spirit to Bagnold's but done in more sophisticated devices (e.g. Craig, Buckholtz & Domoto, 1987), have confirmed the quadratic dependence of the stresses for rapid flows of a variety of dry granular materials and provided an indication of the importance of the boundary conditions for spherical particles. A typical cell provides control over the normal stress and the rate of shear; the thickness of the flow and the magnitude of the shear stress necessary to sustain it are measured. Knowledge of the amount of material in the cell and the thickness of the flow provides an estimate of its average concentration. A cell with a transparent wall permits the study of particle velocities and local concentrations near the wall. The shear cell is a relatively simple apparatus, but its potential has not yet been exhausted. It should continue to be used to quantify the influence of the boundary conditions on the flow for a variety of boundaries and particles. This would be particularly valuable if non-invasive measurement of the local flow properties could be carried out at the same time. The development of such non-invasive measurement techniques should have a high priority. In recent years, several existing techniques have been adapted to rapid granular flows, and local flow properties have been measured by capacitance probes, x-rays, high speed optical cameras, and magnetic resonance imaging. Even if only limited local information can be obtained non-invasively, its use in conjunction with numerical simulations and careful measurements of the overall flow variables might suffice. An example of this is provided by a rapid flow down an inclined chute. Experiments on rapid flows down chutes attempt to establish a region of fully developed flow and to measure its overall properties. Overall properties are taken to be the angle of inclination of the chute, the mass flux through a unit area normal to the flow, and the mass above a unit area of the base. Given the parameters that characterize the collisions between particles and their interactions with the boundaries, we have reason to have confidence in a numerical simulation if it is successful in predicting the overall features of the flow. If, in addition, its predictions agree with several values of local flow parameters measured non-invasively, we seem justified in regarding the numerical simulation as a faithful representation of the flow. The point is that the existence of the numerical simulation may make more than a few non-invasive measurements unnecessary. Of course, in order to be able to make any comparison between the results of numerical simulations and the observations in physical experiments, we must be able to model collisions and to measure the parameters of the model. More experimental work is required here to measure interaction parameters for particles other than spheres. Not until this fundamental link between the physical experiments and the numerical simulations is made can we take advantage of the power of the simulations. Simultaneously, more accurate determination of the real particle interactions will permit the use of realistic collision models in the kinetic theory.

ACKNOWLEDGMENTS

The preparation of this review has been supported by the Granular Flow Advanced Research Objective of the Department of Energy and the Coastal Sciences Program of the Office of Naval Research.

REFERENCES

Bagnold, R. A. (1954) Experiments on a gravity-free dispersion of large solid spheres in a Newtonian fluid under shear. *Proc. Royal Soc. Lond.* A226, 49-63.

Campbell, C. S. and Brennen, C. E. (1985) Computer simulation of granular shear flows. *J. Fluid Mech.* 151, 167-188.

Chapman, S. and Cowling, T. G. (1970) *The Mathematical Theory of Non-Uniform Gases*, Third Edition, Cambridge University Press, Cambridge.

Craig, K., Buckholtz, R. H. and Domoto, G. (1987) Effects of shear surface boundaries on stress for shearing flow of dry metal powders - an experimental study. *J. Tribology* 109, 232-237.

Forester, S. F., Louge, M. Y., Chang, H., and Allia, K. (1994) Measurements of the collision properties of small spheres. *Phys. Fluids* 6, 1108-1115.

Goddard, J., Haff, P., and Walton, O. (1992) Non-linear dynamics of multi-particle systems. In *Research Directions in Computational Mechanics*, (Oden, J. T., Babuska, I., and Belytschko, T., Eds.) NAE.

Hopkins, M. A., and Louge, M. (1991) Inelastic microstructure in rapid granular flows of smooth disks. *Phys. Fluids* A 3, 47-57.

Jenkins, J. T. (1992) Boundary conditions for rapid granular flow: flat, frictional walls. *J. Appl. Mech.* 59, 120-127.

Jenkins, J. T., and Mancini, F. (1989) Kinetic theory for mixtures of smooth, nearly elastic spheres. *Phys. Fluids* A 1, 2050-2057.

Jenkins, J. T., and Savage, S. B. (1983) A theory for the rapid flow of identical, smooth, nearly elastic particles. *J. Fluid Mech.* 130, 187-202.

Jenkins, J. T., and Zhang, C. (1995) Kinetic theory for identical, slightly frictional, nearly elastic spheres. *Phys. Fluids* (in press).

Louge, M. (1994) Computer simulations of rapid granular flows of spheres interacting with a flat, frictional boundary. *Phys. Fluids* 6, 2253-2269.

McPhee, J. D. (1988) The control of nature. I & II. *The New Yorker*, Sept. 26 and Oct. 3, 1988.

Merrow, E. W. (1986) *A Quantitative Assessment of R&D Requirements for Solids Processing Technology.* Rand Corp.: Santa Monica.

Ogawa, S. A., Umemura, A. and Oshima, N. (1980) On the equations of fully fluidized granular materials. *J. Appl. Math. Phys.* 31, 483-493.

Richman, M. W. (1988) Boundary conditions based on a modified Maxwellian distribution for flows of identical, smooth, nearly elastic spheres. *Acta Mech.* 75, 227-240.

Richman, M. W. (1989) The source of second moment in a dilute granular flows of highly inelastic spheres. *J. Rheol.* 33, 1293-1306.

Savage, S. B. (1992) Instability of unbounded uniform flow. *J. Fluid Mech.* 241, 109-123.
Savage, S. B. and Sayed, M. (1982) Stresses developed by dry cohesionless granular materials sheared in an annular shear cell. *J. Fluid Mech.* 42, 391-430.

Sayed, M. and Savage, S. B. (1983) Rapid gravity flow of cohesionless granular materials down inclined chutes. *J. Appl. Math. Phys.* 34, 84-100.

Schmid, P. J., and Kytomaa, H. K. (1994) Transient and asymptotic stability analysis of granular shear flow. *J. Fluid Mech.* 264, 255-275.

Walton, O. R. and Braun, R. L. (1986) Viscosity, granular-temperature, and stress calculations for shearing assemblies of inelastic frictional disks. *J. Rheol.* 30, 949-980.

ACOUSTICS

Edward J. Kerschen
Aerospace and Mechanical Engineering Department
University of Arizona
Tucson, AZ 85721

INTRODUCTION

The processes of sound generation, propagation and attenuation have a significant impact on many aspects of human activities. The development of a technology-based society has, in general, been accompanied by increasing levels of man-made noise. In order to decrease noise pollution, an understanding of acoustic phenomena and an ability to predict these features in specific applications is required. Much progress has been made in noise abatement–perhaps the most dramatic example is the evolution of the commercial jet transport aircraft–but much remains to be done. In addition, possibilities for using acoustic fields to control important aspects of flow systems are now becoming possible, as discussed below.

Although the basic principles of classical acoustics were understood in the last century, many issues that impact the development of current and future technology are not adequately understood. These issues often involve the interrelationship of the acoustics to other features of flow fields, as will be seen in the discussion below. Thus, modern research in acoustics is an integral part of modern research in fluid mechanics in general. Synergistic interactions between research activities in acoustics and those in other areas of fluid mechanics have played a vital role in the development of current technology. Future progress depends even more heavily on such synergistic interactions.

In this section we shall discuss research needs in various fields of acoustics. It is not possible for this list to be all encompassing, and we admit to some bias toward aeronautical and hydronautical applications. However, it is these fields that have spawned the most intense research efforts in acoustics. Furthermore, developments in acoustics that originated in the aeronautical and hydronautical fields have found applications in a broad spectrum of commercial products.

Much acoustic research over the last several decades has focused on developing methods to predict and/or reduce noise fields. Much more needs to be done in this area, especially if the development of a supersonic transport is to be seriously contemplated. However, important research is also being done on the use of acoustic phenomena in other applications. Acoustic detection methods of increasing sophistication are being developed, particularly in ocean applications. Acoustic signatures are also being used to identify important features of various flow fields. Active control of some sound fields using microprocessor based systems appears feasible in the near future. With such control it may well be possible to influence not only the *unsteady* characteristics of various machines, but also their *steady state* characteristics such as overall efficiency, for example. *Thermoacoustic heat transport* is an approach that is mechanically much less complex than conventional systems and does not require CFCs. Finally, as in other aspects of fluid mechanics research, computational methods are playing an increasingly important role. However, some fundamental features of acoustic problems, particularly related to noise generation, that are not encountered in other branches of fluid mechanics pose significant challenges to computational fluid dynamics; these are discussed in the final subsection.

JET NOISE

Aircraft engine jets are an important source of noise for both subsonic and supersonic aircraft. Commercial air transport is currently dominated by subsonic aircraft. However, economic forecasts indicate that trans-Pacific air routes will represent more than 50% of the world market by the year 2010. Due to the long ranges of these routes, supersonic flight offers clear benefits. Economic studies indicate that a supersonic commercial aircraft would be viable in this market. However, the noise generated by the aircraft's supersonic jets during takeoff represents a formidable barrier to successful development. Thus, ways must be found to reduce the noise of supersonic jets without adversely affecting their propulsion efficiency.

Noise prediction techniques for subsonic jets are usually based on some form of Lighthill's acoustic analogy, in which the sound field is viewed and predicted as a weak byproduct of the turbulent flow. In the acoustic analogy the governing equations are rearranged in order to isolate a linear wave operator, the remaining terms being treated as a source which drives the wave motion. The source terms contain the fourth time derivative of the two-point Reynolds stress covariance, a quantity that is difficult to determine by experimental or other means. Thus, in order to utilize the acoustic analogy to predict the characteristics of jet noise, the source terms must be modeled empirically, guided by some combination of experimental data, computations and theoretical concepts. For subsonic jet speeds, noise generated by the fine-scale turbulence usually dominates over that due to the large-scale turbulent structures, and hence the empirical source term models typically ignore the large-scale structures.

In contrast, the large-scale turbulent structures of the jet shear layers play an essential role in noise generation by supersonic jets, particularly when the convection speed of the structures is supersonic relative to the surrounding medium so that the structures radiate Mach waves. If the jet is imperfectly expanded, a shock-cell pattern is present in the mean flow and passage of the large-scale structures through this pattern leads to the radiation of broadband shock-associated noise and, in some situations, jet screech. Theoretical models of these two supersonic jet noise mechanisms have been developed by Tam and his collaborators. His approach is called the wave model, since the sound radiation is calculated directly without recourse to the acoustic analogy.

Various concepts for suppression of supersonic jet noise have been proposed, such as the acoustically treated mixer/ejector and turbine by-pass concepts. Methods which reduce noise through enhanced mixing appear attractive. Among these are shock-free supersonic plumes with non axisymmetric distributions of exit-plane momentum thickness and jets which are excited by oscillating control devices or by feedback of time-dependent disturbances within the initial jet column. Further exploration of such concepts is necessary if supersonic jet noise is to be reduced to acceptable levels.

Supersonic jet geometries that incorporate noise suppression concepts are typically quite complex. They contain three-dimensional, high temperature, free and confined mixing layers. In order to determine the unsteady flow field in such configurations with sufficient accuracy that the noise generation can be understood and predicted, substantial improvements are required in measurement techniques and computational methods for the compressible turbulence in such streams. Noise prediction methods that account for the complexities of realistic geometries also need to be developed.

Research Needs

a. Experimental and computational investigations of methods to generate streamwise vorticity in incompressible and compressible shear layers, and evaluation of the effects on shear-layer growth rates.

b. Development of global three-dimensional velocimetry methods, for high-temperature ducted and unducted supersonic shear layers, that are capable of extracting the turbulence statistics required for jet noise prediction.

c. Development of prediction methods for the absorption of sound by an acoustic liner in the presence of a high-speed, high-temperature flow field.

d. Evaluation of the noise-reduction benefit associated with the tailoring of mean flow fields to reduce growth rates of high convection Mach number turbulent structures in shear layers.

e. Development of actuators and control law algorithms for excitation of supersonic shear layers to control mixing rates.

f. Development of noise prediction methods that are applicable to the complex geometries and flow fields associated with supersonic jet noise suppression concepts.

NOISE OF PROPELLERS, ROTORS AND TURBOMACHINERY

Propellers, rotors, fans and turbomachinery blade rows make significant contributions to the noise of aircraft, rotating machinery and industrial equipment. In order to reduce noise levels or, conversely, increase power while holding noise levels constant, accurate noise prediction methods are required so that design trade offs between noise, aerodynamic performance, structure and weight can be assessed. More sophisticated noise prediction methods also provide new insights concerning noise generation mechanisms, leading to new concepts for lowering noise levels.

The noise generation mechanisms can be separated into two classes. In the first class are noise mechanisms associated with the steady (or nearly steady) aerodynamic loading of the rotor. The acoustic analogy provides an effective means of analyzing noise sources of this type, particularly for propellers and helicopter rotors in which the distortion of the mean flow generally has a negligible effect on the wave propagation. In the acoustic analogy approach, the problem is separated into aerodynamic and acoustic components. An aerodynamic computation is performed to determine the aerodynamic loading on the rotor blades. This aerodynamic loading then provides the "source terms" for calculation of the acoustic field. The acoustic analogy approach is well developed and forms the basis for most of the currently available noise prediction methods. The primary area where improvements are needed is in the aerodynamic computation for the acoustic source terms, especially near blade tips in the presence of separated flow.

The second class of noise generation mechanisms involves the interactions between blade rows, and the interaction of rotors with nonuniform or unsteady inlet flows (wakes, vortices, etc.). These interactions produce broadband as well as harmonic noise components, and are important sources in aircraft turbofans and counter-rotation propfans and in turbomachines in general. For moderate blade row spacings, the most important sources are the impingement of the wake structure of an upstream blade row on the downstream row and interaction of the rotor with endwall boundary layers. At closer spacings the direct interaction of the aerodynamic potential fields of two blade rows in relative motion also becomes an important noise source. If the length scale associated with the blade row interaction (blade chord or spacing) is small compared to the wavelength of the generated sound, the sound source is "acoustically compact" and the acoustic analogy approach of separating the problem into unsteady aerodynamic and acoustic components is very convenient. This limit is particularly useful in hydronautical applications where the small Mach number guarantees that the length scale ratio is small. At the high subsonic Mach numbers typical of aeronautical applications, the length scale ratio for unsteady blade row interactions is typically not small and application of the acoustic analogy requires more care. In this situation acoustic effects, involving phase variations in the pressure distributions on the blade surfaces, become important even in the unsteady aerodynamic computation. A further complication that arises for ducted fans in aeronautical applications is that the propagation of acoustic waves in the region between the rotor and stator blade rows is strongly affected by the swirl present in that region.

Research Needs

Progress in the reduction of propeller, rotor, fan and turbomachinery noise requires theoretical, computational and experimental research on the following topics.

   a. Determination of the acoustic analogy source terms in the vicinity of propeller and rotor blade tips, including the influences of leading-edge and tip vortex flows, flow separation and interaction with endwall boundary layers.

   b. The downstream development of wakes of ducted rotors, including endwall and tip clearance effects, over distances on the order of several blade chords (i.e., at the stator inlet).

   c. Analyses and experiments that examine which of the complicated features of rotor wakes are most relevant to noise generation.

   d. Prediction methods that account for unsteady coupling between the rotor and stator blade rows and for the presence of swirl between the blade rows.

   e. Incorporation in prediction methods of installation effects, including diffraction, refraction and the presence of acoustic liners, for nonaxisymmetric nacelles at angle of attack.

COMBUSTION NOISE AND INSTABILITIES

Combustion noise and instabilities are important in a variety of applications, including aircraft engines and afterburners, rockets, ramjets, gas turbines and furnaces. As in jet noise, theoretical predictions of combustion noise are typically based on acoustic analogy formulations. The theoretical description of the sound generated by specified unsteady heating is well established. There are two main sound sources. The first is the unsteady expansion caused by fluctuations in the rate of heat release. The sound field this generates is influenced by the geometry of the combustor and the propagation path to the external fluid. Variations in the combustion rate also lead to spatial inhomogeneities in the exhaust gases, and additional sound is generated when these are convected through a region of nonuniform flow, as in a nozzle or turbine blade row. The difficulty in applying this theoretical understanding to practical problems, such as the noise generated by the combustor of an aircraft engine, is in obtaining an adequate description of the time-varying rate of combustion. Current prediction schemes resort to using empirical correlations to describe the unsteady heat release. However, computational fluid dynamics (CFD) is being applied to combusting flows of increasing complexity, and it is timely to explore its potential to describe the acoustic sources.

Combustion chambers are often highly resonant acoustic systems, and attempts to increase the density of energy release within the combustor can lead to instability. Combustion instabilities occur because, while unsteady combustion generates sound, the sound waves within an acoustic resonator can be so intense that they lead to further fluctuations in the rate of combustion. The oscillations can become so intense that they cause structural damage. Alternatively, they may enhance the heat transfer, leading to overheating, or the perturbations may simply become so violent that the flame is extinguished. The mean rate of energy release in many of the devices listed above is limited by the onset of damaging combustion oscillations.

It has long been known that high-frequency combustion oscillations can be controlled by the insertion of passive devices such as sound-absorbent liners, but low-frequency oscillations have proved more troublesome. In recent years, active control has emerged as a practical means of reducing these low-frequency oscillations and thereby extending the operating range of a combustor. There has been a number of laboratory-scale demonstrations of the power and versatility of active control of combustion instabilities. The benefits of such a control system have also been demonstrated on the afterburner of a full-scale aircraft engine. Further research and development are required before active instability control can be installed in operational systems. Robust controllers adapting to a range of operating conditions are required.

Of course, it would be preferable to design systems without combustion instabilities and so avoid the need for active control. But accurate prediction schemes have not yet been developed for most practical geometries. The difficulty is that predictions of the onset and frequency of combustion instabilities require knowledge of how the instantaneous rates of combustion are affected by flow unsteadiness. This information is usually not available. The geometries of practical interest are complicated, often involving a turbulent separated flow. Nevertheless, CFD is emerging as a promising means of predicting the unsteady combustion. In addition to its application in practical devices, active control can be used to eliminate naturally occurring instabilities in laboratory experiments, allowing the flow to be driven in a variety of ways to investigate the unsteady response of the flame to forced disturbances.

Research Needs

a. Combustion Noise

   i. Evaluation of CFD as a tool for predicting combustion noise sources.

b. Combustion Instabilities

   i. Investigation of the influence of unsteady flow on the rate of combustion.

   ii. Treatment of combustion oscillations as nonlinear dynamical systems.

   iii. Development of robust adaptive controllers and implementation of active control on practical systems.

ACOUSTIC-SHEAR FLOW RESONANCES

In many applications, geometrical and/or flow features may trigger acoustic feedback which stimulates shear flow instability waves, leading to resonances and the emission of strong acoustic waves of discrete frequencies. Typical of this class of feedback phenomena are jet screech tones, jet impingement tones, cavity resonances, edge tones and Helmholtz resonators. Cavity resonance due to aircraft wheel wells is an important component of airframe noise. Related acoustic-shear flow resonances that occur in transportation vehicles such as automobiles make a significant contribution to the overall noise level in these vehicles. When stronger resonances occur, the intense tones that are generated may cause violent structural vibrations leading to sonic fatigue or other structural failures in addition to an unacceptable noise level.

The discrete frequency sound field is generated by a feedback loop which contains the following elements: shear flow instability waves, one or more geometric surfaces or edges, and acoustic waves propagating upstream either inside or outside the shear flow. The instability waves arise at the upstream boundary of the shear flow (a nozzle lip, for example) and amplify as they convect downstream to a location where an interaction with some geometric or flow feature occurs. This downstream interaction generates an acoustic wave which propagates back to the upstream boundary of the shear flow. The impingement of the acoustic wave at the upstream boundary in turn regenerates the instability wave, closing the feedback loop. In jet screech the downstream interaction consists of the passage of the instability waves through the shock pattern of an imperfectly expanded supersonic jet; in most other cases the downstream interaction takes place at an edge of a solid surface.

Theoretical models based on the feedback loop concept have been reasonably successful in predicting the tone frequencies, at least in cases which involve well defined upstream and downstream boundaries and in which the shear layer development and acoustic feedback loops are easily modeled. However, there is no known way to predict the tone intensity theoretically or empirically. Details of the upstream and downstream geometry are known to significantly influence these resonance phenomena, but are not adequately accounted for in current models.

Research Needs

a. More accurate models for the coupling between the instability and acoustic wave fields at the upstream and downstream boundaries of the feedback loop.

b. Better understanding of the nonlinear behavior of the feedback loop.

c. Development of theoretical and/or computational models to predict the intensity of the radiated sound.

d. Experimental and theoretical study of the sensitivity of feedback resonances and tones to changes in the edge geometry or in the external environment.

PROPAGATION IN NONUNIFORM OR RANDOM MEDIA

Acoustic propagation in nonuniform or random media has long been a topic of vigorous research. Much of this work has been motivated by problems arising in propagation through the atmosphere or the ocean. Issues of acoustic source detection, noise prediction and reduction, and communication are of interest. In direct problems, one assumes that the properties of the medium are known, and examines how these properties affect the propagation of the wave field. Inverse problems, in which propagation phenomena are used to monitor or characterize the medium through which sound has traveled, are also of great interest. Issues related to the propagation of linear acoustic waves are discussed in this section; nonlinear effects are discussed in the following section.

The theoretical analysis of linear acoustic waves in inhomogeneous media is well developed. When the length scale of the inhomogeneity is small compared to the acoustic wavelength, the phenomenon is most conveniently viewed as a scattering problem. In the opposite limit of inhomogeneities whose length scale is large compared to the acoustic wavelength, geometric acoustics (ray theory) is applicable. Ray theory has provided physical understanding and predictive capabilities for a wide range of important phenomena, such as the development of shadow zones in inhomogeneous media, for example. Ray theory is the lowest-order result in a singular perturbation approach and may break down in local regions. For example, local analyses are required to describe the field near a caustic where adjacent rays intersect and ray-tube areas vanish, or in a shadow region where the penetration of the acoustic field involves creeping waves. Guided mode theory, which is applicable to situations involving disparate length scales such as when sound is channeled in layers in the ocean, has also been extensively developed. More recently, the parabolic wave equation approach, which attempts to incorporate both geometric acoustic and diffractive effects, has been utilized for the study of these sorts of features in long range sound propagation.

Methods that incorporate statistical concepts have been developed to analyze propagation in media with random properties. These analyses have been utilized to study the effects of the randomness of the medium on the performance of underwater sensing systems, for example. The propagation of linear waves over relatively short distances through a medium with weak and statistically homogeneous randomness is well understood. However, the statistical properties of the medium are inhomogeneous in many problems of interest, leading to significant complications in the analysis. Application of the theoretical approaches for wave propagation through random media to practical problems is currently severely hampered by a lack of theoretical tools for the simulation of the properties of the medium itself.

Current sensing systems are typically based on geometric acoustics, and therefore require sound with wavelengths much shorter than the scales of the inhomogeneities of interest. With advances in computational acoustics, full wave solutions for the propagation phenomena of interest should be possible, allowing sensing systems to extract additional information through the use of lower frequencies. Developments in instrumentation and data-processing technology are also having a significant impact. In particular, we anticipate new applications that will exploit the emerging feasibility of simultaneously collecting acoustic data in traditional and novel manners at a large number of sites and then processing these data with considerably reduced constraints on data storage and computational complexity. For example, bistatic scattering (including backscatter) of sound could be used to infer properties of turbulent wakes, shear layers and boundary layers. Also possible are tomographic systems that could yield nearly instantaneous spatial patterns of temperature, density, and sound speed.

Research Needs

a. Development of full-wave models for propagation through strongly inhomogeneous media, including analytical and computational methods.

b. Development of sensing systems that make fuller use of the basic principles of fluid mechanics and wave propagation and which utilize simultaneous collection of data at many sites.

c. Application of these models and sensing systems to direct and inverse problems of practical importance.

NONLINEAR ACOUSTICS

Nonlinear effects are important in a variety of situations in acoustics. Among these are the propagation of sound or weak shock waves over large distances, and the saturation of acoustic resonances in enclosed regions. Nonlinear effects lead to the development of harmonics and subharmonics and can also drive acoustic streaming. The fluid velocities associated with acoustic motions are almost always small compared to the speed of sound. Therefore, on the local space and time scales (wavelength and period), the acoustic motion satisfies the linear acoustic equations. However, the nonlinear effects are cumulative, so that the development of the acoustic field over a distance large compared to the wavelength (or a time large compared to the period) is described by a nonlinear equation.

In the last two decades, the development of soliton theory has led to major advances in the understanding of nonlinear waves. However, soliton theory is primarily applicable to situations in which nonlinearity is balanced by frequency dispersion. In contrast, acoustic waves in unbounded media usually propagate with very little frequency dispersion, and nonlinearity is balanced by diffusion or relaxation effects. Due to the lack of dispersion, short-scale components do not propagate away from regions of wave steepening and therefore shocks can form. The shock appears as a discontinuity on the global scale of the waveform; the structure within the shock is influenced by thermoviscous and/or relaxation parameters.

The simplest nonlinear equation expressing a balance between linear evolution, quadratic nonlinearity and thermoviscous diffusion is Burgers' equation, which describes forward propagation in one space dimension. An exact solution to this equation for arbitrary initial conditions can be obtained through the Hopf-Cole transformation. Important nonlinear phenomena, such as a stronger shock overtaking and swallowing a weaker shock, are described by the solutions to Burgers' equation. However, extensions to Burgers' equation are necessary in order to analyze phenomena of practical interest, including the influences of relaxation phenomena and inhomogeneous media. Unfortunately, it is apparently not possible to generalize the Hopf Cole transformation to obtain exact solutions to these extended equations. Investigations of these extended equations have therefore been carried out using asymptotic or computational methods.

Significant progress has been made in nonlinear acoustics, but many questions of practical importance remain to be answered. For example, the nonlinear focusing and defocussing of beams plays a crucial role in underwater applications and is not sufficiently understood. Much information could be gained through studies based on the Zabolotskaya-Khokhlov (ZK) equation, which is the nonlinear version of the parabolic wave equation mentioned in the previous section. The nonlinear propagation of sonic booms through the atmosphere, an inhomogeneous medium with non-uniform mean velocities and turbulent fluctuations, is an issue that has important implications for the development of a new supersonic commercial aircraft. There are some grounds for speculation that the turbulence may occasionally act as a giant lens, creating highly magnified booms at the ground. Nonlinear acoustic phenomena are also important in noise generated by aircraft engines and helicopter rotors and in the operation of internal combustion engines. The influence of real gas effects on the propagation of nonlinear acoustic waves is also not sufficiently understood.

Research Needs

Theoretical (asymptotic and computational) and experimental research on the following issues, especially for propagation in two and three space dimensions, should lead to significant advances.

a. Nonlinear focussing and defocussing of beams; studies of ZK and related equations, including integrable and partially integrable equations, behavior in small-dispersion limits and the embedding of model problems (e.g., ZK) in global descriptions; shock dynamics.

b. Propagation of nonlinear waves through randomly inhomogeneous media, or with random initial data; shock propagation over large ranges including turbulent scattering and thickening effects; nonlinear propagation and evolution of random, broadband signals (e.g., jet noise).

c. Nonlinear effects in the generation of sound fields by bodies in high-speed motion (e.g., propfans, helicopter rotors, etc.) and in the subsequent propagation of these sound fields.

d. Interaction between dispersive, dissipative and nonlinear mechanisms; mixed and other types of nonlinearity, retrograde fluid effects; applications to fluidized beds, geological materials, etc.

e. Nonlinear waves with phase change; evaporation and condensation phenomena; nonlinear acoustics in multiphase media and in gases at very high temperatures ("real gas" effects, dissociation, etc.).

f. Nonlinear propagation in media with special dispersion and absorption characteristics; propagation over walls with linear and nonlinear elastic and damping properties or frequency-selective absorption mechanisms; nonlinear fluid-structure interaction problems.

g. Coupling between gasdynamics and chemistry in exothermically reacting media; linear and nonlinear acoustics coupled to nonlinear chemical models; generation of aero-thermal fields including shock and detonation waves.

FLUID-STRUCTURE INTERACTION

Sound generation mechanisms often involve the motion of solid bodies, as discussed in Section 3. In many cases the density and rigidity of the solid bodies are so large that their motion can be considered as prescribed, independent of the fluid pressure field that exists on the surface of the body. The phenomena discussed in Section 3 usually fall into this category. However, in other situations the density and bulk modulus of the fluid are not negligible compared to the corresponding properties of the solid structure, and the "fluid loading" must be taken into account in analyzing the motion of the structure. In particular, wave propagation within the fluid and the structure is then coupled, leading to a fluid-structure interaction that plays an essential role in physical phenomena of interest.

Fluid-structure interaction is important in underwater noise radiation from ocean vessels, due to the relatively high density and bulk modulus of water. In addition to the direct radiation of noise from the propeller, structural vibrations of the hull can be important sources of noise when scattered at local inhomogeneities (ribs, rivet heads, edges, angles, etc.) that may be far removed from the hydrodynamic source region. With the increasing use of lightweight composite materials, one may anticipate a growing importance of fluid-structure interaction in aeronautical vehicles as well. For example, the flexibility of helicopter main rotor blades suggests that the associated noise generation mechanisms may be strongly influenced by fluid loading effects. In addition to its influence on noise radiation, the coupling of wave motion in the structure and the fluid may result in instabilities that lead to structural damage. The currently available theories do not provide adequate predictions of these phenomena.

Research Needs

Theoretical, experimental and computational research on the following topics is needed.

a. Basic problems of coupling between wave modes (shear, torsional, and compressive in structures; compressive and vortical in fluids) in plates and shells at local inhomogeneities; fluid-structure interaction with slow and rapid variation of structural properties; mean flow effects, absolute and convective instabilities; effects of laminar or turbulent boundary layers; fluid-structure interaction with anisotropic structures (laminates, composites).

b. Fluid-structure interaction for "complex" structures; inclusion of fluid loading in statistical energy analysis and related approaches; long range and short-range fluid coupling.

c. Development of doubly asymptotic approximation and other hybrid asymptotic-computational approaches; development of computational schemes for fluid-structure interaction incorporating structural inhomogeneities and mean flow effects.

ACTIVE CONTROL

The mechanisms by which sound fields are generated and the subsequent propagation of these sound fields are in many cases described by linear equations. The principle of superposition then applies and, at least in principle, it is possible to cancel the existing sound field by introducing an additional sound field of equal amplitude but opposite phase. Although this general concept had been recognized for many years, it is only in the last decade that "anti-sound" has begun to emerge as a useful noise-control technology. In the application of anti-sound to noise generated by combustion instabilities, it was found that the anti-sound field not only eliminated the noise field, but also suppressed the instability process itself. Thus, in addition to cancelling noise fields, low energy acoustic fields can also suppress instabilities in fluid mechanical systems that in many cases limit the performance of these machines. Much of the pioneering work on active control of sound fields and of system instabilities was carried out by Ffowcs Williams and collaborators. The potential benefits of active control are now becoming widely recognized and the research activity in this area is expanding rapidly.

To apply the anti-sound concept for noise control, three elements are required: sensors that measure the existing sound field, acoustic sources that produce the anti-sound, and a controller which utilizes input from the sensors to determine the appropriate output signal to the acoustic sources. The application of anti-sound for noise control is most easily achieved when the noise field has a relatively simple structure. The first successful demonstrations of practical anti-sound systems involved noise propagating in a constant-area tube, at frequencies below the cut-on frequency of the first higher mode. In this situation the sound field consists of nondispersive plane waves propagating along the tube axis, leading to modest requirements for the sensing system, acoustic sources and controller. At higher frequencies, additional acoustic modes can propagate within the duct. These higher modes are also dispersive, further complicating the application of the anti-sound approach. However, it is primarily at the lower frequencies that more effective noise control techniques are needed; passive devices are quite effective in attenuating high frequency noise. Therefore, the combination of active control for low frequency noise and passive devices for high frequency noise is an attractive approach. Anti-sound has also been applied to cancel low frequency noise radiated from industrial stacks and from the tailpipes of gas turbines.

Active stabilization of fluid mechanical systems by the introduction of acoustic fields has the potential for even greater economic benefits. The design and operation of fluid machinery are often limited by instability characteristics of the systems. When permitted to develop, these instability mechanisms produce unacceptable noise and vibration, and may also lead to operational or structural failure of the device. Although the limit state of such oscillations is generally highly nonlinear and difficult to predict or understand in detail, the initial development is in most cases accurately described by linear theory. The linear stability characteristics of the flow depend crucially on the dynamic response of the system boundaries. Thus, a flow system can be stabilized through active control of surface elements, with a control algorithm based on linear theory. Since the active flow stabilizer responds only to unsteady disturbances in the machine, stable flow can be maintained by low-power control systems acting on the initial low amplitude stages of the instability.

Active stabilization has been successfully applied to a number of systems. A flexible wing fluttering in a wind tunnel has been stabilized by activating a wall-mounted loudspeaker. Combustion instability is preventable at small scale with a low-energy audio device, and that knowledge has led to the suppression of buzz instability at flow conditions found in aircraft engine afterburners. The rotating stall instability in axial compressor systems has also been controlled by active methods. The delay of surge by active control has been demonstrated on a gas turbine engine, allowing a significant increase in power beyond the surge boundary for the uncontrolled engine.

Research Needs

a. Development of anti-sound systems for control of low-frequency noise in industrial equipment and transportation vehicles.

b. Exploration of active methods to control the noise generated by turbomachinery blade rows in aircraft engines and other machines.

c. Further development of techniques to increase turbomachinery performance by active control of flow instabilities.

d. Evaluation of the effect of external disturbances on the performance of active flow stabilization systems.

e. Exploration of novel actuators and sensors for active flow stabilization systems.

f. Application of recent developments in control theory (robust, nonlinear control for lumped and distributed parameter systems, neural nets, etc.) to active anti-sound and flow stabilization systems.

THERMOACOUSTIC HEAT TRANSPORT

Thermoacoustic heat transport is a process through which a time-periodic acoustic field generates a steady flow of heat. Like conventional engines, a thermoacoustic engine can be configured as either type of classical heat engine–a heat pump (or refrigerator) or a prime mover. Also like conventional engines, in order to generate work from a thermoacoustic engine it is necessary to provide proper phasing of the processes forming the thermodynamic cycle. However, in this respect, thermoacoustic engines are fundamentally different from conventional engines. In typical conventional engines, which are designed to be most efficient in the reversible limit, considerable mechanical complexity is required to achieve the proper phasing. In contrast, thermoacoustic engines function only when the natural irreversibility of thermal diffusion is present. This inherent difference means that the necessary phasing can be accomplished with few or no moving parts. Hence, thermoacoustic engines are mechanically much less complex than conventional engines. Moreover, the thermophysical properties of inert gases make them excellent working substances for thermoacoustic engines, eliminating the need for CFCs.

Typical thermoacoustic engines consist of a porous element (usually called the stack), with a large ratio of surface area to volume, housed in an acoustic resonator. In thermoacoustic refrigerators, a high-intensity acoustic driver generates a standing wave in the resonator. This standing wave stimulates a net transport of heat over the surface of the stack, resulting in a thermal gradient being established across it. In prime movers, a thermal gradient is imposed across the stack and, at sufficiently high gradients, sound is spontaneously generated in the resonator. Thus, in this case thermal energy is converted into acoustic energy, which, depending on the particular application, can be converted into other forms of energy (e.g., electrical).

Thermoacoustic devices have a broad range of application, including refrigerators, air conditioners, cryocoolers, drivers for pulse tube refrigerators, sound sources, and thermal-to-electrical generators.

Research Needs

Theoretical and experimental research in the following areas should lead to significant advances in the understanding of thermoacoustic heat transport.

a. High-capacity heat exchange in oscillatory flows.

b. Operation of thermoacoustic heat engines in high-amplitude acoustic fields.

c. The role of nonlinear processes in limiting the steady-state acoustic amplitude in prime movers.

d. Enhancement of the performance of thermoacoustic engines using acoustic fields with both standing wave and traveling wave components.

COMPUTATIONAL ACOUSTICS

In the past, direct numerical simulations have played a less important role in acoustics research than in other areas of fluid mechanics, for reasons that are discussed below. Presently available noise prediction methods are still dominated by theoretical approaches consisting of programmed analytical formulae, with empirical input from a data base of experimental results for the relevant "non-acoustic" elements of the flow. However, as computational acoustics matures in the coming decade, it will become an essential tool for understanding and predicting a wide range of acoustic phenomena.

Computational acoustics poses some unique challenges to developers of numerical algorithms. The computational fluid dynamics (CFD) schemes utilized so successfully in aircraft design and performance evaluations typically consider time-independent problems. Thus, these CFD schemes must satisfy only consistency and stability requirements. Time accurate methods have been developed for direct numerical simulations of transitional flows, but these flows have relatively low Reynolds numbers and hence do not contain a very wide range of spatial or temporal scales. Attempts to extend these calculations to higher Reynolds numbers have met with only limited success.

In developing numerical methods for computational acoustics, a number of issues must be addressed. Since acoustic waves are nondispersive, isotropic and nondissipative, the numerical scheme must also have these properties. In contrast, most CFD codes are dispersive and anisotropic and may be highly dissipative, particularly if artificial dissipation terms have been added to improve numerical stability. The outer boundary condition also requires special treatment, since in most situations only the acoustic source region would be contained in the computational domain. Nonreflecting boundary conditions are required to allow acoustic waves to propagate out of the finite computational domain. Thus, computational acoustics requires numerical methods that are quite different than those used for CFD.

Acoustic source mechanisms are often very inefficient, especially at low Mach numbers where the acoustic variables may be five orders of magnitude smaller than the hydrodynamic fluctuations. Thus, the separation of the physical noise field from computational noise due to roundoff and truncation errors is a nontrivial issue. This difficulty is particularly acute for low Mach number computations of quadrupole sources such as jet mixing noise, where a slight violation of mass or momentum conservation in a numerical scheme would introduce a spurious monopole or dipole source whose greater radiation efficiency could lead to significant errors in the far field signal.

Finally, the Reynolds numbers in acoustic problems of practical interest are often very high. In addition, the frequency range of human hearing (50 Hz - 20 kHz) is quite wide. Thus, accurate simulation of problems such as jet noise requires a numerical method that is capable of accurately resolving a wide range of spatial and temporal scales. The wide frequency range also implies the need for quite long run times in order to obtain accurate spectral information.

The methodologies for computational acoustics are presently in an early stage of development and the challenges are formidable. However, the development of numerical methods tailored to the special requirements of computational acoustics, coupled with the continual improvements in computational hardware, will most certainly have a major impact on all areas of acoustics and noise control.

Research Needs

a. Development of stable, nondispersive, nondissipative computational schemes capable of resolving acoustic waves with wave lengths as short as four to five grid mesh spacings. Development of nonreflective boundary and inflow-outflow conditions.

b. Feasibility studies and calculations for model problems in order to develop confidence in the methodologies and their robustness.

c. Direct numerical simulations of complex acoustic problems such as supersonic jet noise and rotor-stator wake interaction noise.

ACKNOWLEDGEMENT

The following individuals generously contributed to the preparation of this section: A.A. Atchley, D.G. Crighton, A.P. Dowling, J.E. Ffowcs Williams, D.B. Hanson, M.S. Howe, A.D. Pierce, J.M. Seiner and C.K.W. Tam. The author gratefully acknowledges support from the Air Force Office of Scientific Research and NASA Lewis Research Center.

REFERENCES

Atassi, H.M. (ed.), (1993) Unsteady Aerodynamics, Aeroacoustics and Aeroelasticity of Turbomachines and Propellers, Springer-Verlag.

Blackstock, D.T. and Hamilton, M. (eds.), (1996) Nonlinear Acoustics, Academic Press.

Candel, S.M. and Poinsot, T.J., (1988) *Interactions between acoustics and combustion*, Proc. Inst. Acoustics, 10, 103-115.

Crighton, D.G., Dowling, A.P., Ffowcs Williams, J.E., Heckl, M. and Leppington, F.G., (1992) Modern Methods in Analytical Acoustics, Springer-Verlag.

Crighton, D.G. and Innes, D., (1984) *The modes, resonances and forced response of elastic structures under heavy fluid loading*, Phil. Trans. R. Soc. Lond. A312, 295-341.

Culick, F.E.C., (1988) *Combustion instabilities in liquid fueled propulsion systems - an overview*, AGARD Conference Proceedings No. 450.

Ffowcs Williams, J.E., *Anti-sound*, (1984) Proc. R. Soc. Lond., A395, 63-88.

Hardin, J.C. and Hussaini, M.Y. (eds.), (1993) Computational Acoustics, Springer Verlag.

H.H. Hubbard (ed.), (1991) Aeroacoustics of Flight Vehicles, Vols. I and II, NASA Reference Publication 1258. Reprinted by the Acoustical Society of America, Woodbury N.Y., (1995).

Jensen, F.B., Kuperman, W.A., Porter, M.B. and Schmidt, H., (1994) Computational Ocean Acoustics, American Institute of Physics, New York.

Lighthill, M.J., (1980) Waves in Fluids, Cambridge University Press.

Munk, W., Worcester, P. and Wunsch, C., (1995) Ocean Acoustic Tomography, Cambridge University Press.

Nelson, P.A. and Elliot, S.J., (1991) Active Control of Sound, Academic Press.

Pierce, A.D., (1989) Acoustics: An Introduction to its Physical Principles and Applications, Acoustical Society of America, Woodbury, N.Y.

Swift, G.W., (1988) *Thermoacoustic engines*, J. Acoust. Soc. Am., 84, 1145-1180.

Tam, C.K.W., (1995) *Supersonic jet noise*, Ann. Rev. Fluid Mech., 27, 17-43.

MOLECULAR DYNAMICS OF FLUID FLOW

Joel Koplik
Levich Institute and Department of Physics
The City College of the City University of New York
New York, NY 10031

INTRODUCTION

Since a fluid is composed of molecules, one always has the option of calculating its static or dynamic properties by computing the motion of these constituents. For most purposes such a procedure is very inefficient, because it provides detailed information at molecular length scales, which are far beneath the usual region of interest for continuum fluid mechanics. There are, however, situations where the microscopic details of a fluid flow are interesting if not crucial. For example, fluids in microscopic geometries or under high stress may exhibit deviations from the continuum equations, and one would like to calculate such effects. Alternatively, in some problems the boundary conditions to be applied to the Navier-Stokes equations are not fully established or are unsatisfactory, as in the presence of moving contact lines or at the edge of a porous medium. The usual way to address such issues is laboratory experiment or statistical mechanical calculation, but these methods have their own limitations, and molecular simulation can provide alternative insights and results. Another relevant class of problems concerns the merger or rupture of fluid interfaces, as in the fission and coalescence of droplets. Here most of the process is a slow viscous free boundary problem, but the breaking of the interface itself occurs at molecular length and time scales, below the resolution of the continuum equations. Likewise, the behavior of very thin liquid films, in particular in spreading on solid surfaces, can involve these small scales and, in addition, the details of the solid-liquid interaction which are hidden in bulk surface tension coefficients. These cases exemplify problems where a microscopic calculation can provide new insights as well as hard-to-obtain quantitative information which complements other techniques.

In this chapter, we first review the technical aspects of MD calculations, then turn to applications in bulk flow simulations, flows in this films and microporous systems, interfacial dynamics, and wetting phenomena, and then consider the somewhat related subject of lattice gas (or cellular automata) simulations. We conclude by listing some promising areas where molecular simulations are likely to provide significant insights in the near future.

Technical Issues

In a typical molecular dynamics (MD) computer simulation, one begins with a set of molecules occupying a region of space in two or three dimensions, with each assigned a random velocity corresponding to a Boltzmann distribution at the temperature of interest. The interaction of the molecules is prescribed, in the form of a two-body potential energy, and the time evolution of the molecular positions is obtained by integrating Newton's equations of motion. (Alternatively, at the expense of realism, the molecules may be treated as rigid discs or spheres, which move freely between elastic collisions.) One integrates for some time period, observes how the molecules have rearranged themselves, and computes averages of density, velocity, stress, temperature, etc., fields. The latter quantities may then be analyzed and used to augment results from experiment or continuum calculations. MD simulations are intrinsically computer-intensive calculations. At present, high-end scientific workstations can handle $O(10,000)$ molecules easily, with CPU times in tens of hours, general-purpose vector supercomputers can increase the speed and deal with somewhat larger systems, and with massively-parallel machines the size limit is $O(10^6)$ particles.

Evidently, the MD method could be applied equally to gases, liquids or solids, provided the interactions are known. Solids are outside the scope of this report, and for a rarefied gas an MD calculation is exceedingly inefficient, because most of the time the molecules are isolated and move along linear trajectories. In gas flows where numerical molecular calculations are of interest, Monte Carlo methods are preferred see Muntz, Rarefied Gas Dynamics, in this volume. In the remainder of this discussion we focus on dense liquids.

Although straightforward in principle, several problems arise in actually carrying out the above procedure. The first is that the molecular scales are so small that typical MD simulations at best involve a region of fluid of $O(100\mathring{A})$ in linear size, over time intervals of $O(10^{-9}sec)$. Fortunately, as discussed below, this seems to be enough for continuum behavior to set in. Secondly, some simplification in the intermolecular interaction is needed so that large systems are accessible. The computation itself is the integration of a large set of coupled ordinary differential equations for molecular positions, whose driving term is the force on one molecule due to the others. ODE algorithms lend themselves to parallelization easily, and so the bulk of the computation is the force calculation. Even the simplest interaction with a physical basis, the Lennard-Jones 6-12 potential for isotropic non-polar liquids, has a long range tail coupling any pair of molecules in the system and leads to computation times $O(N^2)$ for N molecules. In practice, one usually cuts off the potential, which with an interacting-neighbor list or a linked-cell list brings the effort down to O(N) or so. Occasionally one can estimate the correction terms using the two-body correlation function. In more general fluids such as (hydrogen-bonded) water or ionic liquids, the long-range forces are essential and simulations are restricted to smaller systems.

A third general difficulty concerns boundary effects: generally one prefers periodic boundary conditions to improve the rate of convergence to the thermodynamic limit, and to avoid anomalous behavior at the edges of the sample box. However, interesting flows frequently have non-uniform geometries not quite consistent with periodicity, and since the calculations are often at the limit of computer feasibility, one may not have the luxury of adding a buffer region at the system's edges. Fourthly, viscous flows generate heat which can significantly raise the temperature of a small system. The easiest method of temperature control is constant-kinetic-energy equilibration, where the velocity of each molecule is rescaled at regular intervals, but this is rather unphysical. For homogeneous systems, at least, one can alter the equations of motion to couple the molecules to an artificial heat bath. Another alternative, if the simulation involves a solid boundary, is to extract heat from the outer parts of the boundary, imitating the cooling of the apparatus in a laboratory.

Finally, there is the issue of extracting a signal from the noise. Most interesting small-scale fluid phenomena occur at low Reynolds number, which is to say low average molecular velocity. On the other hand, the molecules have a roughly-Brownian thermal motion, which cannot be eliminated because if the temperature is too low, the liquid will solidify. In practice the desired mean (Eulerian) velocity can be substantially smaller than the thermal velocity, so that some significant averaging is required. In steady state flows, one can simply average the molecular velocities in a sampling bin fixed in space over as long a time interval as needed. Practically speaking, the degree of difficulty depends on the number of dimensions in which the velocity varies and the presence of variation in time. In channel flows, two directions can be averaged over, allowing the bin size to be relatively fine in the direction of variation while still containing $O(100)$ molecules. Flows with two-dimensional variation, and $O(10)$ molecules per bin, are often resolvable, particularly when steady, while fully-three-dimensional flows seem inaccessible at present. Calculation of the stress tensor is intrinsically more difficult, because the microscopic stress tensor is computed from the molecular flux of momentum which involves the instantaneous intermolecular force. On the one hand this quantity is a rapidly-varying function of position, and on the other it is intrinsically non-local as it relates to pairs of molecules. Nonetheless, it can be determined with reasonable precision in, e.g., steady channel flows.

Often, the molecular motion itself is restricted to two dimensions, allowing a simulation of a larger system. This procedure is problematical for two reasons. First, there is no strictly two-dimensional Newtonian fluid, because in 2d the viscosity and other transport coefficients diverge in the thermodynamic limit. Secondly, if one is interested in boundaries or interfaces, there is known to be a very sensitive dependence of interface dynamics on spatial dimension, and 2d results can by no means be carried over to 3d.

Bulk flow simulations

One might naively believe that MD cannot provide any new information about simple bulk flows, because of the scales involved. The classical counterexample is the discovery of the power-law decay at long times of the velocity autocorrelation function, discovered numerically by Alder and Wainwright, and subsequently verified by other methods. Furthermore, a number of bulk flow calculations have demonstrating the applicability of the technique despite the limitations cited in the last paragraphs. For example, one can calculate a quantitatively correct Couette flow in a system of only $O(1000)$ molecules, in which the velocity profile is linear, the shear stress is constant, and the ratio of the shear to velocity gradient gives a viscosity consistent with both experiment and the Kubo-Green formula. Likewise, Poiseuille flows through a channel with solid walls have been simulated, using a great variety of model walls ranging from a heat-bath-like thermal boundary to a crystalline molecular solid, and invariably giving the expected flow fields.

Channel flow simulations have provided a firm basis for understanding the familiar no-slip boundary condition. In numerous MD simulations, the average fluid velocity is always observed to vanish within a few molecular diameters of a solid wall, independent of the details of the interactions. The origin of this behavior is that packing considerations tend to produce fluid layers at a wall, whose tangential momentum is absorbed by the solid lattice. Some degree of microscopic slip appears in general, which varies with the details of the solid-fluid interaction and the solid lattice structure, but such deviations are macroscopically irrelevant. Another novel feature of bulk flow calculations is that one can determine the length scale at which agreement with the Navier-Stokes equations breaks down, or in other words, how small a system can still be described as a viscous fluid continuum. In these configurations, as well as in a variety of others which have been studied, the answer seems to be that the continuum behavior sets in when the linear size of the system is about 10 molecules. (In more complicated geometries or at higher Reynolds numbers, where the flow develops structures on larger scales, obviously one would require a larger simulation.)

More generally, one can use MD simulations of bulk flow to examine the limitations of Newtonian behavior, by determining fluid behavior under conditions which are not readily accessible to experiment and where theoretical calculations are uncertain. An early instance was the quantitative study of the corrections to Stokes' law for the drag on a sphere, as its size decreases. More recently a number of simulations have indicated shear ordering in Lennard-Jones fluids at extremely high shear rate, approaching the inverse period of oscillations about the potential minimum. The usual linear constitutive relations survive at lower shear, where most measurements are made. Such results provide some insight into the observed wide range of validity of the linear behavior. Furthermore, they provide a data set for more sophisticated (often non-local) rheological models of fluids. While these phenomena are perhaps of limited practical relevance for simple fluids, they illuminate the underlying statistical mechanics of a liquid and motivate analytic work. Furthermore, MD calculations using more sophisticated molecular structure can provide detailed useful information for the rheology of complex fluids, where the time scales are longer and non-Newtonian behavior more evident. A somewhat related question concerns the assumption of local thermodynamic equilibrium used in the study of shock waves in compressible fluids. The shock itself occupies a small region of space in the flow where fine-scale laboratory measurements are difficult, but the detailed information present in MD simulations both verify the Navier-Stokes modeling assumptions and provide structural information about the fluids near the shock.

In situations where fluctuations are significant for bulk behavior, the intrinsic thermal noise in MD may actually be helpful. For example, various calculations of the onset of instability in Rayleigh-Benard convection for 2-d systems of $O(10000)$ hard discs, show the appearance and slow evolution of 15 convection rolls, depending on system size and boundary conditions. Analogous behavior is seen in simulations of the wake of one or another solid obstacle in a flow. Two points are illustrated by this work: the general agreement of a molecular simulation with a continuum calculation with the same parameters for a non-trivial flow, and the presence and consequences of substantial thermal fluctuations in small systems. The advantage of this calculational method is that while in continuum calculations the fluctuations are generally put in by hand, in an ad hoc manner, in MD they are "physical" and are guaranteed to have the correct form.

Microporous systems and lubrication films

Systematic deviations from continuum behavior occur in microporous systems when the flow channels are only a few molecules in diameter, and several groups have begun to characterize the deviations from bulk behavior by combining MD results with statistical mechanics. A number of general studies have emphasized the effects of the confining geometry on diffusion and viscosity, although further work is needed to understand how the interconnectivity of the pore space influences transport. Similarly, one expects to observe deviations from classical lubrication theory when fluid film thicknesses approach the molecular scale. In fact, in MD simulations of a sphere translating or rotating near a wall, the deviations are found only when the film becomes extremely thin, roughly one molecule across, and would normally be overwhelmed by the roughness of realistic solid surfaces.

A striking related application of MD to liquid film behavior occurs in atomistic simulations of the friction between solid bodies. A monolayer film confined between two regular lattice solids will typically behave as a solid in the absence of bulk motion, whose atoms try to settle into the potential minima formed by the lattice around them. When motion commences, the shear stress builds until the film melts, whereupon the solid can translate across one atom and relax the stress. As the solid continues to move, the stress again builds up, and so on, so the film continually melts and freezes as the solids slide past each other. Although the oscillatory stress can be observed in the lab, MD simulations are invaluable in arriving at a detailed understanding of the mechanism.

Interfacial fusion and fission

In continuum fluid mechanics a phase interface represents a deformable but persistent surface where material properties change and boundary conditions are applied. Microscopically however, there is a smooth transition region in molecular structure, which may appear or disappear, and such interfaces may split or merge. Earlier MD simulations have investigated simple static cases in quantitative detail, for example a drop in equilibrium with vapor, while more recent work has begun to address interface evolution and stability. Recently, the surface tension induced breakup of a liquid cylinders into a sphere, and the fission and fusion of drops in shear flow have been studied in MD simulations.

In general, one finds that the "bulk" behavior of such interfaces is in accord with continuum estimates, to a degree which depends on the amount of fluctuation in the system. A cylinder of several thousand MD atoms surrounded by vapor actually has a rather diffuse boundary, and undergoes appreciable three-dimensional shape fluctuations with only limited damping, whereas a similarly-sized drop surrounded by an immiscible background fluid is comparatively stabilized by the external molecules of the background. The former case shows an order of magnitude agreement between observed rupture times and simple continuum estimates, while in the latter case the agreement for drop shape and rupture characteristics is much better–10-20%. At the same time, attempts to measure velocity and stress fields only display thermal noise. These results not only indicate again that continuum behavior sets in on the 5-10 molecule length scales, but that such behavior appears even when there are no discernible continuum fields. The molecules themselves are affected by interface breaking only in an average sense - in rupture a region of interface is gradually depleted of molecules until the two fluid bodies are held together by thinning necks whose molecules eventually withdraw into the bulk, while in coalescence a thin tendril of molecules slowly forms between drops, which gradually thickens as the drops merge.

In these problems, MD simulation techniques have been shown to be capable of exhibiting the phenomena of interest, but have not as yet produced any new quantitative results. The shortcoming is one of computer power, a front on which great progress is expected in the near future. Aside from fundamental studies, the fact that apparently most macroscopic interfacial phenomena have their quantitative microscopic counterparts can provide complementary information for new methods and materials in microscopic fabrication.

Wetting dynamics

The wetting properties of fluids in contact with solids have a number of difficulties from the point of view of experiment and continuum calculation. In the laboratory, impurity and contaminant effects, coupled to difficulties in the small-scale observation of a two or three phase dynamic system have long plagued experiments, while from a modeling point of view, all of the complications of the subject tend to be buried in a few phenomenological coefficients (surface tensions, contact angles, Hamaker constants, etc.). An MD simulation provides very controlled inter-species interactions along with a complete knowledge of the microscopic state. The difficulty is one of time scales: wetting phenomena are often slow, if not hysteretic, and it is not always possible to reach the asymptotic state of interest. In addition, since the solid-fluid interaction is crucial, it is almost obligatory that the solid be treated as a molecular system as well, and such simulations inevitably require large numbers of particles.

Static MD simulations of wetting have a relatively long history, and provide the dependence of contact angle on interaction parameters, the shape of static drops, and a verification of Young's equation. The latter is a non-trivial accomplishment, since solid-fluid surface tensions are extremely difficult to measure. Molecular systems as complicated as realistic water have been used in this context. More interesting are studies of moving contact lines and drop spreading dynamics. The contact line problem is really the problem of the origin and limitations of the familiar no-slip boundary condition. This condition was originally inferred experimentally in the 19th Century, and supported by simple kinetic theory calculations of Maxwell. MD simulations have indicated its origin in the combination of interaction between liquid and solid particles, coupled to the crowding in a dense liquid. Simulations of a contact line between immiscible fluids moving across a solid, where a standard hydrodynamic analysis predicts a nonintegrable divergent shear stress, show that finiteness is restored by slip on a length scale of a few molecular diameters. Essentially, the fluid tries to "stick" to the solid, but can only do so at the expense of an infinite force, and instead effects a compromise with some slip and a large but finite stress. The boundary condition which should replace no-slip is not known from these simulations, since the high stress also invalidates the Newtonian properties of the fluid locally, but again this could be resolved by more extensive computations.

Dynamic drop spreading simulations have indicated that the full range of spreading regimes observed in the laboratory partial to complete to terraced may be reproduced with simple modifications of a Lennard-Jones interaction. The simulation of the terraced case, in which distinct molecular monolayers advance across a solid surface at different velocities, is particularly informative, since the experiments have very poor resolution within the monolayers and the additional information obtained can be critical in theoretical modeling. The initial studies on this problem (as well as the contact line problem) employed only simple molecules and structurally and chemically perfect crystalline solids. While exhibiting terraced spreading, the layer growth rates disagreed with experiment. Subsequent simulations with chain molecules made by tying Lennard-Jones atoms together, using either an extra confining potential or a fixed bond, were in agreement with experiment. Related additional studies have found similar variation in spreading rate as a function of surface corrugation. The differences in spreading due to molecule and surface structure, as well as the chemical heterogeneity of realistic materials, have only begun to be examined, and MD simulations are ideal because in contrast to real materials simulated parameters may be varied one at a time. Likewise, an understanding of hysteresis in wetting phenomena, in contact angles for example, is ripe for further work.

Lattice-gas hydrodynamics

Usually one thinks of the Navier-Stokes equation as a long and time and wavelength asymptotic limit of a Boltzmann equation, describing molecular constituents with a local momentum and energy-conserving interaction. It is plausible (and the result of a detailed analysis) that the same equation describes a similar limit of a much simpler microscopic system of point particles moving on a lattice, with discrete velocities and with suitable collision rules. Although originally devised as a statistical mechanical model, it was subsequently realized that simple lattice collision rules could be implemented as a computational alternative to standard CFD techniques. In particular, much of the floating point arithmetic of standard numerical simulations could be replaced by fast integer bit-wise operations. The picture is not quite as simple as this in several regards, having to do with extra (unwanted) lattice symmetries, having to complicate the collision rules and slow down the calculation to get a reasonable equation of state, and having to average over the intrinsic fluctuations of the lattice particles. The extra-symmetry problem is particularly severe in 3d, but the recent construction of an appropriate look-up table seems to have this under control.

Despite the difficulties, a number of successful two-dimensional studies have shown that lattice gas models can simulate real flows. Unfortunately, because of the complications, they are not particularly fast. Comparisons between lattice gas and finite element calculations of the same problem tend to give comparable computation times. The prospects for high accuracy in the original lattice gas method is not encouraging either, since statistical fluctuations (roughly Gaussian) must be overcome. Furthermore, estimates of the computational requirements for turbulence simulations, where new methods are most desirable, are quite discouraging. (A recently developed alternative is the lattice-Boltzmann method, where the evolution of the probability distribution function of the Boltzmann equation, rather than the explicit local occupation density field, is computed in a lattice-gas approximation.) Nonetheless, there are problems in which this technique may be very useful. The first involves flow in complicated domains, as arise in porous media and suspension problems at the particle scale. While there is no problem in principle with conventional techniques, the programming can be extremely tedious and the storage requirements can be excessive. In a lattice gas formulation, an irregular solid boundary is easily implemented by a particle reflection rule on certain sites, and previous work has shown that tracer diffusion or even immiscible fluids can be simulated at no great expense. A second potential application involves modeling complex non-Newtonian fluids, by appropriate modification of the lattice rules. Shear-ordering and other non-trivial structural effects in composite fluid systems have been studied in this way. While simulations by this method do not have the realism of standard MD, they are vastly easier to carry out, and should provide semi-quantitative insight in many situations.

Opportunities

There are several exciting avenues for future work on the MD simulations of fluid flows.

- There is the purely computational problem of simulating large enough systems to see velocity and stress variation in three-dimensional time-dependent situations, such as a distorting interface. Presumably, a few orders of magnitude in number of particles is needed to overcome the thermal fluctuations, but it must be remembered that a larger system requires a longer time for a flow to develop. Recently, MD computations with up to 108 particles on massively parallel computers have appeared, although no attempts have yet been made to study non-trivial flows.

- A related partly conceptual problem is that of matching MD simulations to bulk flows. Often, only a small subregion of a flow might entail any molecular scale questions, and it would be desirable to put ones molecules there and match to a continuum flow calculation outside, but the difficulty is the stress. A number of methods exist for controlling the average velocity of the molecules in a region, all based on the fact that the Eulerian velocity is simply the average molecular velocity, but the stress is a nonlocal function of molecular separations and forces and is not readily controlled.

- Problems of wetting are particularly well suited for molecular simulation. The particular advantages of MD are the ability to explicitly control the properties of the fluids and solids involved, as well any possible heterogeneity in the system, and study the evolution of interfacial shapes, local mass transfer, forces exerted on the substrate, and so on. The question of hysteresis in wetting is particularly apt, since this phenomenon is sensitive to random heterogeneities which are difficult to control in the laboratory or address theoretically.

- Flow in microporous systems intrinsically involves small spatial regions, so that MD is directly relevant, and other theoretical methods are limited by the absence of a reliable continuum description at very short distances. Most MD work to date considers a single pore, in effect, but enhancements in computational power will soon make at least small interconnected pore networks accessible to simulation. Aside from viscous transport problems, important issues in selective absorption, hindered diffusion, and modified phase transformations are open.

- Questions in the rheology and flow of non-Newtonian fluids have begun to be subjects of MD simulation. Given the wide variety of constitutive relations found in the literature, and the non-trivial nature of many rheological measurements, an independent source of information is most desirable. Several groups have studied the viscometric properties of chain molecules and the equilibrium dynamics of polymer melts and solutions, where typically, the chains are made of Lennard-Jones monomers with lengths up to about 60, and the solvent is treated as a Brownian force. Outstanding open questions amenable to MD include determining the behavior of singular corner flows, understanding the molecular underpinnings of rod-climbing and other exotic non-Newtonian properties, and perhaps the direct determination of constitutive equations.

GENERAL REFERENCES

M. P. Allen and D. J. Tildesley, (1987) *Computer Simulation of Liquids* (Oxford).

G. Cicotti and W. G. Hoover, eds., (1986) *Molecular Simulation of Statistical Mechanics Systems* (North Holland).

J. Koplik and J. R. Banavar, (1995) Continuum deductions from molecular hydrodynamics, *Annu. Rev. Fluid Mech.* **27**, 257.

M. Mareschal, ed., (1990) *Microscopic Simulation of Complex Flows* (Plenum).

D. Rothman and S. Zaleski, (1994) Lattice gas models of phase separation, *Rev. Mod. Phys.* **66**, 1417.

STABILITY OF FLUID MOTION

Sidney Leibovich
Sibley School of Mechanical and Aerospace Engineering
Cornell University
Ithaca, NY 14853-7501

INTRODUCTION

The study of stability is motivated by the observation that many states of flow, particularly when dissipative effects are small, can be drastically altered by small disturbances. Such alterations may be desirable or unwelcome, and in either event, a knowledge of whether a flow can be easily altered, or of what kind of disturbances to it will produce the greatest desired effect, provides the basis for the decisions required. This sort of question is addressed by characterizing the instabilities to which the flow may be subject. As such, stability issues are fundamental to all areas of fluid mechanics, and are not associated with any particular arena of application. Once an esoteric, perhaps even somewhat mysterious, fluid mechanical specialty, stability investigations have now permeated most application areas.

The progress made in stability theory and experiment has been explosive in the past decade. To a large extent, this may be attributed to the same factors that have affected all areas of fluid mechanics – and science in general – the availability of powerful computing equipment, and the development of new experimental instrumentation and techniques such as noninvasive optical tools, which themselves often are made possible by the existence of powerful computers. This has been important, but major conceptual progress has also been made, and is of comparable importance.

Although the subject is fundamental to all flows, it usually is divided into the kind of "basic" flow whose stability was being subjected to investigation (such as shear flows, rotating flows, thermally unstable flows, and so on), since in each of these, instabilities that may occur are associated with different physical mechanisms, and to a certain extent distinctive theoretical paradigms have developed as the appropriate (or convenient) methods of treatment.

Much progress has been made in the investigation of particular types of flows, but a description of these advances far exceeds the scope of this essay. In this chapter, attention will be restricted to new approaches that have appeared in stability *theory*, and which have proven to be valuable or offer promise for future development. This is followed by suggestions of

129

opportunities for future research, again offered in a problem-independent vein. An attempt will be made to confine references to literature reviews where original research contributions are given their proper due. As a general recommendation, the reader is referred to the most comprehensive collection of literature reviews extant, the *Annual Reviews of Fluid Mechanics.*

GENERAL TRENDS

The primary problem in stability theory devolves to the solution of the linear initial value, boundary value problem (IBVP) for a perturbation given to a prescribed flow state. If perturbations emanating from any physically admissible initial conditions can grow indefinitely large as time (or distance) tends to infinity, instability is declared. If not, the flow is stable. One of the intellectual attractions of the subject is the up or down decision - the flow is either pronounced stable or not. Whether such a determination is of use in interpreting real phenomena is another matter. For that, one needs to know whether the conditions that lead to a declaration of instability can in fact be presented to the system; whether disturbances of finite size may cause the flow to be altered to another flow so that the linearized analysis carried out has no relevance in describing the development of the disturbed state; whether disturbances grow very large for a while but decay asymptotically in time; and indeed whether "asymptotically in time" is meaningful for the problem at hand; and so forth. Many of the newer developments address questions such as these.

A large part, but not all, of the first five items in the list below pertain to boundary layers, mixing layers, and jet flows that have streamlines that are approximately parallel. The crucial feature of these flows, from the viewpoint of stability, is that they are "open flows," meaning that they are most naturally represented in terms of some boundaries (inflow and outflow) that all fluid particles cross. "Closed flows," by contrast, have inpenetrable boundaries (or are assumed to have directions in which the flow is spatially periodic).

With rare exception, early (defined in the roughest kind of way as earlier than about 1980, see the monograph of Drazin and Reid for a survey of this body of work) stability studies mainly dealt with basic flows that were steady and invariant in both the streamwise and cross-stream directions. Linearized instability in these cases can be addressed by restricting attention to "normal modes," which are the Fourier coefficients of the perturbed flow in the invariant directions. Most classical results are based on the examination of normal modes, and most physical instability mechanisms have been identified guided by these analyses. Without difficult additional calculation, a normal mode analysis fails to give insight into the spatial as well as temporal development of disturbances that develop from given perturbations of initial data or time-varying boundary conditions. More fundamentally, a normal mode analysis provides at best an approximate framework in which to examine the instability of basic flow models of virtually all real open flows, which *do* vary in the streamwise direction. These two independent shortcomings, spatio-temporal development and the effects of non-parallel streamlines, have been addressed with notable success.

Spatio-temporal development

The importance of characterizing spatio-temporal disturbance development has been partially addressed by adopting methodology originating in plasma physics that classifies unstable flows into "absolutely unstable" or "convectively unstable" categories. The issue is discussed in the fluid mechanical context by Huerre and Monkewitz. This has been particularly important in settling ambiguities - and errors - in early spatial stability studies.

Non-parallel effects

Accounting for non-parallel effects in boundary layer and shear flows in stability analyses, within a formal rational framework, has been another advance of note, and has required at least partial abandonment of the normal mode disturbance structure. The asymptotic structure required was shown by Smith to be determined by a "triple-deck" structure characteristic of perturbed boundary layers.

New critical layers

The critical layer arising in high Reynolds number, nearly parallel, flows is the cradle of instability in flows without body forces, and it is now understood that the critical layer is not always governed by the balance of linearized inertia and viscous forces of classical theory. The balance may instead involve nonlinear inertial effects, or nonequilibrium (non-steady) effects, or a combination of all of the above. Maslowe covers much of this territory.

Receptivity

Receptivity analysis (see Goldstein and Hultgren) is another important new development connected to stability theory, especially of boundary layers, jets, mixing layers, and other nearly parallel flows. "Receptivity" relates to the way in which disturbances in the environment, which typically are not in the range of unstable wavenumbers, enter the potentially unstable region, and how they are dynamically distorted there to scales that can be destabilized.

Boundary layer transition

Without question, the paramount goal motivating boundary layer stability analysis has been the understanding and prediction of transition to turbulence. This goal may now have been reached, for practical purposes at least. Transition occurs after a nonlinear stage in which linearly unstable Tollmien-Schlicting waves combine in inherently three-dimensional resonant interactions, leading to a short rapid nonlinear growth phase culminating in turbulent flow. The process can now be described and computed, probably with a reasonable measure of confidence, well into the regime of resonant interaction. This is a great achievement, and a comprehensive account of the saga is given by Kachanov; the theoretical elements are more fully described in the monograph by Craik, and in the review by Herbert.

The practical question of transition prediction suitable for design purposes (depending in a minimal way on empiricism) depends on the availability of methods of stability calculations extending through the linear instability region and into the nonlinear growth phases. The procedure must properly account for non-parallel (in fact, three-dimensional) as well as weakly nonlinear effects, and must be fast enough to make extensive parametric surveys feasible. Suitably formulated parabolic partial differential equations (with the streamwise distance the time-like coordinate) promise to account for the necessary physical effects while posing a tractable computational task for design work. Examples of such formulations are given by Hall and by Bertollotti et al.

Post bifurcation behavior

When a flow loses stability as a consequence of alteration of its control parameters (such as its Reynolds number, Rayleigh number, Grashof number, or other controlling dimensionless groups), bifurcation occurs to other flows with qualitatively different features, and the stable members of this new family of flows are the physically realizable candidates to succeed the parent flow. Thus a steady flow may be replaced by a time-dependent flow, or a flow that is invariant in a given spatial direction may be replaced with one with a particular spatial pattern. A major advance has been the widespread consideration of the nonlinear consequences of instability, the post-bifurcation flow states. This kind of effort greatly extends the scope of the subject. It was initiated long ago by Stuart and by Landau, but was not routinely included in stability analyses, nor was there a clear understanding of the *context* of the nonlinear amplitude (Stuart-Landau) equation. In many flows, the post-bifurcation flows have a range of control parameter values in which they are both stable and simple to characterize, and the (secondary) stability of these successor flows can be examined.

Recent flow stability considerations have been informed by the blossoming of dynamical systems theory (the renaissance of which can arguably be traced to Lorenz's 1969 paper on a low-dimensional model of thermal instability). The bulk of the advances in dynamical systems has been made for finite-dimensional systems, not for the infinite-dimensional systems describing mechanics of continuous media. Nevertheless, the complex dynamics in even low order finite-dimensional examples have been invaluable in illustrating the possible connections between flow states, including transitions to chaotic flows. The books and other sources covering this area are legion: Manneville gives an elementary account aimed at physicists interested in fluid mechanics. Insights from low-dimensional dynamical systems have guided numerical studies that have confirmed complex bifurcation sequences culminating in chaotic states in relatively simple fluid dynamical systems (such as thermally unstable flows and their doubly-diffusive cousins). The knowledge generated so far is tightly connected to dynamics of finite-dimensional systems and is therefore limited to temporal complexity, with spatial dependencies remaining orderly. The existing examples consequently do not provide evidence of transition to turbulence. Nevertheless, they have revealed experimentally verifiable examples of complex fluid dynamical phenomena that can be traced and understood within a rational framework. More important, perhaps, is the global view of phenomenology in extended regions of parameter space that is intrinsic to dynamical systems theory. The techniques of bifurcation theory familiar to those working dynamical systems theory, in particular reduction of analyses to a center manifold followed by a normal form analysis are also penetrating the fluid mechanical community, and providing alternatives to the explicitly asymptotic methods that have been traditional. While the center manifold reduction is not limited to finite-dimensional problems, it currently appears to lack the flexibility to treat spatial modulations in unstable systems. Despite this problem, the dynamical systems viewpoint, and particularly its insistence on a global setting, has been conceptually invaluable.

Symmetry and symmetry breaking

Group theoretical considerations in stability and bifurcation theory have been placed in a systematic framework reviewed by Golubitsky et al. and by Crawford and Knobloch. This permits an *a priori* determination of the possible flow states accessible at a bifurcation point and therefore serves as a unifying principle for discussing post-bifurcation flows. Since the number of possibilities usually is quite limited, *and independent of the nature of the physical system*, this approach serves to bring out universal characteristics of bifurcating dynamical systems. As such, it provides a powerful general purpose tool that will likely see much wider application in the future.

Nonlinear stability

Energy stability methods permit the determination of conditions under which a given flow state can never be asymptotically destabilized by any initial disturbance ("global stability"). This approach, which basically requires the presence of dissipation (due to viscosity, heat conductivity, etc.) is masterfully explained by Joseph. More recently, a method due to Arnol'd (see Holm et al.) that shares the same starting point (construction of a Lyapunov functional), but is applicable to flows without dissipation, has been further developed and applied by a substantial number of investigators. The theory currently cannot be described as a general method, since it is not applicable to an arbitrary flow (even if first restricted to be dissipationless) that might be presented for consideration. It is, however, applicable to flows restricted to two space dimensions, and consequently has found popularity in the geophysical fluid dynamics community (see, for example, Mu Mu et al.), where standard models for large scale flows *are* two dimensional and where dissipation is ignored.

Primacy of the initial boundary value problem

Flows, like plane Couette flow, and circular pipe flow, which are linearly stable yet undergo instability and transition to turbulence have always been an affront to theory. It had long been known that disturbances can undergo transient growth in linearly stable flows (and even globally stable flows, see Joseph), before ultimately decaying. The presumption has been that transient disturbances grow large enough, and alter the basic state sufficiently, to excite some isolated attracting flow state, a solution branch not connected to the basic flow. Recently the ability of stable flows to undergo such transient growth, has been placed on firmer ground (Trefethen et al. review these developments). The possible amplifications can be surprisingly large, and the initial disturbance forms leading to the greatest rate of linear growth, or greatest total growth, can be determined (Farrell). These are not of normal mode form, even in flows (like plane Poiseuille and Couette flows) for which a normal mode decomposition is rigorously acceptable. Since the eigenvalue spectrum fails to indicate the extent of such growth, the central position of the linear IBVP in stability questions is reasserted. Large transient linear growth occurs due to the appearance of non-normal operators (characterized by non-orthogonal eigenvectors) in the conventional eigenvalue analysis. The potential for transient growth can be quantified by consideration of the concept of "pseudo-spectra" of non-normal operators promulgated by Trefethen et al. This concept, and the nature of flows evolving from optimal initial disturbance forms, suggest that there is a prospect of identifying attracting isolated solution branches, and of obtaining a clearer understanding of the mysterious by-pass routes to transition.

RESEARCH QUESTIONS

Pauli has said that prediction is risky, especially when it is about the future. This being so, the identification of promising and significant directions for future research may be a fool's exercise. Nevertheless, that is our charge here. With trepidation, then, here is a step into dangerous waters.

1. The role of boundary conditions on perturbations at inflow and outflow boundaries in open flows deserves serious study. This is not an issue confined to stability questions, but is applicable to all theoretical questions. What are the physically required *and enforceable* boundary conditions at artificial inflow and outflow boundaries that typically are placed in the computational treatment of stability (and all other) problems in attempts to capture an unbounded flow domain. Presumably, causality constraints should be fundamental, and are expressible in some sort of generalized radiation conditions applicable to fully three-dimensional, unsteady, and nonlinear, flows.

2. Intensified study of the connections between linear optimal initial conditions and flows emanating from them, and possible detached solution branches in circular pipe flow, and plane Couette flow, and between these linear properties and subcritical nonlinear effects in flows like plane Poiseuille may well prove to be highly productive. The "pseudo-spectra" of non-normal linear partial differential operators of the kind arising in this class of flows presumably is important and further study is warranted.

3. There is a need to better understand the limitations of parabolized stability algorithms, and to seek improvements as needed.

4. Weakly nonlinear analyses should be accompanied in future work — to the extent possible — by sample numerical experiments with computational techniques that do not embed the analytical assumptions. Testing the parametric range of applicability of analytical developments should provide a measure of confidence in their use in the vast number of situations in which direct computation is impossible.

5. Can a version of center manifold analysis be devised that accommodates spatial modulation as well as temporal dynamics in a convenient way?

6. Arnold-type nonlinear stability formulations for flows without dissipation need to be both extended and tested, again against numerical experiment and comparison with independent stability calculations. Are flows declared stable by these variational approaches in two-dimensions

destabilized by three-dimensional perturbations? If so, is this a generic defect in the method?

7. Symmetry considerations should be part of the workaday stability analysis toolbox. It is likely that the powerful insight provided so far in closed flows extend, in some ways, to open flows as well.

REFERENCES

Bertollotti, F.P., Herbert, T., and Spalart, P. (1992) Linear and nonlinear stability of the Blasius boundary layer. *J. Fluid Mech.* **242**, 441-474.

Craik, A.D.D. (1985) *Wave Interactions and Fluid Flows.* Cambridge: Camb. Univ. Press.

Crawford, J.D. and Knobloch, E. (1991) Symmetry and symmetry-breaking bifurcations in fluid mechanics. *Ann. Rev. Fluid Mech.* **23**, 341-387.

Drazin, P.G. and Reid, W.H. (1981) *Hydrodynamic Stability*, Cambridge: Camb. Univ. Press.

Farrell, B.F. (1988) Optimal excitation of perturbations in viscous shear flow. *Phys. Fluids* **31**, 2093-2102.

Goldstein, M.E. and Hultgren, L.S. (1989) Boundary-layer receptivity to long-wave free-stream disturbances. *Ann. Rev. Fluid Mech.* **21**, 137-166.

Golubitsky, M., Stewart, I., and Schaeffer, D.G. (1988) *Singularities and Groups in Bifurcation Theory, Vol. II*, Berlin: Springer-Verlag.

Hall, P. (1983) The linear development of Görtler vortices in growing boundary layers. *J. Fluid Mech.* **130**, 41-58.

Herbert, T. (1988) Secondary instability of boundary layers. *Ann. Rev. Fluid Mech.* **20**, 487-526.

Holm, D.D., Marsden, J.E., Ratiu, T., and Weinstein, A. (1985) Nonlinear stability of fluid and plasma equilibria. *Physics Rep.* **123**, 1-116.

Huerre, P. and Monkevitz, P.A. (1990) Local and global instabilities in spatially developing flow. *Ann. Rev. Fluid Mech.* **22**, 473-537.

Joseph, D.D. (1976) *Stability of Fluid Motions I*, Berlin: Springer-Verlag.

Kachanov, Y.S. (1994) Experiments on boundary-layer transition. *Ann. Rev. Fluid Mech.* **26**, 411-482.

Mannevelle, Paul (1990) *Dissipative Structures and Weak Turbulence.* New York: Academic Press.

Maslowe, S.A. (1986) Critical layers in shear flows. *Ann. Rev. Fluid Mech.* **18**, 405-432.

Mu Mu, Zang Qingcun, Shepherd, T.G., and Liu Yongming (1994) *J. Fluid Mech.* **264**, 165-184.

Smith, F.T. (1979) On the non-parallel flow stability of the Blasius boundary layer. *Proc. R. Soc. Lond. A* **368**, 573-589.

Trefethen, L.N., Trefethen, A. E., Reddy, S.C., and Driscoll, T.A. (1993) Hydrodynamic stability without eigenvalues. *Science* **261**, 578-584.

COMPRESSIBLE AND HYPERSONIC FLOWS

Sanjiva K. Lele
Departments of Aeronautics and Astronautics
and Mechanical Engineering
Stanford University
Stanford, CA 94305-4035

INTRODUCTION

Compressible flows are most common in aeronautical applications involving high speed internal and external flows. A wide range of non-aeronautical applications such as Laser technology, vaccum technology, gas phase reactors, plasma processing of materials, manufacturing processes involving shock waves and the rapidly developing field of micro-electronic flow sensors and actuators also require a fundamental understanding of compressible flow. The development of a new generation of high speed civilian and military aircraft, the development of new aircraft engines using high pressure ratio compressors and turbines and supersonic combustion ramjets for high altitude airbreathing propulsion, and the development of new helicopter concepts - all involve research on compressible flows. Applications involving high altitude flight or operations in the earth orbit or space, entail hypersonic flows. The development of high power gas dynamic lasers and the plasma synthesis of new materials such as diamond films require a study of non-equilibrium processes occurring within a compressible flow. Models of processes occurring in nature such as solar convection, dynamics of cosmic gas clouds, interstellar jets, galectic evolution etc. also involve compressible flows.

Whether fluid compressibility is important can be determined by suitably scaling the governing equations (Thompson, 1988). This scaling reveals three important non-dimensional parameters: Mach number $M = U/C$ which is the ratio of characteristic flow speed U to a representative speed of sound C, Strouhal number $S = \omega L/U$ which the ratio of the convective time scale L/U to the time scale of unsteadiness $1/\omega$ in the flow, and Knudsen number $Kn = \Lambda/L$ which is the ratio molecular mean free path Λ to a characteristic length scale of the flow L. It may be convenient to express the Knudsen number Kn in terms of the Mach number M and the Reynolds number $R = \rho UL/\mu$ as $Kn \approx M/R$ to conclude that for flows with order unity Mach number and Reynolds number in range 10^5 to 10^8, the continuum approximation is reasonable.

If M and S are both small, compressibility in generally unimportant, however, for spatial scales much larger than L, such as acoustic waves generated by a low Mach number unsteady flow, the fluid must be treated as being compressible. The manner in which small linear disturbances propagate in the flow depends locally upon whether $M < 1$ (subsonic) or $M > 1$ (supersonic) when the propagation is confined to Mach cones emanating from the disturbance source. For M close to unity (transonic) nonlinear effects may not be neglected (Landahl, 1989). For high frequency phenomena, i.e. $S \gg 1$, fluid compressibility remains important even if M is small. Fluid motion set up in response to rapid heat release from a chemical reaction as in an explosion or from the absorption of incident radiation pulse in laser pyrolysis are two common examples. For hypersonic flows ($M \gg 1$) the Mach number M is no longer a central parameter and to achieve similitude with respect to the conditions encountered in high altitude flight high enthalpy flows (relative to the dissociation energy of O_2 or N_2 or the ionization energy of the monatomic species depending upon the altitude to be simulated) need to be obtained (Hornung, 1988, Anderson, 1989).

Some Research Directions

An outline of the research issues pertinent to some broad topical areas which involve compressible and hypersonic flows is given below. This discussion highlights the research needs in these areas and is not intended to be comprehensive. Not all topical areas are discussed; this reflects the subjective choices of the present author, not their lack of importance.

Vortex dynamics in a compressible flow

Much of the current understanding of steady and unsteady vortex dominated flows is based upon incompressible vortex dynamics (Saffman, 1992). The dynamics of coherent vortical structures in turbulent flows have also been interpreted using the notions of induced velocity of dominant vortex tubes in the flow, vortex stretching due to the associated strain field, and the vorticity reorganization processes (roll up, pairing/merger, tearing, filamentation and reconnection). These interpretations have yielded new insights for the manipulation/control of the vortical structures and in turn the overall flow properties. Extending the vortex dynamics framework to a compressible flow requires: (1) proper account be taken of the velocity associated with the dilatation field, (2) baroclinic vorticity generation be represented, (3) nonlinear wave steepening and (4) the coupling between the solenoidal and irrotational fields due to physical boundary conditions be treated. A description in terms of interacting hyperbolic (wave-like) and elliptic (vortex-like) components may have conceptual and computational advantages. For low

Mach number flows two distinct approximations have been developed: (a) acoustic analogy approach, describing the far-field acoustics as a by product of near-field nonlinear flow (Lighthill, 1955, Crow 1970), (b) anelastic approximation, describing variable inertia effects but filtering out the acoustic waves (Speigel, 1971, Majda & Sethian 1985). Vortex methods for stratified flows or for low speed combustion problems are based on (b). Extension of such methods to compressible flows may yield new insights. Studies of simple vortex configurations in a compressible flow may help in characterizing how compressibility alters the basic vortex interactions. Studies of the interaction of flames with vorticies and acoustic waves provide simplified paradigms for compressible vortex interactions.

The basic structure of slender compressible vortices, their formation and evolution in response to 'external' changes are also important in view of their relevance to the aerodynamics of delta wings and blade-vortex interaction noise problems. Compressibility is reported to be important at the onset of dynamic stall in a pitching airfoil. Research challenges offered by the development of new helicopters including the research on dynamic stall problem are addressed by McCroskey (1983). Studies of unsteady separating flows and mechanisms responsible for self-sustained oscillations in compressible flows (Meier et al., 1990) may also benefit from a basic understanding of compressible vortex flow phenomena.

Instabilities and transition in a compressible flow

Instabilities of laminar compressible flows rearrange the vorticity in the basic flow increasing its spatial and temporal complexity. Inviscid instabilities rely upon the presence of a generalized inflexion point, i.e. local maximum of $\rho dU/dy$ in a parallel flow $U(y)$ along the x-direction with a local density $\rho(y)$ (Lees and Lin, 1948). The free-stream may be subsonic or supersonic relative to the phase speed of the disturbance. The subsonic disturbances suffer an exponential attenuation in the free-stream while the supersonic disturbances are oscillatory in free-stream and called radiating vorticity mode (Mack, 1984). For subsonic disturbances the presence of an embedded region of supersonic flow relative to the phase velocity provides a sequence of eigen modes - so called acoustic modes - defined say by the zero crossings of the pressure mode shape (Mack, 1984). For confined flows supersonic acoustic modes are also possible (Tam and Hu, 1989).

In boundary layers over adiabatic walls the most unstable mode switches from the compressible Tollmein-Schilichting (TS) mode at low M_∞ to the first subsonic acoustic mode (Mack second mode) for M_∞ above 4.5. The growth rate of TS waves is stabilized by wall cooling but the acoustic modes are destabilized by wall cooling. What mechanism is responsible for this change is not known. Suction appears to stabilize these modes and is a major design factor in achieving laminar flow over a wing. Extended laminar flow at supersonic speeds over swept wings is projected to provide significant drag reduction (Bushnell, 1992). Stability of three-dimensional compressible boundary layers is central to the development of supersonic laminar flow control. Thus crossflow instability and Görtler instability problems relevant to supersonic wings and determining the cost effective methods of suppressing/delaying these instabilities and transition to turbulence are important.

The problems of the receptivity of boundary layers and free shear flows to external disturbances and the impact of non-ideal flow environment (non-uniform wall stiffness, suction holes, local roughness, etc.) on the receptivity are important not just for laminar flow control but also as a means of altering the structure of a turbulent flow. The nonlinear mechanisms operational in compressible flow instabilities are not well understood. They are important to understanding the friction and heat transfer charateristics in the region of transition and its extent. The sensitivity of transition to local and distributed roughness and free-stream non-uniformities and unsteadiness need study. In hypersonic flows the interaction of the instabilities with the curved bow shock and the influence of real gas effects on the instabilities are important research topics. There is already some evidence that the TS waves and Mack second mode are effected in qualitatively different ways (Malik and Anderson, 1991) but no new modes which rely on real gas effects are reported. In the hypersonic conditions the use of the Navier Stokes equations and no slip wall boundary conditions is also uncertain.

Turbulence in a compressible flow

Engineering approaches to predict complex turbulent flows employ turbulence models in one form or another to mimic the effects of turbulence on the gross features of the flow. The predictive ability of such tools is often constrained by the inadequacy of the models or of the modeling assumptions or both. In a supersonic flow the turbulence is subjected to the inviscid effects of mean acceleration/deceleration due to flow turning and suffers bulk dilatation (in an expansion fan) or bulk compression (in a shock wave). Turbulence significantly alters the interaction between a shock wave and the boundary layer and its separation and reattachment. In many circumstances the separation is highly unsteady and interacts with the unsteady shock motion.

It is useful to ask if some elements of the unsteadiness are better regarded as a larger scale quasi-deterministic unsteady flow which is distinct from turbulence, and thus modeled differently. This separation may be vital in developing better models of shock-wave boundary layer interaction phenomena. Studies of idealized shock-turbulence interaction problems are important to answer basic questions on the change in the turbulence as it passes through a shock, the shock jitter generated by the turbulence, the noise radiated by such an interaction and the recovery of the flow after the interaction.

Discriminating between the large scale unsteadiness of the flow and the background turbulence may also be critical to other compressible flows involving massive flow separation. High angle of attack phenomena over aircraft wings and the rotating stall in the jet engine compressors are two prominent examples. The unsteady flow may be three-dimensional and depend significantly upon the geometry and other flow conditions and isolating it should render the background turbulence less sensitive to flow details.

The structural changes in shear flow turbulence due to compressibility, the impact of compressibility on the transport properties of turbulence, on turbulence production, cascade to small scale and dissipation processes need to be characterized. The importance of additional dissipative processes such as eddy-shocklets, the changing role of pressure fluctuations with increasing compressibility and the pathways provide coupling between the wave-like and vortex-like components need study. The effect of compressibility on the mixing and transport carried out by the turbulent flow and mechanisms which can be exploited to increase the mixing or decrease the radiated noise are technologically important.

The effect of variable fluid density, temperature and mean Mach number on the turbulence in the near wall region of a boundary layer, the structure of large eddies in a compressible mixing layer, jet, wake and a boundary layer and how does this change with compressibility need further research. In light of the studies on the eddy structure, it is useful to consider if alternative models which mimic the compressibility effect can be constructed on the basis of the structure of the large eddies.

The issues raised may help identify aspects of turbulent flow that may be attributed to compressibility. To address them a full arsenal of tool is necessary. Detailed experiments (laboratory and computational) aiming to identify the basic processes statistically and deterministically, and their modeling are needed. Reviews given by Lele (1994) and by Spina et al. (1994) on simple turbulent flows can be consulted for a current perspective on compressibility effects in turbulent flows.

Continuum modeling of hypersonic or low density flows

The Navier Stokes equations provide an accurate continuum description of compressible flows as long as the relevant Knudsen number Kn is small (typically $Kn < 0.1$). In gas flows under low pressure such as in low pressure chemical vapor deposition (LPCVD) for growing thin films of electronic or optical materials, and in thrustor nozzles used on satellites for trajectory control and reorientation, as well as in the high altitude flight of the National Aero-space Plane (NASP) the flows encountered are in the continuum transitional regime. In LPCVD the flow speeds are small yielding laminar flows but the Navier Stokes equations may be inadequate. In the other two application areas, mentioned above, the flow speeds are also significantly in excess of the sonic speed. The bow shock waves encoutered in hypersonic flight are strong and extend over a sizable part of the shock standoff distance. The heat transfer to the leading edges is severe and the radiative transfer is important. Two predictive approaches are being developed to deal with such flows: the particulate approach, such as the direct simulation Monte Carlo (DSMC) of Bird (1978) and the continuum approach based on conservation equations (Zhong et al., 1993). Research issues associated with the particulate approach are not discussed here.

The continuum approach is computationally more efficient, provided that the relevant physics can be accurately modeled. This requires that appropriate constitutive equations valid up to higher Knudsen numbers as well as the required boundary conditions may be formulated. To extend the continuum approach to higher Kn higher approximations to the Boltzmann equation may be desired. The next higher approximation than Navier Stokes is the set called Burnett equations. This higher approximation raises several theoretical difficulties. The convergence of the Chapman-Enskog series is not known. It is also unclear that the Burnett equations yield a positive dissipation function to satisfy the second law of thermodynamics. The Burnett equations are not frame invariant to rotating coordinates and there is some uncertainty about terms in the Burnett stress and heat flux involving substantial derivatives (Woods, 1983). The boundary conditions for the Burnett equations are also uncertain, due to their increased order. On the practical side the Burnett equations are unstable to periodic disturbances shorter than some critical length of the order of a mean free path. Recent research, however, shows that the addition of some terms to the Burnett equations can stabilize them and good comparison with particulate based approach have been obtained for a series of hypersonic flows (Zhong et al. 1993). How much of the physics of the continuum transition flows can be incorporated within a continuum model remains a major topic of research. Needless to say that continuum models of the interaction between the flow and surfaces such as catalytic or non-catalytic walls are also required for an overall predictive method. Adequate description of temperature and velocity slip, and chemical processes occurring at the surface is needed. Important research issues pertaining to hypersonic viscous flow are discussed by Cheng (1993), issues pertaining to chemical non-equilibrium in hypersonic flows are discussed by Stalker (1989), and rarefied gas dynamics is reviewed by Muntz (1989).

Compressible flows with chemical reactions

Reacting flows present an additional set of research challenges. For the chemical reactions to occur the reacting species must be mixed at the molecular level. In applications such as supersonic combustion, achieving good mixing within a short flow length is of paramount importance. Achieving high rates of mixing under relative supersonic flow conditions is a major scientific as well as a technological challenge. Processes that govern the entrainment rates and processes that provide the stirring or create a large interfacial area between the reacting species before the molecular mixing and reaction can proceed and the impact of the heat released by the chemical reaction on these processes need study. Similarly the role played by different reaction species, radicals and intermediates in controlling the overall rate of reaction and the composition of the products generated is also critical. Effects of equivalence ratio (fuel air ratio relative to its stoichiometric value), Reynolds number, Damköhler number (ratio of fluid mechanical time to chemical time scale), different Lewis numbers, exothermicity and Mach number on the overall properties need to be understood from first principles. This is a task far more demanding than an engineering prediction of a non-reacting turbulent flow due to the close coupling between the chemical processes and the flow processes. A current review of these issues which emphasizes the role played by the large scale structures occurring in the flow has been given by Dimotakis (1991). A different perspective emphasizing a statistical description of reacting flows has been provided by Pope (1987) and Bilger (1989). Applied aspects of supersonic combustion research are discussed by Billig (1993). Problems of overall combustion stability, ignition, protection of fuel injectors and other elements of the combustor from the high temperature environment without excessive drag penalty, and control over the pollutant formation such as NO_x are also important.

Compressible flows undergoing chemical reactions are central to the processing of thin film materials. The film quality and the crystal structure required depends upon the intended application which range from wear resistant coatings on cutting tools, aircraft turbine blades, fuel injection nozzles and surgical instruments, abrasion resistant coatings on optical lenses, mirrors, optical windows and radomes, microelectronic components and circuitry and opto-electronic devices. The LPCVD and plasma processing applications involve flows under low pressure conditions; the Knudsen number is not small and chemical non-equilibrium species often vital to the desired chemical reactions. To scientifically predict the flow conditions which give the best yield or the best quality in a given application requires the knowledge of specific reaction mechanisms, their rate constants and their dependence on the flow conditions. Such models may also be useful in devising new processes for synthesizing materials with the desired properties. Research on such flows demands a blending of chemistry, quantum mechanics and spectroscopy for flow and species diagnostics and fluid mechanics for understanding the transport processes.

ACKNOWLEDGMENTS

Discussions with Prof. W. C. Reynolds and Prof. D. R. Chapman were helpful in writing this article. The author was supported in part by the NSF-PYI award and grants from AFOSR and ONR.

REFERENCES

Thompson, P. A. (1988) *Compressible Fluid Flow*, formerly published by McGraw-Hill, NY (1972), p. 137-146.

Landahl, M. T. (1989) *Unsteady Transonic Flow*, Cambridge Univ. Press, Cambridge.

Anderson, J. D. (1989) *Hypersonic and High Temperature Gas Dynamics*, McGraw-Hill, NY.

Hornung, H. G. (1988) Experimental real-gas hypersonics, Lanchester Memorial Lecture, *Aeronautical J.* 92:379-389.

Saffman, P. G. (1992) *Vortex Dynamics*, Cambridge Univ. Press, Cambridge.

Lighthill, M. J. (1955) On sound generated aerodynamically, *Proc. R. Soc. London,* Ser. A, 211:564-587.

Crow, S. C. (1970) Aerodynamic sound emission as a singular perturbation problem, *Studies in Appl. Math.* 49:21-44.

Speigel, E. A. (1971) Convection in stars, I. Basic Boussinesq convection, *Ann. Rev. Astron. Astrophys.* 9:323-352.

Majda, A. and Sethian, J. (1985) The derivation and numerical solution of the equations for zero Mach number combustion, *Combustion Sci. Technol.* 42:185-202.

McCroskey, W. J. (1983) Special opportunities in helicopter aerodynamics, in *Recent Advances in Aerodynamics*, eds. Krothapalli, A. & Smith, C. A., 723-752, Springer-Verlag, NY.

Meier, G. E. A., Szumowski, A. P. and Selerowicz, W. C. (1990) Self-excited oscillations in internal transonic flows, *Prog. Aerospace Sci.* 27:145-200.

Lees, L. and Lin, C. C. (1946) Investigation of the stability of the laminar boundary layers in a compressible fluid, *NACA TN* 1115.

Mack, L. M. (1984) Boundary layer linear stability theory, *AGARD Report* 709.

Tam, C. K. W. and Hu, F. Q. (1989) The instability and acoustic wave modes of supersonic mixing layers inside a rectangular channel, *J. Fluid Mech.* 203:51-76.

Bushnell, D. M. (1992) Supersonic laminar flow control, in *Natural Laminar Flow and Laminar Flow Control*, Eds. Barnwell, R. W. and Hussaini, M. Y., Springer Verlag, NY, 233-245.

Malik, M. R. and Anderson, E. C. (1991) Real gas effects on hypersonic boundary layer stability, *Phys. Fluids A* 3:803.

Lele, S. K. (1994) Compressibility effects on turbulence, *Ann. Rev. Fluid Mechanics*, 26:211-254.

Spina, E., Smits, A. J. and Robinson, S. (1994) Supersonic turbulent boundary layers, *Ann. Rev. Fluid Mechanics*, 26: 287-319.

Bird, G. A. (1978) Monte Carlo simulation of gas flows, *Ann. Rev. Fluid Mechanics*, 10:11-13.

Zhong, X., MacCormack, R. W. and Chapman, D. R. (1993) Stabilization of the Burnett equations and application to hypersonic flows, *AIAA J.* 31:1036-1043.

Woods, L. C. (1983) Frame-indifferent kinetic theory, *J. Fluid Mech.* 136:423-433.

Cheng, H. K. (1993) Perspectives on hypersonic viscous flow research, *Ann. Rev. Fluid Mech.* 25:455-484.

Stalker, R. J. (1989) Hypervelocity aerodynamics with chemical nonequilibrium, *Ann. Rev. Fluid Mech.* 21:37-60.

Muntz, E. P. (1989) Rarefied gas dynamics, *Ann. Rev. Fluid Mech.* 21:387-417.

Dimotakis, P. A. (1991) Turbulent free shear layer mixing and combustion, in *High-speed Flight Propulsion Systems*, Prog. in Astronautics and Aeronautics, eds. Murthy, S. N. B. and Curran, E. T., pp. 265-340, Washington, DC: AIAA.

Pope, S. B. (1987) Turbulent premixed flames, *Ann. Rev. Fluid Mechanics* 19:237-270.

Bilger, R. W. (1989) Turbulent diffusion flames, *Ann. Rev. Fluid Mechanics* 21:101-135.

Billig, F. S. (1993) Research on supersonic combustion, *J. Propulsion & Power*, 9:499-514.

ENVIRONMENTAL FLUID MECHANICS

E. John List
Environmental Engineerng Science
California Institute of Technology
Pasadena, CA 91125

INTRODUCTION

This report describes seven basic areas of fluid mechanics that are important to problems of the environment. Clearly, there are major topics in hydrology, oceanography and meteorology which are of extreme importance to the environment in which we all live. These are covered elsewhere. In this section we have covered those topics where human intervention is an important factor at a scale that is less than planetary. Each of the problem areas has some degree of overlap with the others discussed and this will become self-evident in the presentation.

THE ATMOSPHERIC BOUNDARY LAYER

The atmospheric boundary layer is that portion of the lower atmosphere whose behavior is affected by the underlying land or water surface. Although the surface generally takes horizontal momentum from the wind through frictional drag it may be both a source or a sink of heat, moisture, and chemical species. The vertical extent of the boundary layer is defined in a complex way by the surface fluxes of heat, momentum, and moisture together with the local gradients of pressure, temperature, and velocity, and it may range in thickness from tens of meters to several kilometers. In some cases, the boundary layer may not exist, such as in large convective storms, while in others it may be limited in height by a strong temperature inversion, which can inhibit any vertical mixing.

The concept of the boundary layer in meteorology originated with Ekman, who showed that in flow over a surface the effect of boundary friction is confined to a thin layer near the surface - the boundary layer. Outside of this layer, the fluid could be treated as "inviscid", which enables great simplifications in the description of large-scale weather systems, where the flow is given by a balance of the pressure gradient and Coriolis forces due to the rotation of the Earth. However, near the surface, the effect of "friction" becomes important, and a boundary layer occurs. In most atmospheric boundary layers the flow is vigorously turbulent and is usually accompanied by significant vertical fluxes of heat and moisture at the surface, which can add or subtract turbulent energy to the flow through buoyancy effects. The properties of turbulence are very sensitive to buoyancy: heating at the surface adds to turbulence, whereas cooling suppresses it. Hence over land, the boundary layer changes drastically throughout a typical diurnal cycle: it is unstable during the heat of the day, stable at night when the surface cools by radiation to outer space, and neutral (no heating effects) at the transitions. Over the oceans, the diurnal cycle effects are much less due to the large heat capacity of the ocean waters, which maintains the surface temperature reasonably constant night and day. World-wide, the ocean is slightly warmer than the surface air, so the marine boundary layer is typically unstable.

The overall picture that emerges for the atmospheric boundary layer is that of a flow driven by large-scale pressure gradients from weather systems, in which rotational and turbulent stresses are important, and for which thermal (and humidity-)-driven buoyancy effects can alter the turbulence structure and vertical mixing. The pressure gradient can also vary with height and the top of the boundary layer can be a sharp temperature inversion.

The analysis of such flows is of fundamental importance to understanding and predicting air pollution episodes and the outcome of control strategies. It also forms the basis of most climate analysis through the mechanisms of surface transfer of heat, moisture and gases. Large-scale numerical weather models need to concisely parameterize the boundary layer with rather coarse grid spacings, and some studies have shown large sensitivities to boundary-layer representations. In short, boundary layer dynamics are of fundamental importance to understanding many environmental problems that involve the atmosphere.

The atmospheric boundary layer is studied experimentally, theoretically and modeled. The main complication to understanding and predicting such flows is the fact that the motion is almost always turbulent. Landmark experiments which investigated the vertical structure of the boundary layer include the 1968 Air Force Cambridge Research Laboratories surface-layer experiment in the flat fields of Kansas and the 1975 Wangarra experiment in Australia, where the entire boundary layer was measured. For the flow over the ocean, the 1969 Barbados Oceanographic Meteorological Experiment (BOMEX) used the stable spar buoy R/P FLIP for surface-layer measurements. Many other experiments have been conducted since, using a variety of remote sensors (sodars, lidars, radars), and specially-instrumented aircraft are enabling experiments to move from point and line measurements to area and volume measurements.

Theoretical and modeling studies have included the usual turbulence modeling techniques such as eddy viscosity, (k,ϵ) and other closure methods, including "Large Eddy Simulation" (LES). Although these studies are becoming more sophisticated they must still be done in concert with experimental data collection for validation and support.

The boundary layer is also host to other processes which are linked to dynamics and thermodynamics: clouds, radiative heat transfer, and mixing of a variety of pollutant species, many of which are chemically reactive. Particulate transfer, particularly aerosols such as di-methylsulfide, is important in the boundary layer over the ocean in addition to various anthropogenic aerosols over land.

The atmospheric boundary layer therefore presents a fluid-dynamical problem with a wide spectrum of processes and variables that are important to the environment and to the basic understanding of turbulence and turbulent mixing. New results in turbulence may lead to improved LES models and more LES simulations will become possible as computing power rises. However, direct numerical solution of the governing Navier-Stokes equations (with the addition of Coriolis and buoyancy forces) is still not possible for the high Reynolds number of the atmospheric boundary layer, but results at lower Reynolds number may provide insight for the high Reynolds number cases. Experiments will continue to provide more sophisticated measurements in a wide variety of boundary-layer settings that have traditionally uncovered new challenges for theoretical and modeling efforts. Some of the main areas that should be addressed are: examination of the limits of Monin-Obukhov similarity theory and how to address the departures from it; complete examination of the boundary layer in three-dimensions (and time) to compare to LES results, and inclusion of chemical and particulate measurements. Experimental devices are being developed to provide more areal and volumetric measurements through "local remote" sensors such as Lidars and high frequency sounders. However, chemical instruments are slow and usually cannot resolve the spatial and temporal scales needed to understand the basic mixing processes in the boundary layer, and improvement in these is needed.

COASTAL OCEANOGRAPHY

Coastal oceanography deals with processes which take place at the interface between land, ocean, and atmosphere. Because of society's interest in developing resources in the coastal ocean and protecting the environment, issues which arise in this area are of considerable interest to the general population, ranging from recreation to oil drilling on the continental shelf. The list of research topics relevant to coastal oceanography is therefore substantial. However, this report emphasizes only the major topic areas where progress is urgently required.

The motive force behind processes in the coastal ocean is ultimately the stress applied to the ocean surface by the wind, and the accompanying fluxes of heat and moisture between the ocean and the atmosphere. The air-sea interaction processes in the coastal zone are qualitatively and quantitatively different to what occurs far from shore. For example, the presence of a coastal mountain range can have a strong influence on the atmospheric flow, particularly in the marine boundary layer. Moreover, there are examples of coupled air-sea interaction, where the coastal ocean modifies the atmosphere above it in such a way as to intensify transfer of momentum, heat, and moisture between the two media. The problems therefore have all of the complexity of turbulent flows with complex boundary geometry, coupled with the inherent difficulties of interfacial transport processes.

An important application of coastal studies is determining the eventual fate of pollutants released into the ocean. Evaluation of episodic events such as oil spills and chemical releases, together with more steady discharges such as the power plant cooling water, and treated municipal wastewaters, requires a quantitative understanding of the transport processes active over the continental shelf. The transport mechanisms may be advective or dispersive, and the rich spectrum of motions which contribute to coastal circulation, and the consequent spatial and temporal variability, complicate the task of constructing predictive models that can reproduce observed features. This in turn reduces our ability to assess the potential impact of anthropogenic activities in the coastal area.

While progress has been made in assessing how both local and remote atmospheric forcing can contribute to circulation at any site, it is still not possible to predict the transport processes on most of the world's continental shelves, regardless of whether it is advective or dispersive. To date most efforts have concentrated on describing dispersion at specific sites and little effort has been made to relate existing estimates of dispersion to the components of the spectrum of motions.

The primary mechanism by which polluting material is carried to the ocean is small particulate matter. This is true for river and estuary outflows, stormwater discharges, treated wastewater discharges, or atmospheric fallout. Understanding how these particles behave in seawater is a complex problem that requires the interdisciplinary attention of chemists, biologists, and experts in fluid mechanics. Several theories of particle coagulation mechanics have been developed and some have been applied to drilling mud discharges and wastewater outfalls but there has been no comprehensive field study to evaluate the validity of these theories. Although millions of dollars have been allocated to remove such particulate material from wastewater effluents, the understanding of the fate of particulate matter in storm waters and rivers that is discharged to the ocean remains poor.

Coastal and beach erosion processes are also very much of public interest. Previous research has characterized this process as a dynamic river of sand, in which the strength of sources of sediment, the existence of sinks, and the processes which control the flux along the beach interact to determine whether the coastline will erode or not. However, significant research efforts are still necessary to understand the kinematics and dynamics of the fluid-sediment interaction and this long-standing problem is deserving of continued attention.

ESTUARIES

Estuarine flows occur where rivers meet the ocean and most estuary problems are generally associated with contaminant transport, transformation, and sedimentation, and their effects on the ecosystem, which are often interrelated. For example, understanding how metals move from sediments, to benthic organisms, to waterfowl is a formidable problem.

Estuarine flows are generally much more energetic and more strongly stratified than other environmental flows. Whereas a temperate lake might typically destratify twice a year, it is not unusual for large parts of an estuary to be unstratified for a part of every semi-diurnal tidal cycle. Estuarine flows are also strongly nonlinear in that spatial gradients of important properties can be large. Moreover, many of the important properties are the result of weak residuals of tides. The range of dimensions is also significant; fjords may be deep, except where they connect to the ocean, others estuaries, such as San Francisco Bay, are broad and shallow with a deep, narrow entrance channel.

Information needs about estuarine physics are driven almost entirely by regulatory and resource management agencies, who must have the physics to understand the biology and chemistry important to eco-system function. The needs fall into three main areas: basin-scale circulation processes, vertical mixing, and sediment dynamics. At a large scale, the transport and dispersion of particles, dissolved materials (e.g. salt) and passive biota are determined by the combined effects of tidal currents, gravitationally- and wind-driven circulations. The most important problems involve the difference between Lagrangian and Eulerian descriptions of the tidal flow, the origin and evolution of large-scale vortices, the dynamics of hydraulic controls, and the dynamics of fronts. Actual net particle motions may be substantially different from those inferred from Eulerian measurements because of spatial variations in the tidal currents, so that single current moorings cannot be used to predict particle motions accurately. A related issue is the dispersion associated with eddies generated by headlands, channel shoal-shear layers, and islands. With horizontal length-scales of 100 to 1000 meters, these features may have little effect on mean flow, yet can influence dispersion greatly.

Understanding eddy dynamics, especially the modifications by bottom friction and buoyancy is clearly important.

Currents that arise as a result of pressure fields induced by density differences can vary greatly depending on the strength of vertical and horizontal density gradients, which can range from strongly stratified saltwedge flows to vertically homogeneous flows driven by longitudinal density gradients. Flow fields in many estuaries may also be controlled by topographic features which serve as hydraulic controls; i.e. for some part of the tidal cycle, the flow becomes locally supercritical with respect to the propagation speed of possible internal wave modes. Including tidal variability or continuous stratifications into the analysis seems appropriate but has not been done.

Wind stresses can have significant effects on estuarine flows, driving strong flows, increasing vertical mixing, and generating waves that enhance bottom friction and sediment transport. Despite the clear potential importance of wind stresses, this topic has been little explored in estuaries.

Numerical models have helped the understanding of large-scale circulation processes but the problem of poor spatial resolution remains. Eddies, which are an important source of dispersion are typically lumped into a spatially constant horizontal diffusion coefficient. Work on parameterization of subgrid scale processes is important here. Detailed models have at most only 10 to 15 layers of vertical resolution but there is no basis for judging their adequacy to represent vertical mixing. Field work is needed to resolve this issue and to help predict how local vertical fluxes of scalars, particles and momentum depend on mean flow velocity and density structure. The details of variations in turbulent mixing can also significantly affect horizontal residual circulation patterns through ebb-flood asymmetries in longitudinally stratified estuaries. A feature that is unique to estuarine flows is the strength of both the shear and density stratification. For example, shears of 1 $\sec^{-1}$ are not unusual in an estuary but very unlikely in the open ocean. Likewise, estuarine stratifications can be as strong as 1 kg m^{-4} over much of the depth.

The basic problem is one of describing the structure of a very strongly stratified turbulent boundary layer. Recent attempts at modelling this flow using the Mellorλ21/2 level turbulence closure appear promising, although there is no suitable field data set to which the results can be compared. Moreover, traditional closure models do not predict the horizontal structure of the turbulence field, which can depend dramatically on stratification. Thus, a significant need arises for data and methods of prediction for horizontal turbulent mixing.

Wind-wave effects on turbulence and boundary layer structure also need to be considered for estuarine flows. Langmuir cell formation may also give rise to secondary flows that could significantly enhance vertical mixing rates. The importance of wind-driven waves to estuarine flows is largely unexplored.

In terms of ecological impact and practical importance, the dynamics of particles in estuarine flows is probably the single most important fluid mechanics problem that arises in estuarine hydrodynamics. Besides the fact that the behavior of cohesive sediments is less well understood, estuarine sediments are also exposed to changes in salinity that affect flocculation and to wind-waves that can greatly enhance resuspension. Many of the most difficult problems generally involve wind-induced resuspension in shallow water, where wind-generated waves may induce flows in the bottom muds, or where wave-induced pressure fluctuations may modify the rheological behavior of the cohesive sediment in such a way as to greatly increase the likelihood of erosion. In these cases the rheological behavior of the cohesive sediments is a central issue involving both fluid mechanical and geotechnical problems.

The interactions of sediments and phytoplankton is also of concern. Recent work has suggested that flocculation of algal cells and sediment particles may significantly increase the sinking rate of algal cells and thus play an important role in phytoplankton population dynamics. The effect of sediments on the light field depends on the particle size distribution, so that predictions of sediment-light linkages will require the time-space dependence of particle size distributions. Biological processes may also be an important factor for estuarine sediments. For example, San Francisco Bay is carpeted in places with small clams whose fecal pellets may stabilize the sediments, and whose presence as roughness elements may help increase the likelihood of erosion.

VOLCANOES

Explosive eruptions of the volcanos Unzen and Pinatubo in 1991 called the attention of the world to the hazards of explosive volcanism. Periodically, as with Krakatau in 1883, and Mount St. Helens in 1980, such devastating eruptions have received extensive attention. However, there are some fundamental unsolved problems in fluid mechanics whose solution would provide significant additional understanding of volcanic hazards.

The major threats to habitation during explosive eruptions are debris avalanches and pyroclastic flows. Debris avalanches are caused by the collapse of unstable slopes on the edifice of the volcano, and typically consist of blocks of cohesive material as large as hundreds of meters in size carried in a matrix of more highly dispersed material. Pyroclastic flows are high-speed momentum- and gravity-driven flows of hot, dense dusty gases. They are generated by laterally-directed blasts from volcanic vents (as at Mount St. Helens), collapse under gravity of vertically-directed plumes ejected from vents (known as Plinian columns), and collapse of magmatic domes extruded from vents (as with the pyroclastic flows at Unzen in June 1991 in which three volcanologists perished). High energy pyroclastic flows (blasts) can surmount high (600 m) ridges, while relatively low-energy flows are often channeled by existing topography. The particles in pyroclastic flows range in size from micrometers to meters, and the deposits can be massive and uniform or show complex flow structures (e.g. dunes). The average dusty gas density is usually at least 30 kg m^{-3} so the particulate volume ratio is typically more than 1. Therefore, in regions where the flow is nonsteady or accelerating, particles of disparate size interact by collisions and by wake interactions. Consequently, in these flows the particles participate in large-scale cooperative motions and contribute an added viscosity.

Extensive studies of pyroclastic flow deposits during the past few decades have succeeded in defining many properties of the flows by exploiting the analogy with sedimentation. However, the deposits are most useful in yielding information about the waning phases of flow, and provide only secondary information about the waxing phase when much of the destruction occurs. Nevertheless, from the deposits it is clear that pyroclastic flows exhibit a wide range of variability. Flow concentration (density), depth and mobility can vary by several orders of magnitude. The largest debris avalanches and pyroclastic flows (i.e. the deepest and most dense) in prehistoric times propagated at speeds of several hundred meters per second to more than 100 km from the source. Since the average slope of such flows is of order 1, the mechanisms by which they were transported are not well understood. On a smaller scale, pyroclastic flows that have been documented by eyewitness and photography also propagate large distances. Fluidization by evolving and entrained gases is thought to be an important process.

The acoustic disturbance generated by explosive eruptions depends strongly on the rate of energy release and so is not strongly correlated with the total energy of the eruption. Some large explosive eruptions are heard as far away as thousands of kilometers (Krakatau was heard some 20,000 kilometers away in Colombia) while others are silent. Thus atmospheric waves generated by exploding volcanos are not a hazard even in the near field, but they have potential use as a diagnostic tool.

We list here the major challenges in the fluid mechanics of erupting volcanos. The most important unsolved problem in high-speed volcanological fluid mechanics is the mechanism by which liquid magma is fragmented into fine dust near the exit of the volcanic conduit. This expansion process gives explosive eruptions their high energy. Fragmentation apparently occurs in a relatively narrow zone near the surface of the earth at which the hydrostatic pressure has decreased sufficiently that bubbles of supersaturated dissolved gases (principally steam and carbon dioxide) nucleate and grow. The volume of the two-phase fluid expands by a factor of order 100 during fragmentation, and flow velocities of order 100 m sec^{-1} are generated. Mass fluxes across the fragmentation surface where the phase change occurs are of order 3,000 kg m^{-2} sec^{-1}. Thus in the fragmentation region multi-phase flow and transport processes are far from equilibrium. Significant progress can be made by studying model flows undergoing rapid two-phase expansion in the laboratory.

Further expansion can occur in the flow through the exit orifice and, in cases where the pressure release is large, the flow can become supersonic. Nonsteadiness at the source strongly affects the rate of ejection of material, and is the origin of shock waves in eruption columns reported in some eye-witness accounts. The causes of nonsteadiness are not understood and their consequences have not been studied.

When controlled by gravity, pyroclastic flows behave much like gravity currents. However, the size distribution of particles in pyroclastic flows is very large, so they are probably highly stratified, with the largest and most dense particles near the bottom. As with snow avalanches, eyewitness and photographic records give information only about the top dilute layers of the flow. It is expected that flows with large dispersions of particle size will exhibit a rich variety of behavior and are worthy of study in the laboratory.

The major difficulty in formulating experimental and theoretical models of volcanic flows is that reliable flow data from actual eruptions are virtually nonexistent. Time-series measurements of parameters of dynamical interest (density, velocity, etc.) in full-scale pyroclastic flows would have inestimable impact on the understanding of such flows. Though such an undertaking might at first glance appear to be impossible, increasing knowledge about the frequency and direction of eruptions conceivably could be used to predict where instrumentation should be sited. Such an effort would require innovative design of new rugged transducers for flow measurement. Obtaining data to guide the formulation of theoretical models of high-speed volcanic flows will be the greatest challenge of the future for experimentalists in volcanological fluid dynamics.

GROUNDWATER

Most groundwater environments contain high concentrations of extremely small particles (about 1 μm in diameter) called colloids. These colloids affect a broad range of chemical, biological and physical processes that occur in groundwater. Because colloid behavior is controlled by both physical and chemical processes, significant progress in understanding colloid transport in subsurface systems requires strongly interdisciplinary approaches that couple mathematical modeling, chemistry, biology, and fluid mechanics. The state of current knowledge and research needs in this field are summarized below.

Links between the transport of many contaminants in subsurface environments and the transport of suspended inorganic colloids have increasingly been documented. Classes of pollutants with high affinities for mineral surfaces, such as hydrophobic organics, radionuclides, and metals, are particularly subject to mobilization by colloid transport processes. Certain inorganic colloids are contaminants themselves, such as asbestos fibers and toxic metal sulfide precipitates. To predict confidently the mobility of these types of contaminants in aquifers, colloid production and transport processes must be well understood and characterized. Important areas of research must also address the highly coupled nature between colloid mobility and other biogeochemical processes. Microbial activity, for example, can cause changes in the chemical environment of an aquifer, and subsequently lead to the release of mobile, inorganic colloids. Ultimately, colloids or surface-active macromolecules may be used in a remediation sense to transport and remove otherwise immobile contaminants from aquifers.

In addition to inorganic colloids, viruses, bacteria, and eukaryotic microorganisms, collectively termed biocolloids, are also ubiquitous in natural groundwater environments, where they can exist as natural inhabitants or as contaminants from the disposal of human or animal wastes. These microorganisms are of profound environmental interest as mini-reactors that carry-out important biochemical reactions and as agents for the transmission of waterborne human diseases. Two central questions arise in virtually all studies of groundwater microorganisms. First, what microorganisms are present in the groundwater environment of interest? Current technology for detecting microorganisms is often inadequate when applied to groundwater samples. Second, are the microorganisms present in the groundwater bound to the pore matrix or in the fluid phase? The binding of microorganisms to solid surfaces is a controlling variable for many important microbial processes including groundwater mobility, persistence, gene transfer, and biochemical activities.

The emerging picture from colloid field sampling studies suggests that colloids often form in situ, or are released in areas where certain geochemical conditions exist. Low oxygen groundwaters, for example, have been shown to dissolve the oxide coatings that cement other particles to the aquifer matrix. Analogously, changes in pH or alkalinity can cause the dissolution of calcite mineral cementing agents. In studies downgradient from swamp regions, colloid mobility has been attributed to the presence of ubiquitous organic carbon coatings, which effectively stabilize fine particles. Since these types of chemical perturbations often occur in conjunction with waste sites, there is an imperative need to understand the mechanisms by which colloids form.

Filtration through 0.45 μm membrane filters has typically been used to distinguish operationally between dissolved and particulate material in groundwater samples. Recent efforts to develop sampling protocols for smaller inorganic colloids will require abandoning this approach. New analytical techniques, that can be used to characterize polydisperse size distributions and surface chemistry characteristics of particles in field samples are needed. Moreover, the careful development of groundwater extraction methods is critical to the success of field investigations. "Artifact" colloids may be introduced as a result of common well installation practices or by pumping at rates that are sufficient to shear colloids from the aquifer matrix. Once samples are collected, colloids may also be artificially produced by changes in chemical or gas saturation conditions. Research directed towards the development of appropriate sample preservation techniques is therefore necessary. Sampling of microorganisms in groundwater environments is equally challenging. As with inorganic colloids, the very act of sampling the environment can yield results which are not reflective of true in situ conditions. Samples can become contaminated with foreign microorganisms by contact with the drilling equipment and fluids. Also, detection methods may not identify important portions of the microbial community naturally present in subsurface environments, making the microbial populations difficult to estimate and characterize.

Considerable ambiguity still surrounds the most basic questions about the factors that govern colloid transport, particularly surface chemistry effects. Solution and surface chemistry conditions that result in unfavorable particle-collector interaction energies are also those conditions which favor colloid mobility. In the absence of unfavorable conditions, for example, characteristic travel distances for particles in aquifers would rarely exceed a few meters. Consequently, where colloid transport is significant, it is necessary to understand the surface chemistry characteristics of the system. Theoretical models have been developed to incorporate unfavorable particle-collector interactions and predict colloid attachment rates. All existing models, however, dramatically underestimate the colloid deposition rates observed in laboratory experiments. While the colloid deposition rates from laboratory experiments are larger than theoretical predictions, they are still sufficiently small enough to suggest that significant transport is possible when repulsive interactions exist. Very small colloid deposition rates, which would result in significant transport at field scales, are also difficult to measure at laboratory scales; a limitation that underscores the need for improved theoretical models.

Obtaining descriptions of the surface charge characteristics of the aquifer matrix represents a particular challenge. An equal challenge is the task of adequately describing natural particle coatings. Stochastic approaches to solve these heterogeneity-related problems must eventually be incorporated in colloid transport models.

PHYSICAL LIMNOLOGY

Physical limnology is the study of the hydrodynamics of confined bodies of fresh water which are acted upon by the wind and the sun at the surface, by inflowing streams around the perimeter, outflows from specially constructed structures, or natural river overflows, and by the rotation of the earth. In general, the geographical location of a lake and the climate and weather at the location determine the above inputs and, therefore, the hydrodynamic variability of the water in the lake. The larger the temperature difference between summer and winter the more pronounced will be the seasonal thermal stratification of the water in the lake. The stratification is the direct result of the surface cooling, cold water inflow during the winter period, and the heating at the surface in the summer. The range of stratification from one lake to the next may be very large; some lakes rarely stratify and then only weakly, and some remain stratified for most of the year or even over a number of years. Other lakes are stabilized, in addition, by dissolved salts and may remain stratified indefinitely. Density stratification influences all motions in a lake, from basin scale oscillations and circulations to small-scale turbulent mixing.

Many authors have attempted to establish empirical formulae which relate the seasonal stratification in a lake to its latitude, altitude and lake basin shape. The main limitation has been that the climate and weather are not unique functions of these parameters. However, for limnologists such empirical formulae have great diagnostic value. Given the successful predictive capabilities of energy conserving models, an important area of research would be to use these models to derive simple relationships which relate key parameters of the local conditions in a lake to the observed stratification patterns. One such indicator is the Lake Number L_N derived by Imberger and Patterson, which appears to capture the intensity of the mixing in a lake in a single easily computable parameter. It has been shown that the Lake Number L_N can be used as a predictor for various water quality parameters. From an operational point of view, there is an urgent need to develop further water quality control methodologies where simple indicators are used to capture the physical state of the lake. These should then be developed into predictors for both dissolved and particulate water quality parameters. Emphasis must also be directed towards extending indicators, such as the Lake Number, to large lakes where rotation influences the transport processes.

Near the surface of a lake and around the lake bottom, there are turbulent boundary layers where mixing is active. In principle, these layers are similar except that the benthic boundary layer has the added complication that the boundary surface is not horizontal and slopes at a variable angle depending on the geometry of the basin. The energetics of these layers are now reasonably well understood: turbulent kinetic energy introduced at the boundary surface is complemented by shear production at the entrainment surface and this is used to overcome dissipation and buoyancy flux. The sloping wall in the benthic boundary layer causes a net discharge in the boundary layer itself that in turn generates a circulation in the lake as a whole. Careful experimental verification is now required to quantify the connection between the flow in the turbulent boundary layer and the interior circulation. The theoretical models are based on steady flows and need to be extended to incorporate variability in the strength of the turbulence in the benthic boundary layers which appears to be a common factor in similar oceanic observations, and by implication in lakes. As the benthic boundary layers are sites of direct sediment-water interaction, the boundary circulation is extremely important for water quality considerations.

In the surface layer, longitudinal natural convection may take place if there is differential deepening, heating or cooling. Shoreline topography results in wind sheltering over parts of the lake, and leads to both variable momentum transfer and heat transfer across the water surface. Such mechanisms drive strong horizontal flows which play an important role in the exchange between nearshore and offshore water. The cyclic nature of the solar and wind forcing causes further complications in the flow behaviour which has not yet been explored. Processes driven by horizontal density gradients are complicated at very low temperatures where the density maximum at $4°C$ leads to a thermal bar formation. All these processes have been identified in principle only; quantitative predictive capability for the horizontal fluxes and particle transport in these unsteady layers remains to be determined. Equally, the presence or absence of large mixing cells in both the surface and benthic boundary layers are of extreme importance for water quality considerations as they determine the cycling of particles within a variable light and nutrient regime.

The motion of water in the surface and benthic boundary layers also leads to dynamical adjustments in the core of the lake itself. These oscillatory motions are called seiching and they have a frequency ranging from hours to days. In the case of large lakes, these motions are also influenced by the rotation of the earth and the dispersion relationships depend upon both the stratification and the local inertial period. In addition, high frequency internal waves are produced which interact dynamically to produce horizontal intrusions and convective and shear instabilities sustaining a patchy, intermittent turbulent field within the hypolimnion. Separation at topographical features has been shown to produce eddies in a rotational system, adding further energy to the fluctuating motions within the hypolimnion. A major question in physical limnology is what type of turbulent kinetic energy dissipation field do these dynamical interactions produce? The turbulent kinetic energy dissipation produced is important for two reasons. Firstly, it determines the amalgamation and breakdown of biological and chemically active particles. Secondly, the turbulent dissipation rate when combined with a knowledge of the largest turbulent overturning scale, enables the computation of the local vertical mixing efficiency.

Selective withdrawal of fluid from a stratified basin has been the subject of many investigations and recent studies of the combined effect of rotation and stratification have shown that, contrary to common belief, for a long enough time after the initiation of flow, selective withdrawal may be annulled by the influence of rotation in most, if not all, present reservoir situations. These predictions now require careful experimental verification. It is clear, however, that the complete strategy of selective withdrawal be reviewed and that the effect of unsteady selective withdrawal in a rotating stratified fluid and the influence of the geographic location of the withdrawal structure be carefully re-assessed.

Instrumentation in both physical limnology and physical oceanography has radically changed. There are now instruments which measure not only the mean temperature and conductivity fields but also turbulent fluctuations of temperature, conductivity, dissolved oxygen, velocity and fluorescence. Acoustic techniques for velocity measurement and imaging are also reaching maturity, as are remote sensing techniques. It has now become possible for the first time to contemplate an interdisciplinary experiment, where the total flux paths of the various chemical and biological constituents are documented and where a much fuller understanding is obtained of the interaction between the physical advection and mixing processes and the ultimate productivity within the lake.

Water quality in a lake is predominantly determined by the chemical and biological reactions; the water providing the medium in which the reactions take place. However, the bulk of water quality questions concern reactive particles and it is clear that the motion, the amalgamation and the dissociation of inorganic or organic particles, are strongly influenced by the rate of strain at the smallest fluid motions. Furthermore, particles often have the ability to either adjust their buoyancy or to propel themselves. It is essential that future water quality models do not consider particles as simply an amalgamation forming a certain concentration, but rather treat particles individually as chemical reactors moving in a turbulent/laminar fluid with certain coherent motions. Only when water quality models consider the three components, water, dissolved chemical constituents, and particles, as separate entities, will they be able to provide complete predictive capability. These models will be extremely complicated and it is clearly fruitful to ask whether simple causal relationships can be extracted, which relate overall physical features such as the Lake Number with water quality. Such relationships can then be used in water quality controlled strategies, aimed at improving the water quality in lakes and reservoirs.

SPILLS

Accidental releases of toxic or flammable liquids and gases have become a concomitant of industrial development. Slow leaks are ubiquitous. Fortunately, catastrophic failures are relatively rare. However, when they do occur the results can be almost unbelievably devastating, as the recent events in Bhopal, Chernobyl, and Guadalajara have made so graphic. Analysis of such accidents indicates that they are invariably the result of the failure by humans to anticipate, or recognize, the failure of mechanical equipment. Given that such failures will continue to occur, regardless of the best efforts to prevent them, analysis and prediction of the outcome becomes the ultimate resort.

The prediction of the fate of material released in slow leaks, whether to the air or water environment, is relatively straight forward in that two main processes are involved: advection by natural currents and/or dispersion by turbulent mixing. The geometry of the leak is seldom of significance although the pressure and density of the material involved may play a minor initial role in defining initial jet or plume (i.e. buoyancy-driven convective transfer) mixing. The theory of jets and plumes is well defined and validated and has been used to design specific discharges where dilution of the effluent quickly reduces the concentration of the discharge to safe or environmentally acceptable limits. The analysis of dilution beyond the point where initial mixing occurs depends upon local knowledge as to the form and nature of the environment. This may take the form of knowing the prevailing winds, or currents, and the diurnal nature of the atmospheric or oceanic boundary layer. Most of the time these factors are very site specific. The problems are those of the preceding sections where the dynamics of the environment are the overriding concern.

When the release is a catastrophic one the dynamics of the failure mechanism and the thermodynamic properties of the spilled material become of fundamental importance in defining the outcome of the spill. While spills of miscible liquids into water can be well described by combinations of "dambreak" models together with normal advection and dispersion modeling, the situation with volatile liquids is not quite so straightforward. For example, consider the failure of a liquified natural gas or propane storage facility. The initial spill is a super-cooled liquid, which in flowing over a warm surface vaporizes to a cold dense gas. This in turn causes the spill to continue to spread as a gravitationally-driven cloud. Ultimately, and depending upon the specific weight of the vaporized gas, the gas may lift off the surface and create a huge buoyant plume, or it may continue to roll over the surface as a vast gravitationally-driven current. Experimental and theoretical work has shown that such three-dimensional density currents are unstable and break into a series of periodic frontal motions.

Analysis of such flows is made difficult by the boiling two-phase flow and the inherent instabilities that occur in three-dimensional gravity flows. Despite small scale testing that has been done with liquid nitrogen spills the fundamental physics of these processes is still not understood well enough to develop computer codes that adequately represent the mechanics. Similarly, the lift off process where the plume first becomes buoyant is not well categorized.

With flammable gases the problem of flame front acceleration into detonation waves is receiving a lot of attention but the complexity of the chemical reactions involved with mixed gases, even as simple as acetylene and air or hydrogen and air, is quite overwhelming. The chemical kinetics of the reactions involved becomes very complex and the geometry of the flame front is still being studied in depth.

In cases where the spilled liquid is immiscible but slowly volatile (e.g. crude oil) the relative viscosity and density of the two liquids changes with time so that the dynamics are dependent on the history of the spreading and the previous interactions with the atmosphere. Even at this time there is no theory to predict the relative velocity of oil and water in a wind driven current.

When the spill not only spreads over the surface of the ground but in addition soaks into the soil then all of the complexities discussed in Section Groundwater become part of the problem.

Coordinated and edited: John List.

Text by: Carl Friehe, Greg Ivey, Stanley Grant, Jorg Imberger, Terri Olson, John List, Stephen Monismith, Brad Sturtevant, Clinton Winant.

GENERAL REFERENCES

Atmospheric Boundary Layer

Wyngaard, J. C. (1992) Atmospheric turbulence. *Annual Reviews of Fluid Mechanics*, **24**:205-234.

National Research Council, *Coastal Meteorology*, National Academy Press, 99p. (1992).

Coastal Oceanography

Coastal Circulation

Brink, K. H. (1991) Coastal-trapped waves and wind driven currents over the continental shelf. *Annual Reviews of Fluid Mechanics*, **23**:389-412.

Davis, R. E. (1991) Lagrangian ocean studies. *Annual Reviews of Fluid Mechanics*, **23**:434-464.

Coastal Processes

Peregrine, D. H. (1983) Breaking waves on beaches. *Annual Reviews of Fluid Mechanics*, **15**:149-178.

Battjes, J. A. (1983) Surf-zone dynamics. *Annual Reviews of Fluid Mechanics*, **20**:257-293.

Battjes, J. A., Sobey R. A. and Stive, M. J. F. (1980) Nearshore circulation. *The Sea: Ocean Engineering Science*, **9**:467-493.

Estuaries

Bedford, K. W. (1992) The physical effects of the Great Lakes on tributaries and wetlands. *J. Great Lakes.* **18**(4):577-589.

Fernando, H. J. S. (1991) Turbulent mixing in stratified fluids. *Annual Reviews of Fluid Mechanics*, **23**:455-493.

Sherwood, C. R., Jay, D. A., Harvey, R. B., Hamilton, P. and Simenstad, C. A. (1990) Historical changes in the Columbia River estuary. *Prog. Ocean.* **25**(1-4):299-352.

Sullivan, G. D. and List, E. J. (1994) On mixing and transport at a sheared density interface. *J. Fluid Mechanics*, **273**:213-239.

Symposium Papers from the Eleventh Biennial International Estuarine Research Conference - Estuarine Fronts, Hydrodynamics, Sediment Dynamics and Ecology, *Estuaries* **16**(1):1-159 (1993).

Volcanoes

Fisher, R. V. and Schminke, H.-U. (1984) *Pyroclastic Rocks*, Springer-Verlag, New York.

Kieffer, S. W.(1981) Blast dynamics at Mount St. Helens on 18 May 1980, *Nature*, **291**:568.

Sparks, R. S. J. (1986) The dimensions and dynamics of volcanic eruption columns. *Bull. of Volcanology*, **48**:3-15.

Carey, S. and Sparks, R. S. J. (1986) Sparks Quantitative models of the fallout and dispersal of tephra from volcanic eruption columns. *Bull. of Volcanology*, **48**:109-125.

Ground Water

Bedient, P. B. and Rifai, H. S.(1992) Groundwater contaminant modeling for bioremediation - A review. *J. Hazard M.* **32**(2-3): 225-243.

Gelhar, L. W. and Welty, C. (1992) A critical review of data on filed scale dispersion in aquifers. *Water Resources Res.*, **28**(7):1955-1974.

Mangold, D. C. and Tsang, C. F.(1991) A summary of subsurface hydrological and hydrochemical models. *Rev. Geophys.*, **29**(1):51-79.

Sardin, M., Schweich, D., Leij, F. J. and Vangenuchten, M. T. (1991) Modeling the non-equilibrium transport of linearly interacting solutes in porous media - A review. *Water Resources Res.* **27**(9):2287-2307.

Brusseau, M. L. (1994) Transport of reactive contaminants in heterogeneous porous media. *Rev. Geophys.*, **32**(3):285-313.

Physical Limnology

Gregg, M. C. and Kunze, E. (1991) Shear and strain in Santa Monica Basin. *J. Geophys. Res.*, **96**(16):709-16, 719.

Imberger, J. and Patterson, J. C. (1990) Physical Limnology. *Advances in Applied Mechanics*, T. Wu (ed), Academic Press, Boston **27**:303-475.

Imberger, J. and Ivey, G. N. (1993) Boundary mixing in stratified reservoirs. *J. Fluid Mechanics*, **248**:477-491.

Ivey, G. N. and Imberger, J. (1991) The nature of turbulence in a stratified fluid. Part 1: The efficiency of mixing. *J. Phys. Ocean.*, **21**:650-658.

Ledwell, J. R. and Watson, A. J. (1991) The Santa Monica Basin tracer experiment: a study of diapycnal and isopycnal mixing. *J. Geophys. Res.*, **96**:8695-8718.

Straskraba, M. (1980) The effects of physical variables on freshwater production: Analyses based on models. *The Functioning of Freshwater Ecosystems*. (E. D. Le Cren and R. H. McConnell, eds.), pp 13-84. Cambridge University Press, England.

Thorpe, S. A., Hall, P. and White, M. (1990) The variability of mixing at the continental slope. *Phil. Trans. Roy. Soc.*, London, **A331**:183-194.

Spills

Britter, R. E. (1989) Atmospheric dispersion of dense gases. *Annual Reviews of Fluid Mechanics*, **21**:317-344.

Hunt, J. C. R. (1991) Industrial and environmental fluid mechanics. *Annual Reviews of Fluid Mechanics*, **23**:1-42.

TURBULENCE AND TURBULENCE MODELING

John L. Lumley[1]
Sibley School
of
Mechanical and Aerospace Engineering
Cornell University
Ithaca, NY 14853

INTRODUCTION

History tells us that man has been curious about turbulence for at least 500 years, and has been actively trying to understand it for a little over 100. Most flows in the universe are turbulent, and the irregular, unsteady motions of a turbulent flow transport momentum, heat and matter several orders of magnitude more effectively than molecular motion. Consequently, turbulent flow is responsible for most of the energy consumption (through turbulent drag) and most of the heat transfer and matter transport. The more-or-less uniform distribution of oxygen, carbon dioxide and heat in the atmosphere and oceans makes life possible, and is due to turbulent transport.

Rational design of aircraft, automobiles, nuclear reactors and all sorts of industrial mixing and forming processes, as well as design, siting and regulation of power plants, incinerators, and the like, are dependent on an ability to calculate the effects of turbulent transport reliably. Unfortunately, we cannot do that. One hundred years of intense effort have brought us very good qualitative understanding of turbulent flows in nearly all practical respects, (although often our insight is not at a terribly fundamental level) but have not brought us the ability to calculate reliably. Federal agencies have spent large sums (though not large compared to what has been spent on nuclear physics) in hopes of making calculation possible. Program monitors dispair of justifying to their masters the spending of more money, and yet, what choice do we have? Turbulence will not go away; it continues to play its dominant role in flows, and our need to calculate its effects reliably only gets more pressing.

[1]Supported in part by Contract No. F49620-92-J-0287 jointly funded by the U. S. Air Force Office of Scientific Research (Control and Aerospace Programs), and the U. S. Office of Naval Research, in part by Grant No. F49620-92-J-0038, funded by the U. S. Air Force Office of Scientific Research (Aerospace Program), and in part by the Physical Oceanography Programs of the U. S. National Science Foundation (Contract No. OCE-901 7882) and the U. S. Office of Naval Research (Grant No. N00014-92-J-1547).

Public support of science in the United States has always been characterized by a certain inconstancy, a fickleness, a tendency to unsuitable liaisons and the transfer of affection from one pretty face to another at the drop of a hat. This is nowhere more evident than in the support of research in turbulence; the great need for a solution, coupled with the evident slow rate of progress, means that new ideas sometimes attract more attention than they are subsequently found to deserve. The turbulence community itself is somewhat prone to this kind of behavior.

The experience of 100 years should suggest, if nothing else, that turbulence is a difficult problem, that is unlikely to suddenly succumb to our efforts. We should not await sudden breakthroughs and miraculous solutions. Probably no single approach will provide definitive answers. We should expect progress on multiple fronts that is on average slow, and not uniform, with now one front progressing, and then another, with any given front enjoying periods of stasis followed by small forward (though sometimes backward) jumps of random magnitude, like the modern view of evolution, and a little like turbulence itself.

AN OVERVIEW OF THE FIELD

We hope that research in turbulence will produce information of a number of different kinds. Engineers need predictions of average quantities like drag and heat transfer, although some engineering situations require more complex information on turbulence structure, such as panel vibration, radiated noise, and multi-path interference in communications systems. In a particular situation, we can, of course, make experimental measurements to determine the required quantities. At the very least, we hope to be able to extrapolate measurements and interpolate between them, so as to make relatively few measurements do as much work as possible. This requires a certain amount of physical understanding of the turbulence. Hence, this physical understanding is one goal of turbulence research. Further, we would like to be able to calculate the effects of a turbulent flow, at least in situations not qualitatively different from well-documented situations. This requires even greater physical understanding of the flow, permitting the construction of relatively robust computational models of the flow. Ideally, we would like to be able to calculate the effects of a turbulent flow even in situations qualitatively different from our library of well-documented flows. This is the most demanding requirement of all, since it means that our models must behave like turbulence even in situations with which we have no prior experience - it means we must have a deep and comprehensive physical understanding of turbulence physics, which we hope to acquire from turbulence research.

The state of research in turbulence up to 1975 is summarized in Monin and Yaglom (1975), an encyclopaedic two-volume work. This work is currently undergoing revision; until the new edition is published, there is no convenient guide to the literature of the last twenty years. We will try to summarize here the major trends during that period.

Large Eddies / Coherent Structures

The last twenty years have coincided with an enormous explosion of interest in the more-or-less organized part of turbulent flows. To be sure, it has been recognized for some fifty-five years that turbulent flows have both organized and apparently disorganized parts. Liu (Liu 1988) has documented the first appearance of this idea around the outbreak of the second world war. The idea was probably first articulated by Liepmann (Liepmann 1952), and was thoroughly exploited by Townsend (Townsend 1956), but all within the context of the traditional statistical approaches. However, in Brown and Roshko (1971, 1973), is presented visual evidence that the mixing layer, in particular, was dominated by coherent structures, and this captured the imagination of the field, which was ready for a new approach. Within two years, the number of citations in this area had gone up by a factor of four, and within two decades had risen by a factor of ten. A lot of immoderate things have been said about coherent structures, and the statistical approaches that were previously popular; a discussion of this, and a current position on why coherent structures are present in turbulent flows, and why they are present in different strengths in different flows, and how they can be calculated, and when they need to be calculated, and what to do with them once you have them, are all discussed at length in Holmes *et al* (1996). Briefly, the coherent structures appear to be the result of an instability of (what must be an imaginary) flow with turbulence but without coherent structures, and they can be calculated approximately by using stability arguments of various sorts. It is not always necessary to take them into account explicitly, depending on the particular purpose of the calculation; on the other hand, it can be, for some purposes, very profitable to model the flow as coherent structures plus a parameterized turbulent background, and so construct a low-dimensional model of the flow. In the construction of such models, the Proper Orthogonal Decomposition, or Karhunen-Loève decomposition, has been of great use, because it represents the flow with the minimum number of terms (see Holmes *et al*, 1996). Such models can be used whenever an inexpensive surrogate of the flow is needed, and have been very helpful in shedding light on the basic physical mechanisms.

Chaos / Dynamical Systems Theory

During this same period, it was found that certain mechanical systems of three or more degrees of freedom were capable of chaotic behavior, that is, possessed a strange attractor. Chaotic behavior is complex, aperiodic and appears to be random, but is, in fact, deterministic. Mechanical systems that have a strange attractor display an extreme sensitivity to initial conditions, with solutions initially very close together separating exponentially. This has been called in the popular literature the "butterfly effect", and has appeared even in such movies as Michael Crichton's *Jurassic Park*; there is surely no literate person over the age of thirteen who has not heard of chaos and the butterfly effect. It is not difficult to show (using simple models) that this sensitivity to initial conditions is so exquisite that, after a relatively short time, the divergence of the solutions depends on details of the initial conditions that are practically unknowable.

Nearly everything that is known about chaos relates to simple mechanical systems with a few degrees of freedom. Almost none of these ideas has been put in the context of continuous fields, such as the velocity field of a fluid filling a domain. We know only that a fluid in turbulent motion is chaotic (say, as a function of time, measured at a space point), though deterministic, and displays the butterfly effect. According to certain definitions, this is enough to say that there is a strange attractor, although that is not much more than a matter of semantics.

So far, these ideas have been successfully applied only to turbulence near transition, or near a wall, so that a relatively small number of degrees of freedom will have been excited, and the turbulence will be relatively simple, and can be described by a mechanical system of relatively few degrees of freedom (see above Large Eddies/Coherent Structures). As the system moves farther from transition, or from a wall, more and more degrees of freedom are excited, until the structure of the attractor becomes so complex that the system must be treated statistically. The ideas from dynamical systems theory have been successfully used to analyze the low-dimensional models that have been constructed of turbulence near a wall or near transition. There is a separate chapter in this volume which describes this area in greater detail. See also Holmes *et al* (1996), Farge (1992).

Fractals and Multifractals

There is also a separate chapter in this volume describing the work in this area. Very briefly, fractals and multifractals are statistical models which are remarkably effective in describing certain turbulent phenomena, such as the interface between turbulent fluid and non-turbulent fluid. They are related to log-normal distributions, which were first suggested as descriptions for the distribution of dissipation in a turbulent fluid (Monin & Yaglom 1975). Many quantities in turbulent flow display approximate scale similarity, or can be modeled as doing so, if the Reynolds number is high enough; that is, the structure is statistically similar no matter the magnification, and this is the basic requirement for any of these statistical models.

Wavelets

Wavelet decompositions, like Fourier analysis, are an analytical technique that is not confined to turbulence. In some respects it is felt that a wavelet decomposition matches turbulence physics better than a Fourier decomposition, and consequently there has been recent interest in this approach. Probably the first discussion of wavelets in connection with turbulence (though without the name) can be found in Tennekes and Lumley (1972), where an eddy is associated with the Fourier modes between 0.6κ and 1.6κ, where κ is the central wavenumber, which produces a Mexican-hat sort of eddy. The basic idea is, that as the wavenumber increases, Fourier modes cannot possibly be acting independently, but must increasingly act in groups, the more so since eddies in physical space are known to have limited physical extent, unlike Fourier modes which stretch off to infinity. Wavelets can be organized by scale, in families, and distributed over the space (Farge 1992) for a complete representation (Meneveau 1991), at least in homogeneous situations. Many workers in turbulence feel that wavelets are the ideal representation for coherent structures (which occur in inhomogeneous flows), which are always of limited physical extent, and a certain amount of success has been achieved with other equations, such as the Kuramoto-Sivashinsky equation (Myers *et al*, 1994), but there are difficulties in applying wavelets to Navier Stokes systems with inhomogeneous directions, and no notable successes have been achieved at present.

TRADITIONAL PROBLEMS

Of course, the problems on which people worked before the attractive new approaches came on the scene (above) are still with us. Monin and Yaglom (1975) is probably as good a source as any for a survey of such problems, although it does not contain the most recent results. The following list is intended to be illustrative, not exclusive:

- The asymptotic behavior of turbulence for large Reynolds number, which is beyond the reach of experiment or computation, has been a source of fascination (Lumley 1992).

- The asymptotic behavior of jets, wakes and shear layers at great distances from the origin is also beyond the reach of experiment or computation, and exerts a strange attraction - there is evidence that these flows remember their initial conditions far longer than we had suspected. Is there a universal state to which these flows tend (as was once believed), even if slowly, or is there not?

- It has been an article of faith for fifty years that the small scales were more nearly isotropic than the large scales, and while that does appear to be true, we have found that they are often a great deal less isotropic than we might have expected, especially for a scalar. The mechanisms for this are being investigated.

- We have found that there is a certain residual anisotropy in the small scales even at infinite Reynolds number, associated with inter-component energy transfer.

- While it is clear that in three-dimensional turbulence, if averages are taken over a sufficient region or a long enough time, energy is transferred from large scales to small, it is also now clear that for short times, or small regions, energy can flow the other way. This is known as back-scatter, a term that I feel was poorly chosen.

There are many fundamental questions connected with geophysics, for example:

- How does a turbulent / non-turbulent interface propagate in the presence of buoyancy?

- The effects of rotation on turbulence structure are just coming to be known.

- Relatively little of a fundamental nature is known about the dynamics of interacting fields such as momentum, latent and sensible heat and particulates in a gravitational field, and the interaction with index of refraction fluctuations, for example.

- There are many fundamental questions regarding the instabilities of near-surface flows, and the mixing that the resulting secondary motions produce: for example, Langmuir circulations in the ocean surface mixed layer, and their likely terrain-induced counterpart in the terrestrial atmospheric surface mixed layer.

- How do breaking waves generate turbulence?

As a result of the construction of low-dimensional models for the wall region of the turbulent boundary layer (see Sections 2.1 and 2.2), we are coming to have some understanding for the dynamical role of the secondary instabilities that occur in the wall region, which lead to what are called bursts. Some light has been shed on the interaction between these phenomena and adverse and favorable pressure gradients (Stone 1989). The time is now ripe to extend this understanding to flows on concave and convex walls, with secondary rates of strain, and as influenced by a non-inertial rotating frame. This would have interesting applications in axial flow turbomachines.

The fundamental study of turbulent combustion, or even of turbulent reacting flows without heat release, is still in its infancy.

TURBULENCE MODELING

Direct numerical Simulation is restricted to low Reynolds numbers. Large Eddy Simulation avoids this restriction. For similar numbers of grid points, the costs of the two techniques are similar. These costs are still high for general use as a design tool, though in special circumstances LES is being looked at as an interesting possibility. Nevertheless, in the forseeable future there will be a need for model equations that will behave approximately like turbulence at least in restricted situations. These are typically Reynolds averaged equations, which predict mean velocities and pressures, and perhaps component energies and Reynolds stresses, depending on the order of the model. The simpler models are routinely used in, for example, the airframe and gas turbine industries.

In considering accuracy, we should distinguish (as Peter Bradshaw suggests - private communication)

1. calculations that are better than an expert's guess (the latter being the only alternative other than experiment);

2. calculations that give the right trends and can therefore be used, with care, for parameteric studies;

3. calculations that are accurate enough for use in final design.

Industrial users most typically say they want models of class 3, although this may be changing (see below). Current models enter class 3 only for rather simple and undemanding flows. Many people who do turbulence for a living would be quite happy with models which behaved in a qualitatively correct manner, even though quantitatively in error by 20%, that is, in class 2.

In addition, industrial users do not, in general, have anyone in house to trouble-shoot or fine-tune a model, or evaluate models on a comparative basis, or determine reliably if a model is being applied outside its range. There is now in the literature a great variety of models of varying quality which have in general been tested each in a quite different numerical platform, and each on only a small selection of test cases. The user community, quite naturally, would like to see any proposed model tested on some accepted numerical platform, on a comprehensive collection of test flows; they would, in addition, like to see some independent government laboratory test all the proposed models, and rank order them for performance.

In an attempt to address these needs, a consortium of government agencies supported a collaborative effort (Bradshaw *et al*, 1993), in which the community was invited to compute with various models a battery of test flows, and the performance of the various models was compared. Unfortunately, this represented for each collaborator a considerable investment of computer time and personnel, and many did not bother to participate; those who did fell farther and farther behind the organizers' schedule. In any event, the conclusion was that none of the models can reliably compute a variety of flows with the accuracy that industry desires, but can do so only in highly restricted situations where the model has been extensively calibrated. In general, the Reynolds stress transport, or second order, models do better than the simpler models only in the sense that they are able to compute some flows that are quite beyond the simpler models, because the physical mechanism in question was not included in the simpler model. Where both types of models work, the more complex model is not often significantly better than the simpler one, possibly because more terms have to be modeled, and the terms are less familiar, and possibly because both depend on the same inadequate dissipation equation.

Many of the models that exist are quite simplistic, and do not contain a great deal of physics. Parts of the turbulence community have not been terribly enthusiastic about model construction, and the modeling business has not often benefited from the contributions of the most able members of that community. In addition, turbulence is, after all, an extremely complex phenomenon, and although we understand much of it in a crude sort of way, to construct a model that will reliably behave like turbulence to 1% accuracy in a variety of flows is probably beyond our current understanding. It takes a good deal of effort even to produce a model in class 2, and the user community does not always appreciate the difficulties.

There is a serious need for high-quality work to introduce more physics into the models, so that they will behave more like turbulence in a wider variety of situations. Such work is actually quite rewarding, since the effort to make a model that behaves more like turbulence in some sense can often lead to understanding of basic physical mechanisms. A recent case in point is the work of Reynolds on the effects of rotation (Kassinos & Reynolds 1994). However, there is relatively little support available for this kind of work, since the agencies are a little gun-shy; they feel that they have put a lot of money into this area, and it has not produced models which have been very satisfactory from industry's point of view. In addition, work of this sort has become very demanding and labor intensive, since it is now necessary to test and document any proposed model very extensively; the amount of work required now considerably exceeds that appropriate for a Ph. D. In addition, W. C. Reynolds has observed (private communication) that this is not a young person's game - notable contributions have usually been made only by workers in their forties or older. Evidently the construction of this kind of model requires a synthesis of extensive experience.

The outlook of the user community may be changing in such a direction as to improve the situation. Nissan, for example, almost alone in the automotive community, does extensive computations of flow around vehicles during the design phase. For various subtle reasons (Lumley 1993) the computations are not highly accurate quantitatively, although they are qualitatively correct. Nevertheless, the results can be quite helpful to Nissan designers. The US and European automobile industries have lagged far behind in the use of CFD for automotive design because of a demand for accuracy and cost comparable with wind-tunnel testing. As an aside, I believe that industrial users of wind-tunnel testing tend to confuse reproducibility with accuracy, and it is not clear that they take into account the cost of model construction when they calculate comparative cost. In any event, I feel that there is much of value to be obtained from a model in class 2. There are some indications that the commercial airframe industry is beginning to use class 2 models in this way, combining them with other inputs (e.g. - wind tunnel testing).

CLOSURE

In closing, there are two peripheral aspects that we have not touched on, and do not have room to more than mention. The results of turbulence research are not in general accessible to the user community; this has essentially to do with the difficulty of the field. Undergraduate and graduate engineers who go to industry do not generally know enough about turbulence to do anything useful, so far as solving industry's turbulence problems, or interpreting the results of turbulence research, is concerned. The only solution that can be suggested is the use of specialty consultants to make the results of turbulence research available to industry in the form of solutions of industry's practical problems.

The second aspect concerns the current mood in Washington as it affects basic research. The research that is supported by Federal agencies tends to be more and more application-oriented. It is increasingly difficult to support basic research for its own sake, without tying it to some application. That is not, in and of itself, necessarily a bad thing; applications can be very fecund, rich in interesting basic ideas. To maintain our competitive position in international markets, it is important to give attention to the applications. It is absolutely essential, however, that this tilt toward applications not be allowed to drive out basic research altogether. However slow basic research in turbulence has been to trickle down to the level of applications, it is our only hope for the eventual development of robust computational methods that will be useful in design and regulation.

REFERENCES

Bradshaw, P., Launder, B.E. and Lumley, J.L. (1993) Collaborative testing of turbulence models. *Proceedings, Fluids Engineering Division, ASME*, 196:77–.

Brown, G. and Roshko, A. (1971) The effect of density difference on the turbulent mixing layer. In *A.G.A.R.D. Conference on Turbulent Shear Flows, Conf. Proceedings No. 93*, pages 23/1–23/12. NATO Advisory Group for Aerospace Research and Development.

Brown, G.L. and Roshko, A. (1974) On density effects and large structure in turbulent mixing layers. *Journal of Fluid Mechanics*, 64:775–816.

Farge, M. (1992) Wavelet transforms and their applications to turbulence. *Ann. Rev. Flu. Mech.*, 24:395–457.

Holmes, P.J., Berkooz, G. and Lumley, J.L. (1996) *Turbulence, Coherent Structures, Symmetry and Dynamical Systems*. Cambridge University Press.

Kassinos, S.C. and Reynolds, W.C. (1994) A structure-based model for the rapid distortion of homogeneous turbulence. Technical Report No. TF-61, Thermosciences Division, Department of Mechanical Engineering, Stanford University, Stanford, CA 94305, December.

Liepmann, H.W. (1952) Aspects of the turbulence problem. part II. *Z. Angew. Math. Phys.*, 3:407–426.

Liu, J.T.C. (1988) Contributions to the understanding of large-scale coherent structures in developing free turbulent shear flows. *Advances in Applied Mechanics*, 26:183–309.

Lumley, J.L. (1992) Some comments on turbulence. *Physics of Fluids A*, 2:203–211.

Lumley, J.L. (1993) The prospects for accurate numerical simulation of air flow in and around automobiles. Invited Lecture to Société des Ingenieurs de l'Automobile, Ecole Polytechnique, Palaiseau, Paris, February 1993.

Meneveau, C. (1991) Analysis of turbulence in the orthonornal wavelet representation. *Journal of Fluid Mechanics*, 232:469–520.

Monin, A.S. and Yaglom, A.M. (1975) *Statistical fluid mechanics*. The MIT press, Volume I, 1971; Volume II. Translation editor J.L. Lumley.

Myers, M., Holmes, P., Elezgaray, J. and Berkooz, G. (1994) Wavelet projections of the Kuramoto-Sivashinsky equation I: Heteroclinic cycles and modulated traveling waves for short systems. Physica D (submitted).

Rempfer, D. and Fasel, H. (1994) Evolution of three-dimensional coherent structures in a flat-plate boundary layer. *Journal of Fluid Mechanics*, 260:351–375.

Stone, E. (1989) *A Study of Low Dimensional Models for the Wall Region of a Turbulent Layer.* PhD thesis, Cornell University.

Tennekes, H. and Lumley, J.L. (1972) *A first course in turbulence.* The MIT press.

Townsend, A.A. (1956) *The Structure of Turbulent Shear Flow.* Cambridge University Press.

WATER WAVES AND RELATED COASTAL PROCESSES

Chiang C. Mei
Department of Civil Engineering
Massachusetts Institute of Technology
Cambridge, MA 02139

INTRODUCTION

Since coastal regions are usually densely populated, nearshore fluid dynamics is important because it impacts directly on human activities and on natural resources essential to our lives. In contrast to deep-sea oceanography where attention is largely focussed on the global scale, in coastal hydrodynamics one must examine the details of water motion from many kilometers down to the size of a sand particle. Water waves are often generated by distant wind, and must first propagate across the continental shelf towards the coast. Upon entering shoaling waters, they are either refracted by varying depth or current, or diffracted around abrupt bathymetric features such as submarine ridges or valleys, losing part of their energy back to the deep sea. Waves continuing their shoreward march give up some energy by dissipation near the bottom. Nevertheless each crest becomes steeper and mightier, and makes its final display of power by breaking and splashing on the shoreline, and sending countless sand particles afloat, thereby reshaping the shoreline. The role of breaking waves' in mixing and turbulence is also vital to the fertilization, feeding and growth of marine life.

In this section we shall discuss research needs in surface gravity waves and associated coastal processes. For waves in deep water, the need to understand better the effects of nonlinearity will be stressed. We then turn our attention to the nearshore region where interaction with the coastal boundaries is essential. In addition to refraction and diffraction by bathymetry and coastlines, we also include the recent interest in infragravity waves which are caused by nonlinear interactions of short waves within a narrow frequency band. Surf zone dynamics is the most important factor in affecting our shorelines, yet basic understanding is the poorest because of the ubiquitous presence of turbulence and sediments. The final topics are concerned with sediment mechanics of both noncohesive and cohesive kinds.

WAVE PROPAGATION IN DEEP WATER

The apparent randomness of the sea surface has prompted the use of statistics to describe the ocean surface. In particular Pierson and Moskowicz introduced the idea of energy spectrum which has been a standard tool for engineering purposes. Further field experiments have led to JONSWAP spectra for fetch-limited seas and shallow depth. Hasselman and associates have introduced a wave forecasting scheme that formally includes nonlinearity, wind input and disspation. But many of these components must still be dealt with largely empirically because of the lack of physical understanding.

Since surface waves are generated by wind which is invariably turbulent, it has long been held that random eddies in the wind is the dominant source causing randomness on the sea surface. Recent studies simulated by advances in remote sensing show that nonlinearity itself can lead to randomness. For example, in their studies of sideband instability of narrow-banded waves, Yuen and Lake (1982) have given evidence that nonlinear interaction among the carrier wave and its unstable sidebands can lead to chaotic responses once a large number of the sidebands are also unstable. More recently Naciri and Mei (1992) have shown that narrow-banded short waves on periodic long waves of sufficient steepness can be excited to a chaotic state because the latter provides a time-dependent gravity which triggers parametric resonance in the short waves. On the other hand, Janssen (1986) has shown theoretically that a steady wind represented by the statistical mean, can aid the triad resonance of capillarly-gravity waves to period-doubling bifurcation. This phenomenon was first observed in the laboratory by Choi (1977). Since periodic doubling is one of the common paths to chaos (Feingenbaum scenario), it appears likely that even without accounting for randomness of eddies, the averaged flow of a turbulent wind can trigger randomness in sea waves.

Research Needs

Theoretical and experimental research of the following topics should lead to better understanding of the sea spectrum development.

a. Resonant interaction of a triad of capillary-gravity waves under the influence of one or more of the following effects,

    i. Long waves

    ii. Currents

    iii. Wind

    iv. Dissipation near solid boundaries

    v. Dissipation by breaking

b. Resonant interaction of a quartet of gravity waves under the influence of one or more of the effects mentioned above.

In both classes of problems transition to chaos and characterization of the chaotic states are useful for understanding the inherent role of nonlinearity.

REFRACTION AND DIFFRACTION IN SHOALING WATERS

To plan proper defense for beaches and harbors against the perpetual attack of sea waves, engineers need reliable estimates of the wave climate in the vicinity of the project site. Since comprehensive records of direct measurements are rarely available, rational extrapolation must usually be made from wave data or wind information far offshore. Therefore prediction of wave propagation from deepwater to shoaling water is a first step in engineering design. At present the following theoretical methods are available:

a. Ray approximation: This is a method based on linear theory and applicable when the horizontal length scale of variation of depth or current is very large compared to the local wave length. This method reduces the fully three-dimensional problem to a one-dimensional one by first finding the rays and then the amplitude variation along rays. This method fails near a caustic where extra efforts are needed to account for local diffraction or nonlinearity

b. Parabolic approximation: This method extends the ray approximation by adding diffraction in the direction transverse to a wave ray, and has been extended to treat variable depth and current. It has also been extended to include the effects of nonlinearity. The advantage is that the mathematics is similar to solving a one-dimensional diffusion problem. The limitation is that waves must be primarily progressing in one direction.

c. Mild slope equation: By this equation the three-dimensional problem is reduced to a two-dimensional diffraction problem. The computational effort is larger than the preceding methods but can handle complex combined refraction-diffraction problems. So far the nonlinear mild slope equation is limited to the second-order Stokes' wave theory.

Research Needs

a. Extension of mild slope approximation to include nonlinearity, which is applicable in both deep and shallow water.

b. Development of wide-angle models using the parabolic approximation.

c. Development of a wave spectrum propagation model based on the parabolic approximation. The effects of wind, currents, dissipation by bottom friction and by breaking should be included.

NEAR-SHORE TOPOGRAPHY ON THE SEA SPECTRUM

From wind swell records along beaches Munk (1944) and Tucker (1950) observed long waves of periods between 1 to 5 minutes and found strong correlation between the long wave and the swell envelope. For storms of 1-2 days duration the long wave height is roughly one-tenth of the swell amplitude. Munk attributed this correlation to the nonlinear interaction between wind waves and coined the term *surf beats* for them, and suggested that they may in turn excite resonance in a small harbor or bay. Measurements by the Corps of Engineers in the Great Lakes and ocean coast harbors such as Barber's Point Harbor, Hawaii, revealed that long-period oscillations in harbors are ubiquitous despite the absence of long wave energy outside. Recent field experiments by Holman (1981), and by Oltman-Shay and Guza (1987) have also given evidence that long-period edge waves can be caused by short waves incident from offshore.

A related problem in offshore engineering is the slow-drift oscillation of a moored vessel such as a ship or a tension-leg platform. Such a moored system has a long natural period of oscillations and hence low frequency excitation inherent in the incident waves can be detrimental. That long waves can be forced by narrow-banded short waves is known for the open sea and is due to radiation stresses of the short waves (Longuet-Higgins & Stewart 1962). This long wave is called *set down* and propagates at the group velocity of the short waves. Its greatest amplitude is negative and occurs beneath the crest of the short wave envelope. For a number of simple geometries (rectangular shelf, trough or slowing varying bathymetry or current, a harbor of simple form or constant depth), the method of multiple scales has been employed to solve these problems.

Research Needs

Experiments and theories to predict infragravity waves caused by incident short waves with a narrow or broad spectrum:

a. on beaches along a complex coast line,

b. in a harbor of arbitrary basin form and depth variation,

c. over a region of uneven bathymetry.

This is an important step for predicting the change of sea spectrum due to local bathymetry or structures.

SURF-ZONE DYNAMICS

Wind waves break on gentle beaches. The breaking crests are powerful agents of mixing because breaking waves dislodge sediments and throw them into suspension, which can then be readily carried by wave-induced mean currents such as longshore currents or rip currents. The distinguishing feature of the surf zone is that turbulence is due more to wave breaking and less to shear in the flow along the bed.

Prediction of wave as well as current motions in the surf-zone depends critically on a realistic modeling of the local dissipation of wave energy due to breaking. It has been shown by Svendsen & Madsen (1984) that, away from the region of initial breaking, the local wave energy dissipation due to breaking is analogous to that of a hydraulic jump. Therefore, breaking waves can be modeled as propagating bores. The internal motions in the region of initial breaking are still poorly understood. The large-scale vortices are important in the wave transformation, but accurate measurments of velocity and vorticity fields in this region are lacking and hence their quantitative modeling has not yet been developed.

The mechanics of the breaking wave-induced mean current are understood in principle; the concept of radiation stresses was used to calculate longshore current by Longuet-Higgins (1970). Recently, two major advancements have been made in this area. First, it is recognized that the vertical variations of the mean current, which have been ignored in most of two-dimensional depth-averaged current models, play an important role in calculating the mean momentum fluxes in the surf zone and therefore affect the magnitude of current. Secondly, it has been shown that the longshore current can be unstable when the magnitude of the current is large and the curvature of the current profile in the on-offshore direction is significant. This unstable current motion is called "shear wave" because the restoring force is the shear in the longshore current profile (Bowen & Holman 1989). These shear waves have been observed in the field (Oltman-Shay & Guza, 1987; Oltman-Shay *et al.* 1989) and are called "far infragravity waves" due to their long wave periods.

Research Needs

a. Mechanism of breaking of periodic or irregular incident waves on a beach; the location of the breaker line.

b. Measurments of velocity, vorticity and tubulence in the surf zone.

c. Three-dimensional variation of mean flow.

d. Instability of longshore current; transient evolution of long shore current subsequent to instability.

e. Longshore current under the direct influence of overhead wind.

f. Longshore current along a complex shoreline.

TSUNAMI RUNUP

The final phase of the wave propagation is the swash zone dynamics, in which portion of the wave rushes up and down the dry beach face. The flow normally occurs in a thin layer is strongly affected by nonlinearity, bottom friction and the permeability of the beach face. Depending on the beach slope, the run-up front may or may not break. When breaking occurs, the flows become turbulent and basically three-dimensional. The fluid dynamics of the entire run-up processes is complex and many aspects are still not well-understood. Nevertheless, an accurate method for estimating run-up motions is essential for prediction of forces on man-made structures exposed to ocean environments and of the coastal effects of tsunamis and storm surges.

Research Needs

a. Development of modified Boussinesq equations for long wave propagation from intermediate depth to very shallow water.

b. Experiments and theories on three-dimensional run-up. The following effects are to be examined: (1) Bottom friction for both dry bed and wet bed. (2) Beach permeabilty. (3) Turbulence due to breaking.

TRANSPORT OF NONCOHESIVE SEDIMENTS

When the amplitude reaches a certain threshold, a progressive wave over an initially flat sandy bed will force sand grains on the top layer to roll and slide along the bed. When wave amplitude further increases the moving sand particles form ripples at periodic intervals. When rolling grain ripples aquire certain height, stronger waves cause flow separation in the lee of each crest and generate vortices, which dislodge sand particles into suspension. Past and recent studies by solving the Navier-Stokes equations either numerically or analytically for oscillating flows over rigid ripples have revealed a good deal about the fluid motion. So far analytical theories (e.g., Hara and Mei, 1990a) have so far been based on the assumption that the viscous flow is time-periodic. However numerical solution of the initial value problem by Blondeaux and Vittori (1991b) have shown that when the Keulegan-Carpenter number or the Reynolds number is sufficiently large, a series of period-doubling bifurcations occurs and the flow becomes chaotic.

The formation of ripples can be predicted by accounting for the sand motion which can only be described by an empirical relation between the local flow velocity, or shear stress, and sand discharge. Blondeaux and Vittori (1990) couple the hydrodynamic theory with the sand discharge formula and studied the instability of two dimensional ripples and their nonlinear growth to maturity.

Under some circumstances ripples are not perpendicular to the ambient flow but are short-crested. Two mechanisms have been studied so far. One is the Gortler instability of oscillating flow over the ripple crest which amplifies span-wise perturbations. This mechanism has been suggested by Mastunaga and Honji (1980) and a theory of centrifugal instabiltiy has been worked out by Hara and Mei (1990b) who found the associated mass transport on the ripple surface to be in qualitative aggreement with the experiments of Mastunaga and Honji. On the other hand Blondeaux and Vittori (1992) attributed the brick patterned ripple to be due to the span-wise instabilty of the sand motion itself. It is likely that the truth is the combination of both. Much more needs to to be done.

Research Needs

a. Two dimensional oscillatory flow over ripples:

    i. Instability of time-periodic vortex flow over rigid ripples. Bifurcations and chaos.

    ii. Oscillatory turbulent flow over rigid ripples. Although models with a depth-dependent eddy viscosity predicts many mean quantities well, experiments show that the eddy viscosity can be time-dependent. Hence a better closure model is needed.

    iii. Induced streaming in oscillatory boundary layers over rigid ripples. For this one must account for the presence of the free surface and finite wave length

    iv. Three dimensional instability of sand ripples in complex waves.

b. Diffusion and convection of suspended load in the oscillatory boundary layer. Taylor dispersion is expected to be effecitve in assisting the diffusion of suspended sediments.

c. Mechanism of deposition, entrainment and resuspension of particles from a rough bed.

d. Movement of ripples on large-scale sand bars. If the waves are partially reflected, it is well known that variations in induced streaming can cause nonuniform accumulation of particles and form sand bars of wavelength equal to half of the incident wave length. Theories and experiments are needed for this muti-scale phenonmenon of ripples on sand bars in the presence of partially standing waves.

e. Granular mechanics of bedload transport. A basic difficulty in the mechanics of sediment transport is the lack of mathematical model based soundly on microscale mechanics of water/particle interaction. In recent years progresses are being made on the steady flow of dry and colliding granules. Extension of such efforts for granules in flowing water is desirable.

TRANSPORT OF COHESIVE SEDIMENTS

In many esturaries around the world, ultra fine particles of diameters no more than tens of microns are carried to the sea from far upstream of a river. Once mixed with salty sea water they coagulate to form large aggregates and settle down to the seabed. When the mixture of seawater and the settled particles is sufficiently concentrated, it becomes fluid mud whose rheological behavior is no longer Newtonian. In fact in many cases the mixture is jello-like and is well approximated by the Bingham plastic model with a finite yield stress and large viscosity, both of which increase with concentration. As a consequence fluid mud motion can remain laminar even at reasonably high velocities. It can be shown that in tidal flows of very long periods interfacial friction is dominant and the interface between seawater and fluid mud is blurred by continuous variation of suspended particles. However under gravity waves of much shorter period, the pressure gradient is more important which drives the flud mud to flow in bulk, leaving a sharp interface. Bedload transport in bulk is therefore the dominant mode.

Research Needs

a. Theoretical explanation of the non-Newtonian behaviour. Qualitatively it is known that electric charges carried by the tiny particles are responsible for certain peculiar stacking of plate-like particles to form aggregates. The macroscale rheology depends not only on hydrodynamic but also on electrodynamic and other chemical factors.

b. Instability of an inclined layer of fluid mud on the seabed.

c. The effects of density stratification of fluid mud. The non-Newtonian behaviour such as the yield stress depends on the concentration, hence concentration and dynamics are nonlinearly coupled.

d. Wave-induced transport of fluid mud.

e. Mechanism of resuspension of cohesive sediments.

f. Coagulation and breakup of cohesive particles in turbulence.

g. Turbulent diffusion and convection of cohesive suspensions in which coagulation and breakup occur.

These studies are also important in the transport of pollutants in coastal waters

ACKNOWLEDGMENTS

In preparing this report we have benefited from the following workshop reports:

1. Report on the Sate of Nearshore Processes Research (1990), Nearshore Processes Workshop, St. Petersburg, FL, April 24-26, 1989. Report OSU CO90-6 Oregon State University.

2. Tsunami - Research Opportunities (1981), National Science Foundation & National Oceanic Atmospheric Administration.

3. Engineering Fluid Mechanics Workshop Report (1990), ASME CRTD-16, Center for Research & Technology Development.

4. National Hazards and Research Needs in Coastal and Ocean Engineering, Ad hoc Committee Civil & Environmental Division, NSF (1984).

REFERENCES

Blondeaux, P. & Vittori, G. (1991b) A route to chaos in an oscillatory flow: Feingenbaum Scenario. *Phys. Fluids*, **3**, 2492-2495.

Bowen, A.J. & Holman, R.A. (1989) Shear instabilities in the mean longshore current. 1: Theory. *J. Geophys. Res.* **94**, 18023-18030.

Choi, I. (1977) Contributions a l'etude des mecanismes physiques de la generation des ondes de Capillarite-gravite a une interface air-can. Thesis Universiti d'Aix Marseille.

Hara, T. & Mei, C.C. (1990a) Oscillating flows over periodic ripples. *J. Fluid Mech.* **211**,183-209.

Hara, T. & Mei, C.C. (1990b) Centrifugal instability of an oscillatory flow over periodic ripples. *J. Fluid Mech.* **217**, 1-32.

Holman, A.J. (1981) Infragravity energy in the surf zone. *J. Geophys. Res.* **86**, 6442-6450.

Janssen, P.A.E.M. (1986) The period-doubling of gravity-capillary waves. *J. Fluid Mech.* **172**, 531-546.

Longuet-Higgins, M.S. & Stewart, R.W. (1960) Radiation stress and mass transport in gravity wave, with application to "surf beats". *J. Fluid Mech.* **3**, 481-504.

Longuet-Higgins, M.S. (1970) Longshore currents generated by obliquely incident sea waves, 1 and 2. *J. Geophys. Res.* **75**, 6778-6789; 6790-6801.

Matsunaga, N. & Honji, H. (1980) Formation mechanism of brick pattern ripples. *Report Res. Inst. App. Mech.* Kyushu University **28**, 27-38.

Munk, W.H. (1949) Surf beat. *Trans. Am. Geophy. Union* **30**, 849-854.

Naciri, M. & Mei, C.C. (1992) Evolution of a short surface wave on a very long wave of finite amplitude. *J. Fluid Mech.* **235**, 445-452.

Oltman-Shay, J. & Guza, R.T. (1987) Infragravity edge waves observations on two Californian beaches. *J. Phys. Oceano.* **17**, 644-653.

Oltman-Shay, J., Howd, P.A. & Mirkemeier, W.A. (1989) Shear instabilities of the mean longshore current. 2: Field observations. *J. Geophys. Res.* **94**, 18031-18042.

Svendsen, I.A. & Madsen, P.A. (1984) A turbulent bore on a beach. *J. Fluid Mech.* **148**, 73-96.

Tucker, M.J. (1950) Surfbeats: Sea waves of 1-5 minutes period. *Proc. Roy. Soc. Lond. A* **202**, 565-573.

Yuen, H.S. & Lake, B. (1982) Nonlinear dynamics of deep water gravity waves. *Advances in Applied Mech.* **22**, 67-229.

DIRECT AND LARGE EDDY SIMULATION OF TURBULENCE

Parviz Moin
Department of Mechanical Engineering
Stanford University
Stanford, CA 94305

INTRODUCTION

All flows encountered in aerospace technology are at least partly turbulent. The performance of many flow devices is directly determined by turbulence effects, and these must be accurately predicted and, if possible, controlled. Turbulence enhances the rate of surface heat transfer, mixing and chemical reaction in the flow and is responsible for a substantial fraction of skin-friction drag.

Generally, the specific objective of a fluid dynamic calculation and the nature of the flow dictates the accuracy needed for predicting turbulence effects. For example, the effects of turbulence should be accurately accounted for in engineering computational fluid dynamics, CFD, if the object of the computation is to obtain accurate skin friction or surface heat transfer. Moreover, in separated flows where separation occurs due to adverse pressure gradient turbulence affects surface pressure forces and hence lift and form drag.

Although turbulent flows at ordinary temperatures and pressures are governed by the well known Navier Stokes equations, the presence of a wide range of scales of motion has made *direct* computation of turbulence a formidable challenge for computers and numerical methods. The advent of powerful computers has greatly enhanced the ability to directly compute all details of a turbulent flow. All flows encountered in aerospace technology are at least partly turbulent. The performance of many flow devices is directly determined by turbulence effects, and these must be accurately predicted and, if possible, controlled. Turbulence enhances the rate of surface heat transfer, mixing and chemical reaction in the flow and is responsible for a substantial fraction of skin-friction drag.

Generally, the specific objective of a fluid dynamic calculation and the nature of the flow dictates the accuracy needed for predicting turbulence effects. For example, the effects of turbulence should be accurately accounted for in engineering computational fluid dynamics, CFD, if the object of the computation is to obtain accurate skin friction or surface heat transfer. Moreover, in separated flows where separation occurs due to adverse pressure gradient turbulence affects surface pressure forces and hence lift and form drag.

Although turbulent flows at ordinary temperatures and pressures are governed by the well known Navier Stokes equations, the presence of a wide range of scales of motion has made *direct* computation of turbulence a formidable challenge for computers and numerical methods. The advent of powerful computers has greatly enhanced the ability to directly compute all details of a turbulent flow, but the great cost of such computations makes them impractical for engineering applications. Thus, in the applied CFD codes the turbulence itself is not computed, rather its average effect on the mean flow is approximated by a *model* term in the mean flow equations. This statistical description of the effects of turbulence remains the pacing item for computational fluid dynamics.

The hierarchy of methods for turbulence computation, in order of decreasing computational effort, consists of: direct numerical simulation (DNS), where one computes all of the essential turbulence scales of motion and no modeling approximations are made; large eddy simulation (LES), where the effects of the smallest eddies upon the large scale motions are modeled; and the widely used Reynolds-averaged model, which can be viewed as the limit of large eddy simulation in which the computational grid is too coarse to resolve any of the details of turbulence motions. Thus, the accuracy of the turbulence model is most important in the Reynolds averaged approach, the predominant method for engineering CFD, and it is there that improvements in model accuracy have the highest value to the engineering community. In this article we shall discuss only the applications and potential of the DNS and LES techniques. However, as we shall see below one of the most important applications of DNS and (in part LES) is to provide data that are virtually impossible to obtain experimentally for the design and testing of Reynolds averaged turbulence models for engineering applications.

In DNS the computational grid is fine enough to resolve all essential turbulence scales. The unsteady three-dimensional Navier Stokes equations are solved without modeling approximations. Since the number of grid points required is proportional to the 9/4 power of Reynolds number, DNS has been restricted to low Reynolds numbers and very simple flow configurations such as the plane mixing layer, fully developed channel flow and flow over a backward facing step. Even for such flows DNS computations tax the most powerful supercomputers available. In 1992 a typical state of the art DNS computation has about 10 million grid points and may require several hundred CPU hours of a CRAY-YMP processor. The distinction of DNS computations with the largest number of degrees of freedom belongs to two simulations of isotropic turbulence recently completed on massively parallel computers with about 130 million grid points.

The output of a typical direct simulation is an enormous data bank consisting of the instantaneous flow velocity, pressure and temperature fluctuations in three-dimensional space. The primary utility of DNS has been in the use of these databases for the study of the mechanics of turbulent flows. The physical realism of the databases in simple shear flows has been established by very detailed comparison with experimental data. The numerical solutions are also subjected to stringent internal consistency checks, providing considerable confidence in quantities which can not be measured experimentally. In fact, the credibility of simulation data has improved to the point where they have been used to calibrate flow measuring techniques and, in some cases, to check the accuracy of experimental data. DNS has earned its place as a powerful and indispensable research tool in the study of turbulence. In some cases, such as the study of coherent structures in shear flows at low Reynolds numbers, DNS is the preferred tool over experiments. Examination of DNS data has led to a more realistic view of the imbedded coherent structures in shear flows. In the past, it was not uncommon for models of the instantaneous velocity and vorticity vectors to be built from measurements of a single velocity component. Simulations have made possible analysis of turbulence without such extrapolations from incomplete data.

The development of turbulence models for engineering computations in the Reynolds averaged approach has been hampered by a lack of data for the modeled quantities. Experimental measurement of some of these quantities is difficult and for others, impossible. However, significant progress in this area has been made by using direct numerical simulation (DNS) databases. Simulations of simple turbulent shear flows have provided detailed budgets of turbulent stresses which, for the first time, allow direct testing of turbulence models.

The potential benefits of controlling turbulent flows that occur in various engineering applications are known to be significant. A recent and very promising application of DNS is in the development and testing of active and passive turbulence control strategies. It has been shown, for example, that substantial drag reduction can be achieved by active control in fully developed turbulent channel flow. In one case, when the wall boundary condition for the normal velocity was adapted to the instantaneous state of the flow within the boundary layer, drag was reduced by over 20% with negligible (ideal) work input. Although actual hardware implementation of such control schemes is highly impractical at present, DNS has shown what may be possible. As another example, it was known from experiments that longitudinal surface grooves or riblets of the proper size can reduce skin-friction, but the flow mechanisms involved were not well understood. Recent direct numerical simulation of a boundary layer over riblets has helped to elucidate these mechanisms. Again, it was the ability to visualize all turbulence data in three-dimensional space and time that led to both enhanced understanding of the fundamental flow physics and the means to control it. Similar practical application of DNS is expected in the problem of surface heat transfer enhancement.

DNS has recently been extended to compressible and chemically reacting flows with heat release. The effects of compressibility on turbulence have been virtually ignored for some time; there is a unique opportunity for DNS to provide first order information in this increasingly important area in aerospace technology. Simulations have shown that small eddy shocklets occur in turbulent flows and lead to increased dissipation. The interaction between turbulence and a normal shock wave has been simulated, and the results are being used in turbulence modeling. DNS with simple chemical reaction has been performed and inclusion of more complex chemistry with disparate time scales is beginining to appear. DNS is also being used to study the physics of noise generation and the means to control it. The near field acoustic sources can be computed directly, and the contributions of their components on the far field noise can be studied. It has also been shown recently that the actual acoustic field can be accurately computed using highly accurate numerical schemes and proper boundary conditions. These results can be used to test various aero-acoustic analogies.

The future role of DNS in turbulence computation is strongly linked to advances in supercomputer technology, and it appears that massively parallel computers will lead the way in this direction. At present, there is considerable effort toward implementing DNS codes on parallel computers. The structure of current DNS codes allows them to make effective use of massive parallelism.

DNS will not be used in the near future for engineering design, but it will be used with increasing frequency to understand and model turbulence and to develop strategies to control its complex effects in flows of engineering interest. DNS is a powerful but expensive tool for turbulence research and should not be performed frequently; DNS generated data should be archived and made available to many researchers studying diverse aspects of turbulence. Although most researchers do not have the means to perform state of the art DNS, they have sufficient computational resources to probe the data once it has been generated.

In contrast to DNS, the large eddy simulation (LES) technique can potentially become affordable for engineering applications in the near future. The basic philosophy of LES is to explicitly compute only the large-scale motions that are directly affected by the boundary conditions, and to model the small scales (subgrid scale model). The premise of LES is that the small scale motion is more universal and therefore its statistical effect on the large scales can be modeled more easily. LES originated in meteorology and, in the past twenty years, has had limited use in the engineering research community. LES is very attractive for engineering calculations because it is significantly less computer intensive than DNS yet promises to be more accurate and robust than Reynolds averaged models. Unfortunately, subgrid scale models have faced some of the same problems as Reynolds-averaged models including improper asymptotic behavior near walls, variation of the optimal "coefficient" in different flows, and the necessity for ad hoc "intermittency" functions for the computation of transition from laminar to turbulent flow.

A major advance was recently made in subgrid scale modeling for large eddy simulation. The idea is to parameterize the subgrid scale stresses using data from the computed large scale field at several scales, that is, the subgrid scale stresses are extrapolated from the numerical solution during the computation. A particular mathematical formulation of this process leads to an eddy viscosity model with a coefficient that depends on space and time. In retrospect, the notion of an eddy viscosity coefficient that adjusts to the instantaneous local state of the flow is intuitively appealing. It turns out that the resulting dynamic localization model has overcome the above-mentioned deficiencies of previous subgrid scale models and has performed remarkably well in several transitional and turbulent flows without requiring any ad hoc adjustments. The same concept has been extended to compressible flows where it has led to a dynamic localization model for the turbulent Prandtl number. The requisite computer power for LES of flow away from boundaries in turbulent boundary layers scales roughly with Reynolds number to the 0.4 power. Thus, LES may already be feasible for some engineering computations if accurate simulation of the viscous sublayer adjacent to the surface is not required (such as the flow over the windward surface of a fighter aircraft at high angle of attack). However, if, for example, an accurate estimate of skin friction is required, the viscous sublayer must be at least partially resolved which would require much larger computer power. In this case a nested hierarchy of refined grids covering the near wall region must be employed. Even with such carefully constructed nested grids, at present, only component simulations at low Reynolds numbers can be expected (e.g., airfoil at cord Reynolds number of 10^6). However, with computers of a TERAFLOP sustained performance expected to be available by the end of this decade, computation of much higher Reynolds number flows in more complex geometries can be attempted.

The future of LES appears to be bright. The procedure leading to the dynamic localization model may be viewed as an altogether new approach to modeling, and its applications to date have shown that it is indeed very promising. Its inherent robustness should facilitate the use of LES in engineering computations, particularly in turbomachinery where the Reynolds numbers are relatively low.

Research Needs

Progress in LES and DNS technology will benefit from the following:

- *i)* Significant speed and memory improvements in massively parallel computer hardware.

- *ii)* Development of efficient parallel numerical algorithms

- *iii)* Development of numerical algorithms using nested grids to resolve the viscous sublayer.

- *iv)* Validation of the dynamic localization subgrid scale model in complex geometry and fundamental research in parameterization of subgrid scale turbulence.

- *v)* Development of numerical algorithms to capture shocks and/or flame fronts in turbulent flow.

REFERENCES

Chapman, D. R. (1979) Computational Aerodynamics Development and Outlook. *AIAA J.*, **17**, 1293.

Rogallo, R. S. and Moin, P. (1984) Numerical Simulation of Turbulence. *Annual Review of Fluid Dynamics*, **16**, 99.

Germano, M., Piomelli, U., Moin, P. and Cabot, W. (1991) A Dynamic Subgrid Eddy Viscosity Model. *Phys. Fluids A.*, **3**, 1760.

Moin, P., Squires, K., Cabot, W. and Lee, S. (1991) A Dynamic Subgrid Model for Compressible Turbulence and Scalar Transport. *Phys. Fluids A.*, **3**, 2746.

RESEARCH DIRECTIONS IN MHD FOR THE 1990's

René Moreau
Lab. MADYLAM
INPG-ENSHMG - B.P. 95
38402 St. Martin d'Héres, France

INTRODUCTION

Magnetohydrodynamics (MHD) is the science of coupled hydrodynamic and electromagnetic phenomena (Roberts, 1967; Moreau, 1990). This gigantic domain is much vaster than the aggregate of ordinary fluid mechanics and electromagnetism. Indeed, the coupling is so rich that it yields completely original phenomena and mechanisms, which do not occur with purely hydrodynamic or electromagnetic effects (Alfven waves, for instance). The fluids concerned include electrolytes, liquid metals, thermal plasmas $(3,000 - 10,000$ K) and highly ionized gases, while their electrical conductivities vary from 1 to $107\Omega^{-1}m^{-1}$. Density, velocity and magnetic field strength also have extremely different orders of magnitude from one case to another. Any rational classification based on fundamental mechanisms, which would allow one to clearly analyze today's stage of knowledge and most challenging unsolved questions, would also open too many sections for a ten-page report. Therefore, we focus on the principal existing and potential applications of MHD and derive from their needs the most important research directions for the coming decade. Section 2 concerns applications to materials processing, by far the most successfull field in spite of its youth, and section 3 concerns applications to energy conversion, a thirty-year old subject of research and development. Because no applied research can be successfull without the help and the skills of fundamentalists, the goals of basic research in MHD are discussed in section 4.

Mistakes of the past often provide fruitful lessons for the future, so let us recall how MHD research was killed during the 1970's. Open cycle energy conversion with ionized gases at around 4,000 K received big investments during the 1960's. When it became clear that the refractory materials of that time could not survive at such temperature, the rumour that "MHD does not work" spread through scientific circles and the research funding agencies. Soon, most of the groups with MHD expertise were dispersed or reoriented. Today, only a few hundreds American scientists and engineers are involved in MHD research programs. How could such a subcritical community be able to participate in challenging technological developments such as the design of the lithium blanket for fusion nuclear reactors ? How could it respond to the demands from the large community of specialists in the materials sciences who are ready to use MHD effects to improve their technologies (crystal growth, for instance)? Let us discuss the needs and then return to these questions in the conclusions.

MATERIALS PROCESSING

Primary production from ores

Aluminium cells are a classical example of MHD intervention in a primary production process. These cells consume approximately 4% of the electrical energy generated in the USA. The efficiency of the cell is controlled by the fluid flow driven by the Lorentz forces, and by the hydrodynamic instabilities of the interface between the cryolite (molten salt into which alumina dissolves) and the liquid aluminum previously produced. In iron making, electromagnetic smelting is replacing coke-oven production, primarily due to environmental regulations. In both the aluminum and iron processes, the objective is to inject electricity into the bath in order to produce electrolysis or other chemical reactions, to heat by Joule effect, and to stir by the Lorentz forces. Much more knowledge and understanding are needed for the development of good multidisciplinary models to describe the electrodynamics, hydrodynamics, transport phenomena and chemical reactions involved in such processes (Moreau and Evans, 1984). The crucial role of MHD must be stressed because the electromagnetic variables are the only externally controllable parameters, and they can dramatically modify the other phenomena.

Electromagnetic stirring

Electromagnetic stirrers, such as coreless or channel-type induction furnaces, possess many degrees of freedom (frequency, intensity and phase-shift of the coil electric currents, geometry of the inductor and the pool, etc...) and yield a wide variety of electromagnetically driven flows. In spite of the considerable previous research, (Taberlet and Fautrelle, 1985; El Kaddah et al., 1986) several important questions remain. The users of large-scale induction ladles report that mixing and turbulence are weaker than expected. Indeed, the magnetic Reynolds number similarity does not hold, and the AC magnetic field may exert a large-scale damping effect which is similar to that of a DC field and which cannot be observed at laboratory scale. In addition, we need a greater understanding of the effects of the oscillating part of the Lorentz forces in low frequency AC systems.

Electromagnetic steel casting

The development of continuous casting would have been impossible without the use of AC electromagnetic stirrers (Takeuchi, 1992). Various types of stirrers are now used at different locations along the steel strand. Mould stirrers (MEMS) remove slag inclusion, pinholes and blowholes, strand stirrers (SEMS) increase the quality of equiaxed structures, and final stirrers (FEMS) reduce center-line segregation, porosities and cracks. Recent experiments revealed that DC magnetic fields reduce some negative effects to the high-velocity metal jet issuing from the nozzle in curved casters (entrainment of inclusions, shell erosion).

In spite of the important know-how available among steelmakers, which is the basis for today's design of electromagnetic casters, the lack of scientific knowledge is still enormous. Some approximate numerical models of the effects of flow and transport phenomena on the solidified structures have been developed, but there is a critical need for numerical models which would more accurately represent the physical phenomena and for experiments to guide and validate the model development.

Free surface control

High frequency magnetic fields induce currents in a thin skin depth and thus generate surface forces which can balance gravity or surface tension. With these electromagnetic surface forces, levitation melting and electromagnetic shaping or guiding of liquid metal streams are possible (see Moreau, 1990). In addition, a stabilizing effect originates from the surface distribution of electromagnetic forces (Garnier and Moreau, 1983). DC magnetic fields, through their capacity to dissipate the kinetic energy of disturbances, induce a similar anisotropic stabilizing effect. Some aluminum continuous casters use high frequency AC fields to control the shape of the meniscus in the mould and the position of the first solidification line, in order to supress the mould-oscillation marks on the product (Meyer, 1992). The mould itself is eliminated in purely electromagnetic casters, which were developed in Russia (Getselev and Martynov, 1975) and Switzerland (Weber and Sautebin, 1986) and which currently produce more than a million tons of aluminum each year. In addition, a high-intensity DC magnetic field can levitate insulating liquid materials which have a non-zero magnetic susceptibility, such as water.

The scientific problems associated with electromagnetic free surface control are more complex than the usual free-boundary problems because the boundary conditions are non-linear. Magnetostatic approximations (Sneyd and Moffatt, 1982; Brancher et al., 1983), including the zero skin depth assumption, make it possible to predict the equilibrium shape of a given liquid-metal volume located inside a given inductor. Nevertheless, further research is needed to eliminate the errors of these approximations and to include the effects of fluid motion, free-surface waves and instabilities. The most ambitious goal to be reached is the resolution of the inverse free boundary problem, both with AC and DC fields, in order to predict the characteristics of the inductors needed to produce a given stable liquid-metal shape.

Glasses and ceramics

These materials, which are insulating as solids, become weakly conducting at high temperature, so that induction heating is possible with high frequency eddy currents. In practice, the electrical coupling between the material and the coil becomes efficient as soon as the skin depth becomes comparable to the furnace length scale. The Lorentz forces remain small, compared to buoyancy, but the fluid motion is nevertheless electromagnetically controlled (but not driven) since it depends on the Joule power density distribution (Caillault et al., 1989).

While the feasibility of such induction devices has been experimentally demonstrated, no quantitative models currently exist. The strong tempera-ture-dependence of the electrical conductivity leads to an instability in which hot and cold regions experience more and less Joule heating and thus be-come hotter and colder, respectively. Extensive research is needed in order to understand and control this instability and to accurately model internal radiation as well as the effects of stratification on the flow, turbulence and apparent diffusivities.

Solidification under a magnetic field

The transformation of a liquid nutrient or melt into one or more solid phases is of paramount importance because more than 95% of all technolog-ically useful solid materials have undergone this phase change sometime in their history. Their microstructure and chemical composition, which control the material's properties and subsequent behaviour, reflect the melt flow in the vicinity of the solidification front (Garandet et al., 1994). Under normal or reduced gravity, buoyancy and thermocapillary effects are the main driv-ing forces. The use of a properly designed magnetic field (Hurle and Series, 1994) during solidification of electrically conducting liquids is one of the most efficient and promising methods to produce steady convective heat and mass transfer with either enhanced or reduced mixing. Chemical homogeneity and grain size are directly dependent on melt motion and may be improved with an appropriate electromagnetic stirring. On the other hand, the formation of freckles (mesoscale defects) may be inhibited by applying a DC magnetic field to suppress the convective instabilities responsible for their appearance. Ultimately, new ideas and principles leading to controlled and localized mix-ing can be developed. For example, coupling between thermoelectric currents

(Moreau et al., 1993) generated along the solid-liquid non-isothermal interface by the Seebeck effect and an externally applied magnetic field could drive a local fluid flow within the dendritic array in order to directly control the microstructure formation. Such potentialities emphasize the need for further basic and experimental studies, including well instrumented experiments on model systems at room temperature or with transparent materials.

Crystal growth

Single crystals of semiconductors provide the bases for integrated circuits for computers and electronics. The rapid development of ultra small-scale integrated circuits has produced an ever increasing demand for lower microdefect densities and for more uniform dopant and impurity distributions. This demand has led to the use of magnetic fields to control the heat and mass transport to the crystal through control of the melt motion (Hurle and Series, 1994). The electromagnetic suppression of turbulence, surface waves and other disturbances reduces microdefect densities and structural non-uniformities such as striations. In the Czochralski process, which currently produces about 70% of the silicon for electronics, the magnetic field can be tailored (Khine and Walker, 1994; Gelfgat and Gorbunov, 1994) to reduce the concentration of impurities and to homogenize the distribution of both impurities and dopants. In the vertical Bridgman and float-zone processes, uniform magnetic fields can produce purely diffusive crystallization. For crystal growth in microgravity, thermocapillary forces dominate and MHD provide an excellent method to control the thermocapillary convection in microgravity (Bojarevics, 1994). A major effort involving the close coordination of experimental, numerical and analytical studies is needed in order to fully realize the benefits of MHD in crystal growth.

ENERGY CONVERSION

Open-cycle systems (cold plasmas)

The concept of an open-cycle MHD power system (Petrick and Shumyatsky, 1978; Messerle, 1988) working with a cold plasma at $2500 - 3000$ K has been accepted by the two largest national MHD power generation programs in the USA and Russia. Fuel combustion gases, seeded with alkali metal powder to enhance ionization and electrical conductivity, flow across a magnetic field (4-5 T) inducing DC electric power extracted by means of electrodes. This MHD generator, which converts heat directly into electricity, is able to extract more than 10% of the gas enthalpy (topping cycle). The rest of the available enthalpy is utilized in a conventional steam plant (bottoming cycle).

In recent years, the economic crisis has curtailed the Russian program, while the USA program has concentrated on coal-fired MHD plants. Coal-fired power plants with MHD topping cycles should ultimately achieve ultra-high efficiencies of 50-60%, while 40% is an optimistic projection for future conventional steam turbines plants. The MHD coal-fired plant also offers the important environmental benefits of significantly reduced pollution levels (SO_2, NO_X, CO_2, etc...).

The USA program is now in the "Proof of Concept" stage for subsystems of the topping and bottoming cycles and for the seed recovery process. The final conceptual design of an integrated system is in progress. A number of scientific and engineering problems must be resolved, and large-scale and long-duration experiments must be performed. This program has also resulted in a number of valuable spin-offs, such as a pulsed high-energy MHD facility developed in Russia which was successfully employed in a series of geophysical research experiments, a pulsed MHD power source developed in the USA for military applications in space, and a number of valuable developments in combustion and in high-temperature materials.

Closed-cycle systems (gas-liquid metal)

In a liquid metal closed-cycle MHD power system (Petrick and Branover, 1985) there are two separate fluids the thermodynamic fluid (gas or vapor) and the electrically conductive fluid (liquid metal). The two closed loops for the two fluids share a common two-phase section with bubbles of gas or vapor dispersed in the liquid metal. As the bubbles propel the liquid metal, their thermal energy is converted into mechanical energy. The temperature of the bubbles is essentially constant because they are surrounded by a hot liquid metal having a much higher specific heat. Intensive heat transfer from the metal to the bubbles occurs. Thus the thermodynamic fluid expands nearly isothermally during the energy conversion process, and this unconventional thermodynamic cycle has a substantially increased conversion efficiency. Therefore a liquid metal MHD power system has a higher efficiency than a steam turbine system operating at the same temperatures.

A number of specific concepts of liquid metal MHD systems for terrestial and space applications have been suggested. Several pilot plants have been built, and thorough testing has demonstrated viability for the cogeneration of steam and electricity. For the 1 to 20 MW_e range, the cost of electrical energy from these MHD systems is significantly less than that for conventional power plants. The liquid metal MHD system also offers a number of advantages for the protection of the environment. Construction of demonstration plants of an industrial scale is the immediate next step. While most necessary scientific and engineering data are available, further extensive studies in materials engineering and in controlling the two-phase flow are needed.

Liquid-metal blankets of fusion reactors

Deuterium-tritium fusion requires a lithium "blanket" around the plasma to breed the required tritium. A self-cooled blanket design (Malang et al., 1991), in which flowing liquid lithium carries the energy and tritium to external heat exchangers and tritium separators, is less complex than the alternate design, but its price is the potentially large pressure drop needed to drive the liquid metal through the strong magnetic fields which confine the fusing plasma. This may result in significant parasitic pumping power and large stresses in the firts wall which separates the liquid metal from the vacuum around the plasma. A self-cooled liquid-lithium blanket represents a very unusual MHD flow because of the extremely large values of the interaction parameter and Hartmann number ($10^3 - 10^5$). Detailed measurements of velocities, pressures and voltages in such liquid-metal flows have only been obtained during the last decade. This intensive experimental effort and the complementary effort to develop numerical methods to accurately predict velocity distribution in complex geometries have greatly advanced the state of the art and have indicated that there are several ways to reduce the required pumping power and first-wall stresses and thus to make self-cooled blankets a more attractive alternative. One new concept is flow tailoring which uses the strong electromagnetic body force to concentrate the flow where it is most needed for energy removal. Another new concept is to optimize the heat transfer enhancement due to periodic flow fluctuations which have recently been observed experimentally under reactor-relevant conditions for the first time.

The alternate design is the water-cooled blanket (Giancarli et al., 1994) in which the liquid metal (Li-Pb eutectic alloy) flows at an extremely low velocity (a few mm/s) only to extract the tritium. The energy from the plasma is transferred by conduction through the liquid metal to an array of water tubes which are distributed within the Li-Pb ducts and which extract the energy by forced convection. With this small flow rate of the electrically conducting fluid, the MHD pressure drop and required pumping power are much smaller than in the self-cooled design. However, buoyant convective motions in the liquid metal may lead to stagnant regions with undesirable tritium accumulations or other adverse effects. Again, scientists are faced with unusual MHD problems, since the current knowledge about free convection in the presence of a magnetic field is very limited, even for moderate magnetic field strengths.

During the next decade, one blanket design will be chosen for future fusion reactors, most probably within the worldwide ITER (International Thermonuclear Experimental Reactor) program. Intense research is needed to provide a reliable basis for a rational choice. There is a critical need for new knowledge of these unusual MHD flows with large interaction parameters and large Hartmann numbers, of flow instabilities and of the specific properties of turbulence under such conditions.

Sea water propulsion

Conventional propellers produce acoustical signatures. Submarine propulsion by the electromagnetic pumping of sea water has been proposed as a possible way to reduce the acoustical signature and to increase the stealth characteristics of submarines (Meng, 1991; Motora et al., 1994). The electrolysis due to an electric current through sea water generate a large volume of gaseous hydrogen, and the hydrogen bubbles may produce a greater acoustical signature than conventional propellers. Since the electrical conductivity of sea water is very low, even a modest electric current density produces a large Joulean heating loss. For practical channel sizes, the propulsive efficiencies of MHD systems are far lower than those of conventional propellers. Improving efficiencies for practical sized ducts is a very challenging problem.

Liquid metal sliding electrical contacts

One effort to develop highly efficient electro-mechanical energy converters focuses on machines with a typical DC electric current density of 107 and A/m^2 accross each electrical contact between rotating and fixed solid metal surfaces. State of the art solid brush-slip ring systems experience excessive wear at current densities above 7×106 A/m^2. On the other hand, liquid metal sliding electrical contacts have been experimentally demonstrated to run for extended periods of time with current densities in excess of 3×107 A/m^2. A liquid metal fills the small radial gap between the outer perimeter of a rotating copper disk (rotor) and a static, grooved, copper surface (stator) which shrouds the rotor tip. The local magnetic field varies from 0.2 to 6.0 Tesla, and transition from laminar to turbulent flow occurs in the middle of the typical operating range. The design objectives are : (1) to minimize the voltage difference between the rotor and stator for a given electric current between them, (2) to minimize the Joulean heating and viscous dissipation in the liquid metal and (3) to avoid a free-surface instability leading to liquid metal ejection for the entire operating range.

FUNDAMENTAL RESEARCH

Polyphasic fluids

There are many fundamental problems related to the two-phase liquid metal-gas (or vapor) flows occurring in closed-cycle liquid metal MHD power generating systems. The two-phase flow may occur in the magnetic field which usually is transverse to the flow (two-phase flow MHD generators) or

outside the magnetic field region (single-phase flow MHD generators). In
the latter case the gas is used to accelerate the liquid metal or to increase
its pressure and is separated from the metal just upstream of the MHD
generator. In the former case a satisfactory design will require answers to
many major questions, including: a) What is the two-phase flow pattern in
the presence of the magnetic field and the induced electric current interacting
with the field; b) What is the ratio between the velocities of the gas bubbles
and of the liquid metal; and c) What is the effective electrical conductivity
of the two-phase mixture under the described conditions. Although a large
number of studies have been performed and substantial progress has been
made, there is currently insufficient knowledge to give satisfactory answers
to these questions.

Even when the gas is separated from the metal before the generator, further studies are needed. The important question of the gas to liquid velocity
ratio has been resolved empirically for vertical flows, but the investigation
of a means to decrease this ratio (which, is important for the improvement
of preformance of power systems) using surface-active additives is still in a
very preliminary stage. The process of boiling of a volatile liquid in direct
contact with a hot liquid metal is practically even more important, but very
little is yet known about the phenomenon or the benefits of applying an
external magnetic field. There are still other very interesting classes of phenomena related to the MHD of polyphasic fluids, such as pseudo-alloys of two
immiscible metals solidified in conditions of artificial micro-gravity created
by a magnetic field and removal of suspended oxide inclusions in solidifying
melts.

Thermal plasmas produced by DC arcs

The widespread application of plasma systems in materials processing has
stimulated considerable interest in the quantitative representation of fluid
flow and transport phenomena under thermal plasma conditions. Two types
of models have been developed: a) plasma jets where electromagnetic effects
are usually not taken into account but where recirculating flows and jets impinging on a solid surface are treated, and b) plasma columns where electromagnetic effects must be taken into account and where interactions between
the plasma and the electrode surfaces, with strong, local non-equilibrium effects, must be addressed. Many peculiarities of these plasma flows and the
lack of precise or well interpreted measurements make modelling very difficult. Within the arc, the temperature may vary by an order of magnitude,
with corresponding variations in density, specific heat, molecular viscosity
and thermal conductivity. Very steep temperature and velocity gradients are
encountered in the jet fringes (up to 4000 K/mm and 400-500 m/s/mm). The

plasma core is characterized by temperatures higher than 800 K for classical plasma gases $(Ar, He, N_2, O_2,$ Air and $Co_2)$ with no metal vapor, a low density (0.03 to 0.1 times that of the cold gas) and a rather high viscosity (3 to 10 times that the cold gas). It is surrounded by a cold flow (3000 K) of high density. Thus the plasma core flow is usually laminar while the flow in the arc fringes is fully turbulent. The mixing of a cold gas to promote chemical reactions produces heating and quenching rates up to $10^7 K/s$, strong coupling of the fluid mechanics and chemical kinetics, and large differences between the electron and heavy species temperatures. Close to the electrodes, the non-equilibrium effects must be taken into account as well as the strong coupling between the fluid dynamics and magnetic forces at the arc attachment. The tiny attachment of an arc column (at the anode or at a cold cathode) may fluctuate at frequencies from 1 to 50 kH_z. When a plasma jet exists a nozzle, it encounters a steep laminar shear at the outer edge of the jet. The large velocity difference (typical plasma flow velocities even in subsonic conditions are in the range 800-1500 m/s) produces a succession of vortex rings, which result in a large-scale entrainment of the surrounding atmosphere with almost no mixing because of the density difference.

Fusion and space physics MHD

The densities and temperatures of the ionized gases in fusion plasmas and in space or astrophysical situations are very different from those of liquid metals and of generator or arc plasmas. These ionized gases exhibit very different kinds of MHD behaviour. There are four different Reynolds-like numbers of significance, with either a flow of Alfven speed in the numerator and either a kinematic viscosity of magnetic diffusivity in the denominator. At present, there is no comprehensive understanding of the roles of these often large Reynolds-like numbers in the onset or properties of turbulence. MHD turbulence in fusion confinement devices is largely undiagnosed. In the same way, the understanding of the roles of compressibility and variable density is far from complete.

The mean-free-paths for charged particle collisions are often a much larger fraction of the macroscopic length scale for fusion or space plasmas than they are for liquid metals. Microscopic plasma kinetic theory plays a non-tivial role in determining what expressions to use for resistivities and viscosities (which are, in general, tensors). This microscopic kinetic theory may be significantly modified by turbulence. Theoretical calculations of transport coefficients in fusion or space plasmas are as inaccurate as those for water (Balescu, 1988). While careful laboratory measurements have provided accurate values for the viscosity of water, no such measurements have been performed on fusion or space plasmas, and the diffusivities are often uncertain by orders of magnitude.

MHD for fusion and space plasmas can without exaggeration be said to face now many of the same problems that hydrodynamics faced over a century ago. The equations of motion are in doubt, the boundary conditions are in dispute, and the physical constants are highly uncertain. Most of the important tasks which have supposedly motivated the development on the subjects (e.g., production of a breakeven controlled D-T reaction or the prediction of solar flares and magnetic substorms) remain unaccomplished after decades of effort. It is important to initiate programs for fundamental investigations and measurements of MHD effects in ionized gases that are independent of the mission-oriented agencies that have controlled them up to now. There needs to be an academic branch as well as a practical branch of space and fusion MHD, as there is in ordinay fluid mechanics.

Stability and turbulence

The great variety of MHD flows makes the study of their instabilities seems hopeless. Indeed one can easily devote a lifetime to the instabilities in a single MHD flow. However, if we restrict ourselves to flows driven by pressure gradients or electric currents in strong, steady magnetic fields, then velocity gradients are localized in boundary and internal shear layers or jets, and instabilities are expected to appear here first. Thus, whatever the complex geometry or magnetic field configuration, the instability problems reduce to a few classes of nearly shear flows. The techniques for solving such problems are now well known from other branches of hydrodynamics, but have been applied to very few MHD flows to date. Precise predictions for complex flows are in principle much easier in such conditions than in ordinary fluid dynamics. The asymptotic conditions of strong fields apply in many practical applications, particularly in fusion reactor blankets. A major obstacle to fast progress is the lack of communication between the basic and applied research communities. There is also a need to study the many instability mechanisms involving thermo-convective effects, AC fields, thermoelectric effects, free surfaces, strong electric currents and dynamo effects.

For the small magnetic Reynolds numbers arising in most engineering situations, the first effects of a magnetic field on turbulence are the production of a strong anisotropy and a significant damping by the Joule effect. This damping is not as effective as it might appear from simple dimensional analysis. The turbulent eddies are elongated along the magnetic field direction, which considerably reduces their damping. The cascade of energy toward small scales is then suppressed and the energy spectra change dramatically. In channel flows, turbulence may become two-dimensional and has a very weak energy dissipation. In spite of the previous experimental and theoretical studies (Lielausis, 1975; Sommeria and Moreau, 1982), many questions are still unclear, such as the possibility to enhance heat and mass transport in such anisotropic turbulence. And, when this will be clarified, much more work will still have to be done before practical utilization of the phenomenon. The possibility to use MHD turbulence in laboratory for simulation of two-dimensional turbulence in the atmosphere and oceans still increases the importance of this topic. Another reason that turbulence can persist in a strong magnetic field is the confinement of the flow to narrow and strongly unstable jets as the result of entrance effects in a channel. The observed fluctuations become periodic in very strong fields and should be treated with non-linear stability theory rather than with statistical methods. Complex magnetic field configurations can provide other mechanisms for turbulence generation and the associated enhanced transport due to chaotic advection, but such possibilities have not yet been explored. In spite of its importance in astrophysics and in many industrial devices, turbulence in thermo-convection has not been studied, even in the standard Rayleigh-Benard geometry.

Self-sustained magnetic field

The spontaneous generation of magnetic fields (Moffatt, 1978) from purely hydrodynamic motions in the MHD phenomenon responsible for the magnetic fields of the earth, stars and other celestial bodies. The dynamo effect requires the large magnetic Reynolds numbers arising in many celestial systems, but seldom occurring in terrestial devices. The magnetic field grows from infinitesimal perturbations by an instability occurring for large magnetic Reynolds number just as turbulence arises for large Reynolds numbers. The principal problem is to understand how energy is organized in large-scale, complex, turbulent flows with magnetic fields, but such energy organization is poorly understood without magnetic fields. With the development of telescopes with very high resolution and the increased capability of numerical computations, astrophysicists are beginning to address such complex fluid dynamics problems. The dynamics of the liquid metal earth core is increasingly recognized as an important aspect in the evolution of the earth and even of life, since the magnetic field provides protection against cosmic radiation. The earth core flow seems to be driven by the segregation of the metallic components with different densities during the slow solidification of the inner core, with obvious analogies to metallurgical melts. While labotratory simulations of geo-dynamic flows are not feasible, convection experiments in an imposed magnetic field can give hints, especially if small magnetic field perturbations due to eddy currents are measured. There are clear harbingers of important breakthroughs in geophysical and astrophysical MHD for the next decade, guided by experiments and applied studies.

Dynamo effects may also adversely reduce flow rates and produce corrosion in the sodium cooling circuits of breeder reactors. The spontaneous self-generation of magnetic fields has been measured in the Russian BN 600 reactor, but may arise from thermoelectric rather than dynamo effects. Breeder reactors represent a unique terrestial opportunity to observe and measure dynamo effects if administrative obstacles to such tests can be overcome.

CONCLUSIONS

Most applications of MHD in materials processing have been developed by materials engineers who have extended ordinary fluid dynamic numerical codes and have done some laboratory experiments to validate the numerical predictions. It is clear that this approach has now reached its limit. New progress requires systematic studies. In particular, the analyses of disturbances, including turbulence, and the associated heat and mass transports are now among the top priorities. The same conclusion holds for applications to energy conversion, with special needs for flows with very strong magnetic fields. Other challenges for the next decades include the behaviour of two-phase fluids (liquid-gas) in a magnetic field and the properties of plasmas with very steep temperature and velocity gradients. For the highly ionized gases in fusion and space, a special effort is needed to transform the current primitive state of knowledge into a modern branch of fluid dynamics, with validated equations and constitutive laws.

All the contributors to this chapter agree that MHD research has suffered a disaster during the last two decades, especially in the USA. They recommend launching a major recovery program to create a strong community with meetings, journals and other demonstrations of activity and capable of responding to the ever increasing demands from the materials processing, energy conversion and other allied technological communities.

ACKNOWLEDGMENTS

This paper has been written from contributions to the different subsections proposed by H.H. Branover (Beer-Sheva, Isral), P. Fauchais (Limoges, France), Y. Fautrelle (Grenoble, France), J.J. Favier (Grenoble, France), M. Garnier (Grenoble, France), D. Montgomery (Hanover, NH, USA), J. Sommria (Lyon, France), and J.S. Walker (Urbana-Champaign, Ill., USA). The author is deeply indebted to them for their extremely valuable help.

REFERENCES

Balescu, R., (1988) Transport Processes in Plasmas, Vol. 1: *Classical Transport*, North-Holland.

Brancher, J.P., Etay, J. and Serro-Guillaume, O., (1983) *J. de Méc. Théor. et Appl.*, vol. 2, n°6, p. 977.

Bojarevics, A., Gelfgat, Y., and Gerberth, G., (1994) *2nd Intern. Conf. on Energy Transfer in MHD flows*, Aussois, France, p. 117 (to appear in Magnetohydrodynamics).

Caillault, B., Fautrelle, Y., Perrier, R. and Aubert, J.J., (1989) *Liquid Metal Magnetohydrodynamics*, eds. Lielpeteris, J. and Moreau, R., Kluwer Acad. Pub., p. 241.

El Kaddah, N., Szekely, J., Taberlet, E. and Fautrele, Y., (1986) *Met. trans. B*, vol. 17B, p. 687.

Garandet, J.P., Favier, J.J. and Camel, D., (1994) Handbook of Crystal Growth, vol. 2b, *Bulk Crystal Growth, Growth Mechanisms and Dynamics*, ed. Hurle, D.T.J., North-Holland, p. 659.

Garnier, M. and Moreau, R. (1983) *J. Fluid Mech.*, vol. 127, p. 365.

Gelfgat, Y. and Gorbunov, L.A., *2nd Intern. Conf. on Energy Transfer in MHD Flows*, Aussois, France, p. 1 (to appear in Magnetohydrodynamics).

Getselev, Z.N. and Martynov, G.I., (1975) Magnitnaya Gidrodinamika, no. 4, p. 144 (in Russian).

Giancarli, L., Baraer, L., Bielak, B., Eid, M., Futterer, M., Nardi, C., Proust, E., Petrizzi, L., Quintric-Bossy, J., Salavy, J.F. and Severi, Y., 1994, *Proc. of the 11th TMTFE*, New Orleans, USA.

Hurle, D.T.J. and Series, R.W., (1994) Handbook of Crystal Growth, vol. *2a, Bulk Crystal Growth, Basic Techniques*, ed. Hurle, D.T.J., North-Holland, p. 259.

Khine, Y.Y. and Walker, J.S., (1994) *2nd Intern. Conf. on Energy Transfer in MHD Flows*, Aussois, France, p. 57 (to appear in Magnetohydrodynamics).

Lielausis, O. (1975) *Atomic Energy Review*, vol. 13, p. 527.

Malang, S., Reimann, J. and Sebening, H., (1991) *DEMO-relevant test blanket for NET/ITER, Part 1: Self-cooled Liquid Metal Breeder Blanket*, vol. 1 and 2, KfK 4907 and 4908.

Meng, J.C., (1991) *Ist Conf. on Energy Transfer in MHD Flows*, cadarache, France, p. 3.

Messerle, H.K., (1988) *AIAA Progress in Astronautics and Aeronautics*, vol. III, p. 361.

Meyer, J.L., (1992) *Magnetohydrodynamics in Process Metallurgy*, eds. Szekely, J., Evans, J.W., Blazek, K. and El-Kaddah, N., p. 127.

Moffatt, H.K., (1978) *Magnetic Field Generation in Electrically Conducting Fluids*, Cambridge Univ. Press.

Moreau, R. and Evans, J.W., (1984) *J. Electrochem. Soc.: Electrochemical Science and Technology*, vol. 131, p. 2251.

Moreau, R., (1990) *Magnetohydrodynamics*, Kuwer Acad. Pub.

Moreau, R., Laskar, O. and Tanaka, M., (1993) *Materials Science and Eng.*, A173, p. 93.

Motora, S. and Takezawa, S. *2nd Intern. Conf. on Energy Transfer in MHD Flows*, Aussois, France, p. 501 (to appear in Magnetohydrodynamics).

Petrick, M. and Shumyatsky, S. Ya., (1978) *Argonne National Laboratory*, Argonne, 11, p. 707.

Petrick, M. and Branover, H.H., (1985) *AIAA Progress in Astronautics and Aeronautics*, vol. 100, p. 371.

Roberts, P. H., (1967) *An Introduction to Magnetohydrodynamics*, Longmans.

Somméria, J. and Moreau, R., (1982) *J. Fluid Mech.*, vol. 118, p. 507.

Sneyd, A.D. and Moffatt, H.K., (1982) *J. Fluid Mech.*, vol. 117, p. 45.

Taberlet, E. and Fautrelle, Y. (1985) *J. Fluid Mech.*, vol. 159, p. 409.

Takeuchi, E., Zeze, M., Toh, T. and Mizoguchi, S. (1992) *Magnetohydrodynamics in Process Metallurgy*, eds. Szekely, J., Evans, J.W., Blazek, K. and El-Kaddah, N., p. 189.

Weber, J.C. and Sautebin, R., (1986) *Light Metals*, ed. Miller, R.E., p. 869.

RAREFIED GAS DYNAMICS

E. P. Muntz

Department of Aerospace Engineering
University of Southern California
University Park
Los Angeles, CA 90089-1191

INTRODUCTION

Rarefied gas dynamics originated from an interest in flow phenomena associated with very high altitude flight as initially discussed by Zahm (1934) and Tsien (1946). The subject has roots in the kinetic theory of gas flows (Chapman & Cowling 1952) and in early molecular flow studies (Knudsen 1934). Over the intervening years rarefied gas dynamics has evolved from the initial emphasis on high altitude flight to an eclectic range of applications; all woven together by the theme of highly nonequilibrium gas flow physics. The subject encompasses, for example; high-altitude hypersonic flow fields, the reflective and reactive characteristics of gases interacting with solid and liquid surfaces, energy-transfer phenomena in molecular collisions, aerosol dynamics, cluster formation and topology, flows induced by evaporation and condensation, upper-atmospheric dynamics, and the attainment of milli-Kelvin temperatures by flow cooling techniques. Other subjects in the field include vacuum-pump performance, spacecraft contamination, a variety of interactions due to thruster plumes, spacecraft charging, and gas and isotope separations. Underlying all of these subjects is the central theme of the field of rarefied gas dynamics: *the study of gas flow phenomena in which the discrete molecular nature of the gas cannot be safely ignored.* The field has a rich heritage of analysis applied to the study of flows where concepts and techniques related to nonequilibrium statistical mechanics are important. An equally respected tradition is the development and application of instrumentation techniques that can be used to study the details of molecular motion in gas flows, as well as to study flow-generated populations of internal energy states and the characteristics of gases after surface encounters.

In this summary, promising directions for research in rarefied gas dynamics are reviewed based on several application areas discussed in the following section. Future research directions are primarily defined by the growing availability of powerful computational facilities. The introduction of teraflop, massively parallel computers will finalize the transition of rarefied gas dynamics from an interesting exercise in applied mathematics to a predictive powerhouse for use in new developments. Major contributions will be made to, among others, the processing of designer structural materials, the development of micromechanical devices, and the rapid deposition of refractory and other high performance coatings. The applications all require the development of exquisitely precise predictive capabilities for highly nonequilibrium gas flows and gas-surface interaction phenomena, in order to contribute to either the development of efficient, high throughput manufacturing processes or new devices. These are essential issues in a technologically competitive world.

BACKGROUND

The production of high performance structural materials and coatings is frequently done using chemically reacting, partially ionized gas mixtures. The development of processes for materials processing up to this time has been largely empirical (cf., Manos and Flamm 1989). In many cases these processes depend on complicated nonequilibrium phenomena both in the gas phase and at surfaces. Although the original thrust was predominantly in the area of electronic materials there is a growing interest in the synthesis of designer structural materials and coatings. The ability to model and predict the performance of processes leading to materials synthesis will make a valuable contribution to the country's commerical competitiveness.

The recent interest in microelectromechanical devices (Gabriel et al. 1988) has stimulated the study of gas dynamic processes at very small scales. The size scaling of forces (Trimmer 1989) inexorably leads one to the conclusion that gas dynamic pressure forces will be important in microdevices. In fact a new class of transient pressure driven microdevices that can provide GHz mechanical switching is currently being studied using rarefied gas dynamics techniques (Wadsworth 1993). The application of microelectromechanical devices is expected to revolutionize sensor, instrumentation and control technology in the next decade (Wise 1991). It turns out that the generation of transient pressure pulses in typical microdevices involves highly nonequilibrium gas flows because of the rapid changes in gas properties that are encountered. This field is expected to become a fertile development area, requiring extensive application of rarefied gas dynamic calculations, including a careful accounting of surface interaction effects.

During the flights of Space Shuttle's STS-3 and STS-4 a low-light-level television camera on the orbiters recorded a gaseous glow above the windward tail surfaces of the vehicle (Banks et al. 1983, Mende et al. 1983). In the same period it was noticed that thermal insulation blankets returned from orbit had been severely eroded (Whitaker 1983, Peters et al. 1983). In the early 1980s it also became evident that modern satellite computer systems frequently required re-booting due to upsets as a result of discharges traced to differential charging on satellite surfaces. The upsets were most frequent as the vehicles passed through the midnight to dawn quadrant of their orbits (Garrett 1980, Fennell et al. 1983). All of these phenomena indicated that in addition to the effects of the magnetosphere's radiation belts, there are interactions between local space environments and satellite systems of practical, measurable significance. As extended satellite lifetimes are realized, the contamination of optical components and thermal control surfaces becomes increasingly important. Major sources of contamination are the flow fields of control- and station-keeping thrusters. Optical contamination as a result of the ambient high-speed atmosphere interacting with exhaust gases has also been recognized as a potentially serious space-system problem. These occurrences represent phenomena that have been studied within the field of rarefied gas dynamics. Generally, increasingly frequent visits to space, as well as the interest in permanently manned low-Earth-orbit space stations and to the longer range goal of establishing permanent lunar or other bases, leads one to expect numerous rarefied-gas-dynamic problems associated with activities such as the mining and processing of native resources in space. Even the old idea of space scoops (Berner & Camac 1961) has recently been rejuvenated.

The reliable prediction of observables associated with high altitude exhaust plumes is an important and continuing problem in defense technology (Weaver 1993). Several highly nonequilibrium phenomena combine to make the problem intractable using conventional computational fluid dynamic approaches, much more detailed rarefied gas dynamics based predictions are required. At high altitude missile exhaust plumes can extend for many kilometers and produce observable radiation both directly from the hot propellant flowing from the exhaust nozzle and from interaction of the exhaust gases with the gases in the upper atmosphere.

FOUR GRAND CHALLENGES

In this section four developments that require a deep understanding of highly nonequilibrium phenomena coupled to gas or vapor flows are identified. There are of course numerous other applications of rarefied gas dynamics, the ones chosen here were selected for their timeliness and relevance.

- Since commercial technologically important processes depend on the transport of materials—the coupled involvement of nonequilibrium phenomena with flows or transport is important. Interesting new designer bulk materials and coatings require significant activation energy to create. As demonstrated by plasma enhanced vapor deposition it is frequently advantageous to employ nonequilibrium phenomena in order to increase efficiency. The first Grand Challenge covers the general area of materials processing; the Challenge is to develop validated prediction techniques that will enable the efficient design of new, nonequilibrium assisted processes for materials synthesis.

- The application of rarefied gas dynamics to the design of new mechanical technologies is not limited to space vehicle and missile systems. A rapidly developing application is to microelectromechanical systems. In this case very small dimensions (on the order of 10^{-4} cm and smaller) requires the application of rarefied gas dynamic techniques for predictions of gas transport and the production of mechanical motion. The use of time dependent, three-dimensional, rarefied flow predictions to assist in the design of micromechanical devices is a well defined Grand Challenge.

- The fascination of students with the subject of space flight and space exploration suggests that the country will have a long term interest in problems associated with space sciences and astronautics. A Grand Challenge appears in predicting the details of the interactions between space vehicles and diverse space environments in the presence of electric and magnetic fields.

- A technologically vigilant defense posture must be part of our future. One of the critical areas needing attention is theatre defense against intermediate range ballistic missiles (IRBM). The increasingly rapid detection and tracking of IRBMs depends on accurate knowledge of high altitude plume optical signatures in a variety of spectral bands. Current prediction techniques are demonstrably inadequate for this highly nonequilibrium flow problem. A Grand Challenge for defense technology is the successful prediction of the observable characteristics of high altitude missile plumes.

COMPUTATIONAL RAREFIED GAS DYNAMICS AND ITS EXPERIMENTAL VALIDATION–A TOOL FOR THE GRAND CHALLENGES

The dominant predictive tool in rarefied gas dynamics for the past decade at least has been the direct simulation Monte Carlo (DSMC) technique. This approach, which was introduced in 1963 and 1964 by Graeme Bird (Bird 1963, 1965), has been developed, nurtured, and brought to an impressive level of productive capability by Bird and others in the intervening years. The work reported by Moss (1986) using the DSMC method to predict species concentrations and radiation in very energetic flow fields indicates the power of the technique. The basic technique is described by Bird in his book (Bird 1976); its use in chemically reacting flows has been presented by Koura (1973) and

Bird (1976, 1979). Recently, the technique has been extended to radiating flows (Bird 1987). A short review of the DSMC technique, as well as the Hicks-Yen-Nordsieck method (Nordsieck & Hicks 1967, Yen 1971) and the molecular-dynamics method, has been given by Bird (1978, 1989; see also Yen 1984). The Hicks-Yen-Nordsieck (HYN) approach is a Monte Carlo method, but in this case the collision integral is solved by a Monte Carlo sampling technique; the remainder of the Boltzmann equations is solved using standard finite-difference methods. Appropriate implementations of the DSMC technique have been shown by Bird (1976) and by Nanbu in a series of papers (Nanbu 1986) to be in principle an exact solution of the Boltzmann equations, although as Bird (1989) argues, it is not entirely clear that such a connection is necessary.

In the DSMC method a large number of simulated molecules are followed simultaneously (Bird 1976, 1989). Collisions are handled on a probabilistic basis using the molecules found in a small geometric cell after each computational times step. The computation is started from some initial condition and followed step by step in time; steady flow is the condition that is reached at large times. The computational cell network is in physical space, and the time steps can be directly related to physical times. A most important feature of the technique is that it can be applied so that the computation time is proportional to the first power of the number of simulated molecules.

The DSMC technique is a pure form of computational fluid dynamics. In principle it can contain all of the physics needed for any problem without the necessity of nonequilibrium thermodynamic assumptions that are required in nonequilibrium continuum-flow calculations. In practice the technique is computationally intensive compared with its continuum counterparts. However, calculations that overlap continuum calculations (with added slip effects) for STS orbiter flow fields have been accomplished (Moss 1986). The success of the technique depends on computational performance, so it is easy to anticipate further advances in its use. A number of research centers are now looking at efficiently matching DSMC and continuum techniques in a computational hybrid approach (Cheng & Wong 1988) as well as the excitation of continuum models based on kinetic theory (Cheng et al. 1991). As indicated by Bird (1985), hypersonic flow fields can show widely different degrees of rarefaction at different locations, which makes a hybrid approach and its improvement attractive for such flow fields. On the same subject, it was noted by Yen (1984) that the HYN approach offers the possibility of not having to match different numerical techniques for solving mixed continuum and rarefied-flow problems.

Since in many situations where extreme nonequilibrium effects exist, such as in shock transition zones around sharp hypersonic leading edges or in materials processing flows, the DSMC technique is the only realistic method for obtaining solutions, its validity is important. In some sense its validation may well serve as a prototype for many other computational fluid-dynamic techniques. The usual situation, and the one that applies here, is that for flow fields in which one really wants to use the computational technique such as in micromachines or hypersonic flight, it is very difficult to check it directly. The status of DSMC experimental validation is reviewed in the following section.

The success of the DSMC technique, as well as a continued desire to extend calculations to lower Knudsen numbers, has revived interest in discrete-velocity gas models. In these models, gases are allowed to have molecular velocities with only a limited set of possible values (cf., Gatignol 1991). It is surprising how successful the very simplest velocity model of a gas can be in predicting approximate transport properties and the equation of state. These have yet to be extended to chemically reacting flows.

A major triumph of the past 5 to 10 years has been the development of the DSMC technique to provide predictions for highly nonequilibrium, radiating and reacting flows. The DSMC, or any similar particle treatment of a gas where classes of particles are followed (in this case only in a statistical sense), intrinsically is able to provide a representative reflection of nature and can be correct in highly nonequilibrium situations. This of course is only true if all important elastic and inelastic collision cross sections are known as a function of energy. However, the possibility is there, whereas in continuum treatments reliance must be made on the assumptions of nonequilibrium thermodynamics. In the past decade, Graeme Bird and his associates at the Langley Research Center have been able to develop in "an engineering context" (Bird 1985) the DSMC technique to provide predictions for highly nonequilibrium real air flow fields including radiation (Moss et al. 1988). The "engineering context" refers to the numerous necessary approximations that must be made in describing the collision processes and energy transfers during collisions. The computational resources and physical information needed to avoid these approximations were lacking. The exciting point is that the engineering calculations are possible and provide reasonable agreement with the very few experimental results that are available (Bird 1987). It is also clear that the agreement may be fortuitous, since there are a very large number of interacting approximations and not very well known collision cross sections involved.

The possibility of making flow-field predictions in the near future for mul-tidimensional, reacting and radiating flows (also of course viscous) with the DSMC technique implies a clear direction for future work. Significant effort in rarefied gas dynamics will be aimed at developing inelastic collision models and establishing their validity in situations that are less complicated than a nonequilibrium air chemistry flow field. As an example, the standard way of treating inelastic collisions with transfer of rotational energy in the DSMC technique is based on the model of Borgnakke & Larsen (1975). In the model a certain fraction of collisions are considered inelastic; for these collisions, new translational and rotational energies are sampled from the distribution of these quantities that would occur in an equilibrium gas with a specific energy that is the same as the specific energy available in the collisions. This of course is not very realistic, but if the fraction of these inelastic collisions is chosen so that the rotational relaxation rate in the gas is matched, it seems to work quite satisfactorily in an "engineering context". Vibration and electron excitation have been handled by analogous techniques. Experiments can be done quite easily to investigate models of, say, rotational energy transfer [for instance, rotational-level population distributions in a shock wave are already available (Robben & Talbot 1966)]. It is to be expected that different inelastic collision models (e.g., Boyd 1992a) will appear, and that experimental studies in simple situations will be undertaken to validate their predictions.

EXPERIMENTAL VALIDATION OF COMPUTATIONAL RAREFIED GAS DYNAMICS

There has been a significant effort in the past several years to validate experimentally the DSMC technique for monatomic and diatomic gas flows involving essentially only rotational energy levels. The fundamental accuracy of the technique has been verified (cf., Davis et al. 1983, Erwin et al. 1991, Pham-Van-Diep et al. 1989). What remains to be established is the suitability of physical models for vibration, dissociation, electronic excitation, free electrons and ions, and radiation. There are several issues here that involve the availability of basic physical information on momentum transfer collision cross sections, state to state transition probabilities in collisions and chemical reaction probabilities (Kunc 1989). There is indeed potentially far more detailed information required than can be supplied in the forseeable future if the problem is addressed in a direct frontal assault by simply trying to include all interactions in complete detail. Additionally such detail would overwhelm even teraflop computers. Thus, for the next decade at least a reasonable amount of physical modeling will need to be used in computational rarefied gas dynamics. The validity of these models will have to be tested experimentally. The search for appropriate experiments and instrumentation for model verification will be an important aspect of research in rarefied gas dynamics for the next decade.

It is interesting to note that the Grand Challenges used as examples here present situations that are difficult to investigate in satisfactory investigative detail. A micromachine for example, with a few microns characteristic dimension is not an easy place to study the detailed properties of flow fields. Thus, experimental validation will take place on simple surrogate situations that make similar demands on the prediction techniques as in actual applications but are easier to study experimentally. An example is the iodine vapor flow facility being used by Pham-Van-Diep (1992) to study vibrational population distributions in a nonequilibrium dissociating gas. The iodine behaves as a surrogate dissociating diatomic gas but at relatively low temperatures, permitting long experiment times, making careful detailed measurements possible.

Research Directions

The Grand Challenge opportunities mentioned above demand extraordinary efforts either in the development of the DSMC technique or some suitable alternative. In order to contribute meaningfully to the applications suggested in the Grand Challenges extensive research must be done. Inclusion of inelastic collisions; electronic, vibrational and rotational state transitions in collisions; satisfactory approaches to account for radiation, electric and magnetic fields and very importantly, detailed interaction with surfaces. One very interesting problem is to be able to predict trace species that can be important in nonequilibrium flow processes during the synthesis of materials. Over the next decade research in rarefied gas dynamics will include, amongst a large group of additional directions, the following thrusts.

- Effective implementation of computational rarefied gas dynamics on massively parallel computer architecture must be studied. Since even with teraflop machines it will not be possible with present algorithms to account for all of the detailed physics and chemistry in a typical three dimensional computation (Boyd 1992b); a certain amount of modeling will be required along with algorithm and computer architecture development.

- Extensive experimental validation of the modeling and the algorithms will have to be undertaken.

- Theoretical analysis needs to continue despite the emphasis on computational techniques because theoretical guidelines for the computations are invaluable. For instance, the nature of the Grand Challenges require the self-consistent inclusion of electric and magnetic fields. To do so in computationally efficient ways will require significant analysis.

REFERENCES

Banks, P. M., Williamson, P. R., and Raitt, W. J. (1983) Space shuttle glow observations. *Geophys. Res. Lett.* 10: 118–21.

Berner, F., and Camac, M. (1961) Air scooping vehicle. *Planet. Space Sci.* 4: 159–83.

Bird, G. A. (1963) Approach to translational equilibrium in a rigid sphere gas. *Phys. Fluids* 6: 1518–19.

Bird, G. A. (1965) Shock wave structure in a rigid sphere gas. In *Rarefied Gas Dynamics* ed. J. J. deLeeuw, 216–21. New York: Academic.

Bird, G. A. (1976) *Molecular Gas Dynamics*. Oxford: Clarendon.

Bird, G. A. (1978) Monte Carlo simulation of gas flows. *Ann. Rev. Fluid Mech.* 10: 11–31.

Bird, G. A. (1979) Simulation of multi-dimensional and chemically reacting flows. In *Rarefied Gas Dynamics*, ed. R. Campargue, 365–88. Paris: Commissariat a l'Energie Atomique.

Bird, G. A. (1985) Low density aerothermodynamics. *AIAA Paper No. 85-0994*.

Bird, G. A. (1987) Nonequilibrium radiation during re-entry at 10 km/s. *AIAA Paper No. 87-1543*.

Bird, G. A. (1989) The perception of numerical methods in rarefied gas dynamics. In *Rarefied Gas Dynamics*, ed. E. P. Muntz, D. Weaver, and D. Campbell, Washington: AIAA.

Borgnakke, C., and Larsen, P. S. (1975) Statistical collision model for Monte Carlo simulations of polyatomic gas mixtures. *J. Comput. Phys.* 18: 405–20.

Boyd, I. (1992a) Analysis of vibration-dissociation-recombination processes behind strong shock waves of nitrogen. *Phys. Fluids A* 4(1): 178–185.

Boyd, I. (1992b) Private communication.

Chapman, S., and Cowling, T. G. (1952) *The Mathematical Theory of Non-Uniform Gases*. Cambridge: Cambridge University Press.

Cheng, H. K., and Wong, E. (1988) Fluid dynamics modeling and numerical simulation of low-density hypersonic flows. *AIAA Paper No. 88-2731*.

Cheng, H. K., Wong, E., and Dogra, V. K. (1991) A shock-layer theory based on thirteen-moment equations and DSMC calculations of rarefied hypersonic flows. *AIAA Paper No. 91-0783*

Davis, J., Doring, R. G., Harvey, J. K., and Macrossan, N. (1983) An evaluation of some collision models used for Monte Carlo calculation of diatomic rarefied hypersonic flows. *J. Fluid Mech.*, 135: 355–371.

Erwin, D. A., Pham-Van-Diep, G. C., and Muntz, E. P. (1991) Nonequilibrium gas flows 1: A detailed validation of Monte Carlo direct simulation for monatmic gases. *Phys. Fluids A* 3(4): 697–705.

Fennell, J. F., Koons, H. C., Leung, M. S., and Mizera, P. F. (1983) A review of SCATHA satellite results: charging and discharging. *Proc. ESLAB Symp. Spacecr. Plasma Interactions and Their Influence on Field and Part. Meas., 17th ESA SP-198*, 3–11.

Gabriel, K., Jarvis, J., and Trimmer, W. S. (1988) Small machines, large opportunities: A report on the emerging field of microdynamics. National Science Foundation, Washington, D. C., Technical report, 1988.

Garrett, H. B. (1980) Spacecraft charging: a review. In *Space Systems and Their Interaction with the Space Environment. Progress in Astronautics and Aeronautics*, ed. H. B. Garrett and C. P. Pike, 71: 167–226. New York: AIAA

Gatignol, R. (1991) Constitutive laws for discrete velocity gases. In *Rarefied Gas Dynamics*, ed. A. Beylich, 819–29. Basel: VCH.

Knudsen, (1934) *The Kinetic Theory of Gases*. London: Methuen & Co.

Koura, K. (1973) Nonequilibrium velocity distributions and reaction rates in fast highly exothermic reactions. *J. Chem. Phys.* 59: 691–97.

Kunc, J. (1989) Review of efficient semi-classical methods for determination of atomic and molecular interactions. In *Rarefied Gas Dynamics*, ed. E. P. Muntz, D. Weaver, and D. Campbell. Washington: AIAA.

Manos, D. M., and Flamm, D. C. (1989) *Plasma Etching*. San Diego, Academic Press.

Mende, S. B., Garriott, O. K., and Banks, P. M. (1983) Observations of optical emissions on STS-4. *Geophys. Res. Lett.* 10: 122–25.

Moss, J. N. (1986) Direct simulation of hypersonic transitional flow. In *Rarefied Gas Dynamics*, ed. V. Boffi and C. Cercignani, 384–99. Stuttgart: B. G. Teubner.

Moss, J. N., Bird, G., and Dogra, V. K. (1988) Nonequilibrium thermal radiation for an aeroassist flight experiment. *AIAA Paper No. 88–0081*.

Nanbu, K. (1986) Theoretical basis of the direct simulation Monte Carlo method. In *Rarefied Gas Dynamics*, ed. V. Boffi and C. Cercignani, 369–83. Stuttgart: B. G. Teubner.

Nordsieck, A., and Hicks, B. L. (1967) Monte Carlo evaluation of the Boltzmann collision integral. In *Rarefied Gas Dynamics*, ed. C. L. Brundin, 695–710. London: Academic.

Peters, P. N., Linton, R. C., and Miller, E. R. (1983) Results of apparent atomic oxygen reaction on Ag, C, and Os exposed during the shuttle STS-4 orbits. *Geophys. Res. Lett.* 10: 569–71.

Pham-Van-Diep, G. C., Erwin, D. A., and Muntz, E. P. (1989) Nonequilibrium molecular motion in a hypersonic shock wave. *Science* 245: 624–626.

Pham-Van-Diep, G. C., Muntz, E. P., Weaver, D. P., DeWitt, T. G., Bradley, M. K., Erwin, D. A., and Kunc, J. A. (1992) An iodine hypersonic wind tunnel for the study of nonequilibrium reaction flows. *AIAA Paper No. 92-0566.*

Robben, F., and Talbot, L. (1966) Measurements of rotational temperatures in a low-density wind tunnel. *Phys. Fluids* 9(4): 644–62.

Trimmer, W. S. (1989) Microrobots and micromechanical systems. *Sensors and Actuators* 19: 267–287.

Tsien, H. S. (1946) Superaerodynamics, mechanics of rarefied gases. *J. Aeronaut. Sci.* 13: 653–64.

Wadsworth, D. (1993) Microscale Gas Dynamics, Ph.D. Dissertation, USC.

Weaver, D. (1993) Private communication.

Whitaker, A. (1983) LEO effects on spacecraft materials. *AIAA Paper No. 83-2632.*

Wise, K. D. (1991) Micromechanical sensors, actuators and systems. In *Micromechanical Sensors, Actuators and Systems* Winter Annual Meeting of the American Society of Mechanical Engineering, 32: 1–14.

Yen, S. M. (1971) Monte Carlo solutions of nonlinear Boltzmann equation for problems of heat transfer in rarefied gases. *Int. J. Heat Mass Transfer* 14: 1865–69.

Yen, S. M. (1984) Numerical solution of the nonlinear Boltzmann equation for nonequilibrium gas flow problems. *Ann. Rev. Fluid Mech.* 16: 67–97.

Zahm, A. F. (1934) Superaerodynamics. *J. Franklin Inst.* 217: 153–66.

HYDRODYNAMICS OF SHIPS
AND OFFSHORE PLATFORMS

J. N. Newman
Department of Ocean Engineering
Massachusetts Institute of Technology
Cambridge, MA 02139

INTRODUCTION

The oceans of the world comprise 70% of its surface. A similar portion of the boundaries of the United States is formed by the Atlantic and Pacific, the Gulf of Mexico, and the Great Lakes. While most people live and work away from these coastlines, the oceans are vital to our international trade, military defense, energy supply, and environmental health. Ships, and the platforms used to exploit offshore petroleum resources, are the primary tools of these pursuits.

Ships have been designed for a millennium, but the field of naval architecture is constantly pressed for improved technology with increasing concerns for safety, economy, and international competition. The technology of offshore platforms has developed over a relatively short period, starting with the first oil wells in shallow waters of the Gulf of Mexico. Ocean engineers now are challenged by the progression of ever-deeper oil fields, in the Gulf of Mexico and in more demanding environments such as the North Sea.

Ships and offshore platforms share common or similar hydrodynamic problems relating to their interaction with the surrounding fluid. For ships the primary problems relate to the minimization of the hydrodynamic drag force, the efficient development of propulsion to overcome this drag, and the effects of ocean waves on the ship's motions and structural loads. Other requirements include the capability to maneuver effectively and safely, and to minimize vibrations, noise, and cavitation damage. For offshore platforms the primary concerns involve the structural loads due to waves, and the effects of waves and currents on ancillary equipment such as risers, moorings, and supply vessels. Additional hydrodynamic problems are associated with the construction, transportation, and installation of the platforms.

The traditional approach to these topics has been to perform experiments in special facilities including towing tanks, wave basins, and closed water tunnels. Fundamental conflicts between the diverse scaling laws of hydrodynamic similitude (Reynolds number for viscous effects, Froude number for

220

waves, and cavitation number) make it impossible to precisely model practical problems, and the need to minimize such conflicts has led to an evolution of ever-larger facilities. The most recent example in the United States is the 'Large Cavitation Channel' constructed by the Navy in Memphis, to permit experimental investigations of propeller noise accounting for the interaction between the propeller and ship hull (see Figure 1).

Figure 1: The flow past a destroyer propeller, including cavitation on the propeller blades, trailing vortex sheets, and on the rudder to the right of the propeller. This experiment was conducted with a forty-foot long model of the destroyer, in the Large Cavitation Channel at Memphis. The ship's hull is to the left and above the picture. (Courtesy of the David Taylor Model Basin, US Navy Department)

Theoretical knowledge has always been useful to design and interpret experiments. Aided by parallel advances in computational facilities and numerical analysis, the theoretical approach has assumed a more useful role in recent years. Where it is valid, this approach has distinct advantages over experiments, including economies of cost and time, and the ability to examine local details of the flow field. Ultimately, in the engineering workstation environment, answers to practical problems will be obtained readily by designers.

In two specific topics, propeller design for ships and wave loads on offshore platforms, computer programs now permit reliable engineering predictions to be made for a significant range of design parameters and operating regimes. Outside of these limits, and for other types of problems, substantial further progress is required to replace experimental methods by theory and computations. This objective is necessary not only to ensure economical design, when experimental costs are inflating and computational costs are decreasing, but also to address engineering problems where scale effects are of concern without continuing the progression of ever-larger experimental facilities.

Two specific problems are described in §§2-3 to illustrate the status of this field: the steady-state calm-water prediction of a ship's performance, and the unsteady loads on an offshore platform in waves. A more general list of research needs and opportunities in the broader context of the field is outlined in §4.

STEADY-STATE SHIP RESISTANCE AND PROPULSION

Ships moving on the ocean surface rely upon the development of a propulsive force or 'thrust' to balance the drag force or 'resistance'. The resistance is composed of two principal components, one due to the viscous or 'frictional' forces acting along the wetted surface of the hull and the second due to the generation of the waves which are left behind the ship. It is customary in both experiments and theory to assume that these two components are independent. Thus the viscous effects, which depend on the Reynolds number, are considered separately from wave effects (Froude number). From the experimental standpoint this leads to Froude's hypothesis, that experiments can be conducted with small scale models preserving the correct full-scale value of the Froude number to ensure the correct representation of the waves; the measured resistance is then corrected for the difference between the model and full-scale frictional drags, this difference being determined from a fictitious flat-plate with the same length and wetted surface area as the ship. From the theoretical standpoint wave effects can be analyzed as a potential-flow phenomenon, and frictional drag can be considered in the context of boundary-layer approximations.

In fact there are interactions between viscous and wave effects, and important situations exist where the viscous flow is not confined to a thin boundary layer. These are most prevalent near the stern of ships with relatively full form. Separated flow can be expected in the absence of the ship's propeller. A promising computational model described by Larsson et al (1990) is based on three zonal solutions (a) in a thin boundary layer close to the ship's surface and upstream of the stern; (b) outside this boundary layer where the wave effects are considered without viscosity; and (c) in a zone near the stern where the full Navier-Stokes solution is computed, accounting for viscous effects but not (as yet) for waves. This is an appropriate decomposition insofar as we cannot hope to solve the complete exterior flow problem with all effects included simultaneously. Further work is required to consider the most important interactions, particularly between waves and the rotational flow near the stern.

Most ships are propelled by screw propellers which function in a manner fundamentally similar to an airplane wing, moving through the surrounding fluid at a small angle of attack and generating a lift force on each blade with an axial component in the forward direction. In the simplest description the propeller is assumed to operate in an undisturbed uniform stream. In reality the presence of the ship's hull upstream of the propeller modifies the inflow significantly. Conversely, the suction effect of the propeller has a beneficial effect on the flow past the hull, reducing or eliminating the zone of separated flow.

Cavitation is a primary factor which limits the velocity of the propeller blades, and hence their thrust. Cavitation leads to rapid erosion of the propeller blades, and to high levels of acoustic radiation and vibrational loadings on the ship. Most experimental investigations of propellers are conducted in water tunnels where the ambient pressure can be reduced to simulate the correct cavitation number. Except in a few very large tunnels it is impractical to include the ship's hull, and wake screens are introduced to represent the nonuniform inflow. (One reason for the Large Cavitation Tunnel in Memphis is the possibility to include models of the hull of a ship or submarine, instead of just the propeller by itself.)

Computations based on potential theory have supplanted many routine experiments in this field. Current programs use distributions of singularities to represent the propeller blades, hub, and the vortex sheet in the wake. Some programs can reproduce unsteady effects due to rotation of the propeller in a nonuniform wake, as well as the formation and collapse of sheet cavitation on the blades. The details of such flows are important, and careful numerical analysis is required particularly in the vicinity of the propeller tips and trailing edges. Ducts surrounding the propellers can be included, and the next generation of such programs may include interactions with the flow past the hull, and with the wave field on the free surface.

OFFSHORE PLATFORMS IN WAVES

The loads imposed by waves on offshore platforms are important from the standpoint of structural design, the operability of platforms, and ultimately their safety. The most frequent waves, of moderate height, are important from the standpoint of fatigue. Storm waves have a more obvious importance, and as the waves become larger the relevant fluid mechanics of their interactions with the platform become more complex.

Waves of small amplitude, propagating on the ocean surface, are well described by linear potential theory. The complicated irregular surface observed in nature can be separated into spectral components, each a regular sinusoidal wave with specified direction of propagation, wavelength, and period. The constantly changing irregular patterns on the ocean surface are a consequence of dispersion, with long waves travelling faster than short waves.

With spectral analysis it is straightforward to consider the simplest sinusoidal waves separately, in the frequency domain.

The interactions of such waves with platforms are analyzed from two complementary approaches. Morison's formula, which ignores diffraction effects but accounts in a semi-empirical manner for viscous-drag and inertial forces, is relevant to older platform designs where the supporting structure consists of cylindrical elements with relatively small diameter compared to the wave height (see Figure 2).

Figure 2: Waves interacting with a platform in the North Sea during a severe winter storm. The wave in the foreground has a height between 63 and 66OB feet. (The main deck, which separates the diagonal braced sub-structure from the accomodation module in the upper half of the platform, is 96 feet above the mean sea level.) (Courtesy of the Shell Development Company, Houston)

From the fluid dynamic standpoint, this is the regime of large Keulegan-Carpenter numbers, where the trajectory of the fluid particles is large compared to the cross-sections of the structure. Researchers continue to seek a more fundamental basis for this regime, whereas practical designers believe that Morison's formula can be used with confidence provided the drag and inertia coefficients are selected judiciously.

Platforms with larger cross-sectional dimensions, which are prevalent in deeper water and thus of greater current importance, require the consideration of wave diffraction. Computer programs capable of reproducing these interactions have been developed and used extensively in the offshore industry within the past decade. Corresponding experiments can be performed in wave tanks equipped with suitable wavemakers and absorbing beaches, with measurements performed in a realistic irregular wave spectrum or with the separate sinusoidal components.

Before considering nonlinear wave effects in their most general context, it is appropriate to note the significance of second-order wave effects on offshore structures. Various quadratic interactions exist which give rise to mean and second-harmonic wave forces in regular waves, and to long period (difference-frequency) and short period (sum-frequency) time variations in a spectrum. These are small in magnitude, proportional to the square of the wave height, but of singular importance to structures which are sensitive dynamically to such long- or short-period loads. Tension-leg-platforms, which are used increasingly in deep portions of the Gulf of Mexico and North Sea, are important examples. These are platforms with excess buoyancy, moored to the sea floor by vertical tendons. They are dynamically resonant at long periods in a similar manner to an inverted pendulum, and at short periods due to the tendon elasticity. The long-period motion affects the platform's ability to maintain position, and the short-period vertical motion affects the fatigue life of the tendons.

Since their magnitudes are small in relation to the dominant first-order waves, second-order effects are difficult to measure. Sophisticated experimental equipment is required in conjunction with a well-calibrated wave tank. Conventional narrow wave tanks are unsuitable due to the reflection of waves from the side walls. A large wave basin has recently been constructed, primarily for this purpose, at the Offshore Technology Research Center in Texas.

Computations of second-order loads for practical offshore structures require extensive higher-order perturbation expansions in the theoretical development, care in the resultant numerical analysis and computer programs, and hundreds of hours on Cray-level supercomputers to describe the relevant combinations of wave frequencies and headings. The development of more efficient algorithms for performing these computations is a current research goal.

In extreme sea states, or 'survival conditions', linear theory is not sufficient to describe important effects including peak wave loads, and local 'slamming' of waves with high impact upon the structure. The mathematical description of such problems is another topic of current research, which has been greatly facilitated by supercomputer access. Massively parallel computers have much to offer here, at the expense of special programming efforts. Most computations have been performed with idealized two-dimensional problems. Practical design computations for three-dimensional structures must await substantial further progress both on numerical analysis and program development, and on computers with greater capacity.

Another complication in extreme sea states is the role of viscosity and separation. The Keulegan-Carpenter number is too large to ignore viscous effects, and the scale of the structural elements in relation to the wavelength is too large to use Morison's formula, which ignores diffraction. Thus, for contemporary platforms in deep water, there are major uncertainties regarding the appropriate representation of both viscous and diffraction effects in the most severe conditions of extreme waves.

Nonlinear effects are easier to model in a wave tank, simply by increasing the magnitude of the generated waves, but attention is required to consider the sources of nonlinear errors. Representative time-scenarios of ocean waves can be generated, but linear superposition and spectral analysis are no longer valid. Since some of the most important nonlinear effects are infrequent, long experimental runs are often required which place demands on the absorption of reflected waves from the structure itself and from the extremities of the basin. Since the Reynolds number is reduced at model scale, there is uncertainty concerning the role of viscous effects.

RESEARCH NEEDS AND OPPORTUNITIES

a. integrated theories and computational solutions for the flow past a ship hull including the effects of the propeller and viscous wake near the stern

b. integrated theories and computational solutions for the flow past a ship hull including the effects of the propeller and free-surface wave effects

c. computational solutions for the motions and structural loads on ships in waves, for the nonlinear regime of extreme sea states

d. hydrodynamic effects for ships maneuvering in unrestricted and restricted waters

e. improved understanding of high-frequency wave loads on offshore structures due to nonlinear effects

f. wave loads on offshore structures in the regime where both diffraction and separated flow occur

g. wake flows including separation and vorticity, in the presence of a free surface

h. vortex induced vibrations on risers, towing cables, and other compliant structures

i. separated flows past arrays of interacting structural elements

j. more efficient and complete computational solutions for predicting second-order wave loads

k. loads due to the impact of breaking waves on ships and platforms

BIBLIOGRAPHY

Blake, W. K., Meyne, K., Kerwin, J. E., Weitendorf, E., & Friesch, J. (1990) 'Design of APL C-10 propeller with full-scale measurements and observations under service conditions,' *Transactions, Society of Naval Architects and Marine Engineers*, New York, **98**, 77-111.

Faltinsen, O. M. (1990), '*Sea loads on ships and offshore structures*,' Cambridge University Press, Cambridge, UK.

de Kat, J. O., & Paulling, J. R. (1989), 'The simulation of ship motions and capsizing in severe seas,' *Transactions, Society of Naval Architects and Marine Engineers*, New York, **97**, 139-168.

Larsson, L., Broberg, L., Kim, K.-J., & Zhang, D.-H. (1990) 'A method for resistance and flow prediction in ship design,' *Transactions, Society of Naval Architects and Marine Engineers*, New York, **98**, 495-535.

Larsson, L., Patel, V. C., & Dyne, G., eds. (1991) '*1990 SSPA-CTH-IIHR workshop in ship boundary layers*,' FLOWTECH Research Report, Gothenburg, Sweden.

Lin, W.-M., & Yue, D. (1990), 'Numerical solutions for large-amplitude ship motions in the time domain,' *Proceedings, 18th Symposium on Naval Hydrodynamics*, Ann Arbor, 41-66.

Nakos, D., & Sclavounos, P. (1990), 'Ship motions by a three-dimensional Rankine panel method,' *Proceedings, 18th Symposium on Naval Hydrodynamics*, Ann Arbor, 21-40.

Newman, J. N. (1992), 'Panel methods in marine hydrodynamics,' *Eleventh Australasian Fluid Mechanics Conference*, Hobart, Australia.

Price, W. G., ed. (1991), '*The dynamics of ships*,' Proceedings of a Royal Society Discussion Meeting, The Royal Society, London, UK.

Sarpkaya, T. (1992), 'Fluid loading research – future directions,' *Conference on the Behaviour of Offshore Structures* (BOSS '92), Imperial College of Science Technology & Medicine, London, UK.

REACTING FLOWS AND COMBUSTION

Stephen B. Pope
Mechanical and Aerospace Engineering Department
Cornell University
Upson Hall
Ithaca, NY 14853-7501

INTRODUCTION

The practical importance of combustion is matched by its difficulty as a scientific discipline. The next three paragraphs illustrate the importance of combustion in energy conversion, fires and explosions.

Oil or coal-fired power plants, aircraft and automobile engines all depend on the combustion of fossil fuels. Vast amounts of fuel are consumed annually, with commensurate emissions of atmospheric pollutants. A major goal of combustion science and engineering is to understand the fundamental combustion processes better, so that the performance of these energy-conversion devices can be improved, leading to reduced fuel consumption and pollutant emissions.

Fires are responsible for upwards of five thousand fatalities and tens of billions of dollars in losses annually in the United States alone. They also represent a significant environmental hazard, particularly in connection with industrial accidents such as those associated with the extraction, transport and processing of petroleum products. Finally, the oil field fires in Kuwait demonstrate that fires on an enormous scale can pose public health and climate impact problems on a regional and perhaps global scale.

The study of detonative combustion goes back to the 19th century, driven by the desire to understand and thereby prevent coal mine explosions. Explosions still plague, not only mining operations, but also arise in grain elevators, fuel handling facilities and many other industrial operations and are particularly destructive when they involve detonations. There is also recent concern regarding detonative explosions within space vehicles or during the launch of liquid hydrogen-oxygen rockets, and in the case of the Three Mile Island accident there was the possibility that the hydrogen generated during core meltdown might detonate and cause rupture of the reactor containment.

Combustion is an extremely difficult phenomenon to study for several different reasons. While the gas phase is usually dominant both solid and liquid phases can be present for example in the form of coal particles, oil droplets, and soot. Even in the gas phase, the chemistry is complex: typically there are about 50 chemically-significant species participating in perhaps 200 significant reactions. These coupled reactions have time scales varying from $10^{-12}s$ to $10^2 s$.

The coupling of the thermochemistry with the fluid mechanics can be strong, and leads to further complications. The high temperatures involved (e.g. 2,000 K) lead to property variations (e.g. density and viscosity) of a factor of ten or more. The short reaction time scales in conjunction with the fluid mechanics leads to short length scales, for example flame thicknesses of a tenth of a millimeter. In most practical applications the flows are turbulent, leading to the phenomenon of turbulent combustion–the intersection of two non-linear problems with multiple scales. The short length and time scales involved, together with the high temperatures present a significant challenge to experimental investigations.

In spite of these considerable difficulties, in the last ten years great strides have been made in understanding combustion, both experimentally (largely through the development of laser diagnostics) and theoretically (largely by the exploitation of more powerful computers).

It appears that the available tools are now a match for the complexities of combustion, and that significant progress can be expected in the next decade.

RESEARCH

Combustion is a broad discipline, with a wide variety of applications. The present exposition cannot be comprehensive. The following sub-sections illustrate just some of the currently-active research areas.

Laminar Flame Studies

Emphasis on improving the efficiency of power generating units combined with the modeling of turbulent reacting flows and environmental issues dealing with the production of nitrogen based pollutants helps to motivate the study of laminar flames. As an example, most of the oxides of nitrogen are formed during combustion when part of the oxygen combines with atmospheric nitrogen rather than with the fuel. As a result, the burning of hydrocarbon fuels can produce large quantities of NO and NO_2 (nitric oxide and nitrogen dioxide, respectively). Both compounds are considered toxic and nitrogen oxide is related to the formation of photochemical smog. Laminar flames can help elucidate the processes by which NOx is produced in combustion systems.

In the late 1970's and early 1980's, there were definitive computational studies of one-dimensional laminar flames. The conservation equations were solved numerically with detailed combustion chemistry and transport properties, and the results were shown to be in accord with experimental observations.

The study of the interaction of heat and mass transfer and chemical reaction in practical combustion systems requires a multidimensional study. Three-dimensional models combining both fluid dynamical effects with finite rate chemistry are as yet computationally infeasible. As a result, the modeling of chemically reacting flows has generally proceeded along two independent paths. In one case chemistry was given priority over fluid mechanical effects and these models were used to assess the important elementary reaction paths in, for example, hydrocarbon fuels. In the other case, multidimensional fluid dynamical effects were emphasized with chemistry receiving little priority. It is only within the last couple of years that the modeling of laminar flames has begun to consider the complex interaction between the fluid dynamic and thermochemistry solution fields.

The relatively recent movement into multidimensional systems is due in large part to the size of the systems that must be solved. In computational fluid dynamics (CFD) computations one often solves the Navier-Stokes and continuity equations for the velocities and pressure as a function of the independent spatial coordinates. In computational combustion the fluid dynamic equations are coupled to the energy and species balance equations. While CFD computations rarely solve for more than five unknowns, this is not the case in reacting flow computations. In particular, for laminar hydrocarbon flames one often solves for as many as 30 chemical species in addition to the temperature and the fluid dynamic variables.

In addition to the accurate prediction of the basic combustion process in multidimensional systems, the concept of fluid dynamical control of the combustion process is an extremely important issue for a variety of combustion systems. As an example, the lean extinction limit of premixed methane-air flames as measured in the laboratory at atmospheric pressure occurs at a stoichiometric ratio of about 0.5. However, when counterflowing streams of premixed methane and air are counter rotated, a compression zone develops near the stagnation point. This effectively reduces the straining on the system and the flame is able to sustain combustion at lower equivalence ratios than if no rotation were present. In particular a 10-15% reduction in the lean extinction limit can be obtainable in a two-dimensional counterflow configuration. Application of this concept to practical burner systems has not yet been undertaken but the potential is clearly significant from the point of view of lowering CO and NOx levels for gas turbines and home furnaces and in decreasing the thermal stress on system components.

Fluid dynamic considerations are clearly of importance in the modeling of the laminar combustion system. The ability to predict accurate fluid and chemical fields in multidimensional flames will inevitably lead to a better understanding of the processes by which pollutants are formed, in improving engine efficiency and in modeling turbulent reacting flows.

Optical Diagnostics

In a combustion environment, the measurement of fluid velocity, temperature or chemical species concentration is complicated by changes in density, due to heat release, concomitant with mixing of two or more streams. The high temperatures melt delicate probes; probes that do not melt are too large to resolve the fluctuations in the flow. Nonintrusive optical diagnostics are called for. Laser Raman scattering from a single microsecond laser pulse, has produced time resolved measurements of scalars in turbulent flows. These new measurements are most successful when the experiment and the diagnostic are designed in concert. In turbulent nonpremixed flames, high mixing rates lead to flame extinction. It is easy to see the effects of local flame extinction with pulsed Raman scattering. Thus, the laser here is a binary diagnostic that tells us when the mixing rates have exceeded a critical value. By systematically changing fuel and the mixing geometry, the binary signal from the laser Raman becomes a powerful tool for interpretation of turbulent mixing rates. Flame extinction is an abrupt process whose nonlinearities will challenge models in the extreme. As an intermediate challenge, we have explored mixing rates and chemical kinetic rates, without flame extinction. Fuels that contain an oxidizer have been used to explore the competition between mixing (dilution in coflowing air, typically) and reaction. At high mixing rates, the oxidizer in the fuel has little time to convert fuel on the 'rich side' to products, so the oxidizer is a passive scalar. At lower mixing rates, the oxidizer in the fuel has time enough to attack the fuel and so a change is observed using the laser Raman scattering probe. This combination of laser diagnostic and judicious choice of fuel and geometry will continue to be a fruitful avenue of research.

A common assumption in models of turbulent combustion is that all species, and heat mix and diffuse at the same rate. This assumption produces great simplification in the modeling; yet we should have some notion of effects of differential diffusion. Only recently have we been able to isolate the different effects. Laser probing of nonreacting mixing of a jet of propane and hydrogen mixing with air have shown local separations of hydrogen from propane. These point laser measurements quantify differential diffusion but also lead us to consider the nature of large scale mixing as it effects the molecular level.

In the immediate future we can expect further evolution from single point measurements to two dimensional planar images of laser Raman and planar laser induced fluorescence, PLIF. PLIF images will improve our qualitative view of turbulent mixing. It is reasonable to expect a PLIF sequence of images. For combustion flows, the rapid time scales lead us to imagine that the images will be acquired through a combination of multiple cameras and lasers. For nonreacting turbulent flows, such images are already appearing. In addition, the slower time scales associated with laboratory liquid mixing flows allow the planar images to be scanned in space, yielding a three dimensional image of a mixing flow. By collecting many of these three dimensional sets in sequence, a full three dimensional plus time measure is now emerging. These measurements are the experimental analog of direct numerical simulation. As before, much will be learned from these liquid injection experiments that will be extrapolated to gas phase flows and extrapolated further to combustion flows.

Much interest in combustion is due to the formation of pollutants and toxics at a heretofore acceptably small level. These small concentrations are weakly linked to the major reactions in combustion. The numerical prediction of such pollutants, such as nitric oxide and carbon monoxide, is a challenging, but urgent, area of research. Analogously, the measurement of part-per-million species in the surrounding environment provided by turbulent combustion is a challenging, but urgent, area of research. An emerging diagnostic has the intimidating name of Degenerate Four Wave Mixing (DFWM). Suffice to say, research on laboratory laminar flows is showing great sensitivity to part-per-million species with the challenges of a signal that is nonlinear with respect to laser power and species concentration.

Computer Modelling of Turbulent Combustion

In recent years, turbulent combustion models have seen increased use in the design and development of combustion devices including: gas-turbine combustors, furnaces, boilers and internal combustion engines. The aim of these models is to determine the performance of a proposed design, where "performance" may include heat-transfer and temperature-field characteristics, pollutant emissions, and interactions with mechanical components.

Models currently employed are found to be useful even though they are limited both in accuracy and in scope. Typically they are finite-volume codes in which the Reynolds-averaged conservation equations are solved in conjunction with the $k-\epsilon$ turbulence model. Combustion may be assumed to be mixing-limited, in which case finite-rate kinetic effects are not represented.

To meet all practical design needs, current turbulent combustion models need to be improved in the following respects:

i. generality,

ii. finite-rate kinetics,

iii. accuracy of turbulence modelling,

iv. incorporation of additional effects.

Most models are applicable either to diffusion flames (with a single fuel stream and a single air stream) or to premixed flames (in which the reactants are thermochemically uniform). Many important combustion processes are not well approximated by these idealizations. Examples are: staged combustion, piloted diffusion flames, stratified charge engines, and inhomogeneously premixed flames (as may occur in two-stroke engines).

The major improvement currently being sought is the ability to handle realistic finite-rate chemical kinetics. There are two motivations. The first is to calculate the effects of "slow" reactions, such as the production of NO_x and soot, or the post-flame oxidation of CO. The second and much more difficult objective is to calculate ignition, extinction and related phenomena.

There is considerable current research to develop more advanced models that can meet these needs. One approach is PDF methods in which a modelled transport equation is solved for the (one-point, one-time) joint probability density function (pdf) of the fluid properties. A major attraction of this method is that the direct effects of reaction appear exactly in the pdf equation, and hence require no modelling assumptions.

More comprehensive models (such as the pdf method) inevitably result in a computational problem that is one or more orders of magnitude greater than that generated by currently used turbulent combustion models. However, with the rapid advances in computer hardware (especially parallel processing), it is likely that hardware will not be a limiting factor. Algorithm development will be needed, however, to combine more comprehensive models with parallel machines to produce a useful design tool.

The above discussion pertains to turbulent combustion models with a direct engineering objective. Another valuable approach with a different objective is Direct Numerical Simulation (DNS). Here the idea is to solve the conservation equations without modelling assumptions, with the scientific objective of gaining a better understanding of the basic physical and chemical processes involved. Compared to DNS of non-reactive flows, this is a very difficult task–arguably one that is infeasible with current computers.

In practice, to yield a tractable computation, severe modelling assumptions are made–not for the turbulence, but for the thermochemistry. For example, a one-step reaction and equal species diffusivities may be assumed.

Spray Combustion

Sprays and spray combustion are complex and challenging because they involve numerous unresolved problems from a variety of scientific disciplines;

for example, liquid breakup through interface instability, dispersed multiphase flow, turbulent mixing, thermal radiation in participating media, interphase transport in a reactive environment and the chemical kinetics of flames, among others. Thus, although scientific study of spray combustion began with single drop combustion studies nearly fifty years ago, current understanding of spray combustion is still very limited. This is unfortunate because sprays and spray combustion are critical technologies for energy conversion and pollutant production–major concerns that will become even more important in the future as fuel supplies dwindle and sources of pollution multiply.

Research accomplishments and needs for combusting sprays are discussed in the following, emphasizing fluid mechanics issues. The discussion begins with single drop processes and concludes with sprays.

The vaporization and combustion of single drops have received considerable attention because they are well-defined and important unit processes for sprays. Thus, the general mechanisms of drop vaporization, ignition, combustion and extinction are understood and there is empirical information available to estimate drop transport rates for engineering purposes at moderate pressures. Nevertheless, understanding of drop transport properties for conditions relevant to practical combusting sprays is far from complete. The main difficulties are that drops are fluid rather than solid particles, and that drop transport occurs in a turbulent environment.

Current knowledge about drop transport is based on concepts derived from solid spherical particles. In actual practice, however, drops deform, have internal motions and are subject to secondary breakup. These real processes are particularly important for practical combustion devices. For example, at the high pressures of many practical combusting sprays, drop surfaces approach the thermodynamic critical point where reduced surface tension causes effects of deformation and internal motion that have received little attention. Additionally, recent studies show that drops formed by primary breakup at liquid surfaces are intrinsically unstable to secondary breakup, yet the dynamics and outcome of this rate controlling process are unknown.

Drops in combusting sprays are in a turbulent environment where they experience high relative turbulent intensities and scalar property fluctuations (e.g., velocity fluctuations comparable to the mean relative velocity between the phases and temperature fluctuations on the order of 500 K). This causes two effects that are significant for transport in sprays: turbulent dispersion of drops, which is an important turbulent mixing mechanism, and enhancement of drop transport rates by temporally and spatially varying surroundings. Problems of turbulent dispersion of drops are receiving increasing attention, which should be encouraged due to the importance of this multiphase mixing mechanism. Drop transport rates in turbulent environments have received some attention in the past, but better understanding of this process should result from application of modern experimental and computational diagnostics.

Sprays are frequently divided into two categories: dense sprays, which involve the high liquid volume fraction region near the injector where processes of liquid breakup and the presence of irregular liquid elements (e.g., ligaments) are dominant features; and dilute sprays, which involve the low liquid volume fraction region more remote from the injector where drops are more or less spherical and processes of drop transport and turbulent dispersion are dominant features. Thus far, there have been relatively few studies of combusting sprays for either regime because the complexity of these flows already make nonevaporating and noncombusting sprays difficult to measure or analyze. One can sympathize with the desire to reduce complexity by avoiding combustion: however, measurements of the structure of combusting sprays are feasible using contemporary instrumentation and experiments along these lines are needed to highlight unit processes that merit additional study and to gain a better fundamental understanding of combusting sprays.

The main uncertainties about sprays are associated with the dense spray region. This region is crucial because it controls the properties of drops entering the dilute spray portion of the flow. In spite of its importance, however, dense sprays have not been studied very much due to problems of penetrating dense drop clouds with experimental diagnostics. Lack of experimental observations has correspondingly impeded the development of theory. Recent developments, however, have changed this picture appreciably and there is now significant potential for observations of dense sprays that should yield a period of rapid growth in understanding this flow. The main new tools have been the development of holography techniques capable of penetrating dense sprays and even observing processes of primary breakup along the liquid surface, as well as exiplex techniques for resolving liquid and vapor concentrations. Results thus far have shown that existing theories of primary breakup based on stability theory are ineffective and have disclosed a new turbulent breakup mechanism that is active in some instances; therefore, much remains to be done to gain a reasonable understanding of this first step in any spray process. Additionally, secondary breakup processes in dense sprays appear to involve mechanisms of ligament stretching and turbulent distortion that have not been considered in classical shock tube studies of the breakup of round drops.

Another facet of sprays, relevant to both dilute and dense sprays, is the effect of turbulence/dispersed-phase interactions on the turbulence properties of the continuous phase. Two interactions are important in principle: turbulence generation by the stirring action of drops moving with a velocity relative to the continuous phase; and turbulence modulation (damping) by interactions between the velocity fluctuations of the drops and the continuous phase. Turbulence modulation has received greatest attention in the literature, however, this effect is more significant for sedimentation than sprays. On the other hand, turbulence generation by drops creates the turbulence field of dense sprays and this type of turbulence has distinctly different properties from conventional single-phase turbulence. These differences will no doubt attract study of turbulence generation by drops (particles) in the future, which might provide a useful perspective for a better understanding of conventional turbulence as a by-product.

The other side of turbulence/dispersed-phase interactions is turbulent dispersion of drops which was mentioned earlier in connection with studies of individual drops. However, recognition that turbulence within dense sprays differs from conventional turbulence implies a need to study turbulent dispersion in spray environments as well. Recent work has disclosed features of turbulent dispersion that merit particular attention: self-generated dispersion where eddy shedding from drops causes them to disperse laterally even in quiescent gases, and effects of unsteady turbulent flow fields on drop dynamics and transport due to effects analogous to virtual mass, Magnus, Basset, Saffman lift, etc., forces but for Reynolds numbers higher than the Stokes regime that are more representative of drops in sprays.

To summarize: issues for single drops include effects of deformation and motion of drop liquid on interphase transport, and the effects of a strongly turbulent environment on transport to droplets and turbulent dispersion of drops. Issues for sprays include the mechanism of primary breakup for non-turbulent and turbulent liquids, the relevant secondary breakup mechanisms for dense sprays, turbulence generation by drops in dense sprays, and the dynamics of interactions between drops and turbulence representative of practical spray combustion processes. Progress toward gaining an understanding of these problems has been slow in the past, however, recent development of instrumentation capable of resolving processes within sprays, and continued growth of computational capabilities, offer exciting prospects for rapid advances in the future.

Fire Research

The fluid mechanics of fires is of interest for two distinct reasons. First, the combustion processes in most fires are controlled by the turbulent mixing of the gas field fuel and air into a buoyant plume. A portion of the chemical energy released is then fed back to the condensed fuel surfaces by a combination of convective and radiative transport to sustain the fire. Second, the buoyant plume generated by the fire acts as a giant pump, distributing heat and hazardous combustion products over a much broader domain outside the active combustion zone. The transport and dispersion of smoke and hot gases is controlled by the interaction of buoyancy driven flows with the geometric and ventilation constraints imposed by building environments or the meteorological and topographical conditions in the vicinity of outdoor fires.

The above division of interests is reflected in the advances made over the past decade. Most of the progress in fire related combustion processes concerns the structure of free standing fire plumes with known burning rates. The time-averaged velocity, temperature and plume width in purely buoyant plumes inside and outside the combustion zone has been measured as a function of position and correlated with the global heat release rate over a four order of magnitude range of fire strengths. The utility of the mixture fraction (the fraction of material at a given point in the flow that is or was fuel) as a correlator of temperature and major species in both buoyant jets and plumes has been demonstrated for a variety of simple fuels. Finally, the existence of large scale structures in the form of reasonably periodic toroidal pulsations of the entire plume has been demonstrated and plausible correlations of frequency with fire size developed. Indeed, the state of knowledge with respect to fire plumes at present is quite analogous to that of turbulent mixing layers shortly after the famous Brown-Roshko experiments in the early 1970's.

The understanding of plume structure together with other advances in fire driven flow have led to the development of fairly comprehensive mathematical models of smoke and toxic gas transport in complex building structures. These are lumped parameter models with an individual room divided into a hot upper layer, a cold lower layer, and fire plumes where required by the scenario under consideration. Separate studies of flows through doorways, convective and radiative heat transfer to boundaries and catalogs of burning rates of common furnishings have helped these "zone models", as they are known in the fire research community, to become practical engineering tools. More recently, commercially available computational fluid dynamics codes have been adapted to study fire induced flows in enclosures of complex shape, and large eddy simulations based on approximations to the basic fluid dynamics equations in simple geometries have also been performed. The "nuclear winter" scenario has led to a revival of interest in the properties of large fires. The first models intended to explicitly account for some of the fluid mechanical issues raised by such fires, together with the beginnings of a relevant field scale experimental data base, have begun to appear.

The level of sophistication of most of the fluid mechanics related to fire research is still primitive compared with that in other branches of fluid mechanics. The fire plume is as central a fluid mechanical entity to this subject as the mixing layer is in other branches of fluid mechanics and deserves a comparable level of effort. Studies of burning surfaces are either focused on the solid phase phenomena or conducted at a scale sufficiently small for stable laminar flow to exist. More realistic investigations of the coupling between gas and condensed phase phenomena are in their infancy. The application of "zone" models to very large buildings awaits the development of usable, experimentally validated theories of smoke movement down long corridors and in vertical shafts. The role of thermal radiation in almost any aspect of the fluid mechanics of fires is almost entirely unknown, although its effect on the thermal characteristics of fires is the subject of many investigations. The study of large fires is hampered by the difficulty of getting good experimental data, as well as conceptual uncertainty about the intellectual boundary between large scale fire phenomena and micrometeorology. Despite three thousand years of interest in the subject, it is still not possible to quantitatively explain how a fire burns.

Detonation Research

As mentioned in the Introduction, safety considerations continue to be a major factor driving detonation research. But detonations also play an important role in such applications as supersonic combustion ram jets, hypervelocity drivers, detonative manufacturing processes, automotive engine knock, and fuel-air explosions, among others. Detonations are also central to the technology of solid explosives. The detonation of solids is a field by itself involving not only fluid mechanics and chemistry but structural mechanics as well. The detonation of solids is not considered here, but many of the results developed for gaseous and heterogeneous detonations are applicable.

The factors governing the initiation of detonations either by the sudden release of energy or by transition from deflagration to detonation (DDT), the steady state propagation characteristics, and the detonability of various fuel-oxidizer mixtures are the main targets of most detonation research. Early work has shown that the steady propagation velocity or Chapman-Jouguet (CJ) velocity of a lossless detonation in a gaseous fuel oxidizer mixture can be determined from an essentially equilibrium thermodynamic calculation without any consideration of the kinetics of the combustion reactions. Establishing the physics of initiation and of the effect of losses due to the presence of walls, obstacles or bounding compressible media requires consideration of the structure of the detonation front. The earliest proposed structure due to Zeldovitch, Neumann, and Doering (ZND) assumes that the detonation consists of a leading shock wave followed by a one-dimensional reaction zone whose structure depends on the chemical kinetics of the fuel-oxidizer mixture under consideration. Particularly for hydrocarbon fuels, the reaction zone is often dominated by an induction region during which properties remain almost constant followed by a narrow reaction zone in which the heat releasing reactions take place.

While the one-dimensional ZND model has provided a basis for estimating the effect of wall losses and the minimum energy required for direct initiation, analysis has shown that this structure is inherently unstable. The spinning and galloping detonations observed under limiting conditions certainly are not one-dimensional, and the instability of the ZND structure is supported by the experimental observation that the structure of all gaseous detonations is inherently unstable and governed by the the dynamics of the continual interaction of transverse waves propagating across the main detonation front. This structure is referred to as "cellular structure" because these interactions trace out a sequence of cellular patterns on soot covered foils placed on the walls of detonation tubes.

Deflagration transition has been observed to involve highly accelerated combustion fronts which induce shock waves ahead of them. The resultant "processing" sensitizes the combustible mixture ahead of the flame and transition to detonation occurs when a small disturbance due to turbulence or some other phenomenon causes this sensitized mixture to explode. During this initial "explosion within explosion" phase pressures are generated which are far in excess of the pressure behind a steadily propagating CJ wave. While there have been many observations and analytical studies and numerical simulations of this process, DDT is still not completely predictable or understood. A striking feature of DDT is the appearance of combustion fronts with velocities as high as 700 - 800 m/s.

The diffraction of detonation waves by obstacles, which is often involved in DDT, and the interactions between different detonating explosive media or between a detonation and a bounding inert material involve oblique detonations and other two- and three-dimensional reactive shock configurations. Such phenomena also will be involved in supersonic combustion ram jets, and are central to certain types of hypervelocity drivers and the proposed oblique detonation ram jet. It has been possible to observe some of these two- and three-dimensional interactions using Schlieren photography with exposure times as short as 10 nanoseconds and to measure the pressure signatures generated on the containing walls. There also has been some success in simulating these phenomena numerically on a qualitative basis but so far there have not been one on one quantitative comparisons with experiment.

The following items are promising research topics:

Schlieren photography and the measurement of surface pressures are the main diagnostic tools available for the study of gaseous and heterogeneous detonations. High speed optical diagnostics, which are now available, should be applied to resolve the details of gaseous detonation structure, of detonative interactions and diffraction past obstacles, and of Deflagration-Detonation Transition.

The reaction zone structure, initiation, and the propagation character-istics of spray, dust, and film detonations for various materials should be investigated in detail. Detonability limits should be established. The role of cellular structure, if any, in heterogeneous detonations should be explored.

The role of turbulence in the transition to, and propagation of, detona-tions in gaseous and heterogeneous mixtures should be explored in detail. The structure of the turbulence present under various test conditions needs to be carefully characterized.

The high speed combustion fronts which have been observed in DDT need to be studied in detail. The usual mechanisms of turbulent flame propagation do not provide an explanation for the high flame velocities of hundreds of meters per second observed in practice.

Heterogeneous detonations involve the ignition and subsequent combus-tion of fuel droplets or dust particles. Because of the sparsity of data in the microsecond range involved, data on shock induced particle and droplet igni-tion and combustion is lacking. These processes should be tracked in detail experimentally.

Properties of both gaseous and heterogeneous detonations, particularly in marginal cases, are strongly dependent on scale. Both large scale and lab-oratory scale facilities are available for studying detonation behavior. Thus, coal-mine sized facilities are available in both the US and Poland, and inter-mediate sized test facilities exist in many laboratories throughout the world. A coordinated experimental program, which takes advantage of all of these facilities, should be designed and carried out to determine the effect of scale on detonation characteristics under conditions which are otherwise compa-rable.

Although the cellular and dynamic structure of gaseous detonations has been simulated numerically using both two- and three-dimensional models, many questions remain, e.g., in some cases the results are very sensitive to the grid size used. Numerical simulation studies of the dynamic detonation structure should be continued using the more advanced computer technology now available.

Development of simulations of large scale interactions, that is on a scale larger than that of the individual detonation cells, should be continued. Such simulations should be extended to smaller grid sizes and to three-dimensional interactions.

In order to make the computations tractable most simulations of det-onations and other reactive flows use simplified models for the chemistry involved. In the case of detonations a two step model is often used consisting of an induction zone followed by an almost discontinuous reaction front.

Efforts should be made to develop algorithms using full chemistry, or at least some of the partial or approximate reaction schemes currently under investigation.

Heterogeneous detonations can involve mixtures of fuel drops or dust particles with air or oxygen, of gaseous fuel-oxidizer mixtures and combustible or inert particles, or of fuel slurries and a gaseous oxidizer. Most models of such heterogeneous detonations assume a one-dimensional ZND reaction zone structure, and monodisperse droplets or particles. Such models can only serve as a rough approximation. Hence there is a strong need for extensive theoretical and numerical studies of such heterogeneous detonations.

Although DDT has been the subject of research for many years, this process is still not well understood and many uncertainties remain. Recent advances in algorithms and in computing power should be applied to the development of comprehensive simulations of DDT in both all gaseous and heterogeneous media.

Heterogeneous detonations depend on mixing of fuel in solid or liquid form with the gaseous oxidizer then followed by ignition and combustion. The physics of these processes, involving the interaction between the high speed flow behind the leading shock of the detonation and droplets or particles of various materials, is not fully understood. Modeling studies of particle flow interactions are therefore required to provide proper interpretations of experimental measurements and appropriate input for the simulation of heterogeneous detonations.

While there have been some studies of the effects of wall boundary layers on detonation propagation, many analytical and numerical studies of detonations propagating in tubes ignore this effect, a fact which may account for the divergence between theory and experiment in some cases. Boundary layers will be especially important in the propagation of detonations in small tubes and in DDT. The propagation of detonations through the small spaces and gaps in internal combustion engines is thought to be responsible for much of the damage caused by engine knock, and there viscous flows may be a dominant effect. The effects of boundary layers and other viscous phenomena on the propagation of detonations should therefore be investigated, both analytically and numerically particularly in the more complicated two and three dimensional configurations.

In conclusion, the study of detonations, particularly when heterogeneous detonations are included, requires the simultaneous consideration of high speed compressible reactive flow, two-phase flows, turbulence, boundary layers, gas-particle or gas-droplet interactions, and heterogeneous or gas-phase particle or droplet combustion. The study of detonations thus encompasses a wide range of fluid dynamic phenomena.

GENERAL REFERENCES

Bachalo, W.D. (1994) "Injection dispersion and combustion of liquid fuels" *Twenty-Fifth Symp. (Int'l) on Combust.*, The Combustion Institute, pp. 333-344.

Dixon-Lewis, A. (1992) "Structure of laminar flames", *Twenty-Fifth Symp. (Int'l) on Combust.*, The Combustion Institute, pp. 305-324.

Maly, R.R. (1994) "State of the art and future needs in S.I. engine combustion," *Twenty-Fifth Symp. (Int'l.) on Combust.*, The Combustion Institute, pp. 111-124.

Pope, S.B. (1992) "Computation of turbulent combustion: progress and challenges", *Twenty-Fifth Symp. (Int'l) on Combust.*, The Combustion Institute, pp. 591-612.

Prather, M.J. and Logan, J.A. (1994) "Combustion's impact on the global atmosphere", *Twenty-Fifth Symp. (Int'l) on Combust.*, The Combustion Institute, pp. 1513-1527.

MULTI-PHASE FLOW, CAVITATION, AND BUBBLES

A. Prosperetti
Department of Mechanical Engineering
Johns Hopkins University
Baltimore, MD 21218

INTRODUCTION

The words "multi-phase flow" denote a flow in which more than one phase occurs simultaneously. Examples involve mixtures of gas and liquid (e.g., in pipelines and vapor generators), gas and solid (e.g., in fluidized bed combustors), liquid and solid, or two immiscible liquids (both occurring, e.g., in chemical reactors). Mixture of three phases, such as gas, liquid, and solid, are also common, e.g., in liquid-gas chemical reactors where the solid might be the catalyst. These flows are so ubiquitous that it is estimated that about two thirds of all industrial chemical compounds involve them at some step of their manufacturing process.

Many major industries therefore rely on, and are affected by, the ability to design for multi-phase flow. Examples are the oil industry, chemical and bio-chemical manufacturing, synthetic fuels production, conventional, nuclear, and geothermal power generation, paper manufacturing, food industry, pneumatic and hydraulic transport of solids, waste treatment, and many others.

Multi-phase flows also include cavitation, in which highly destructive bubbles spontaneously form in a flowing liquid. This process significantly impacts, e.g., hydroelectric plants regardless of their scale, ship propulsion and design, lubrication systems, pumps, turbines, and bio-medical systems.

Despite this wide occurrence of multi-phase systems and flows, our ability to analyze, and effectively design for, them is in a very primitive state. The main reason is that, contrary to most other situations in Fluid Mechanics, it is nearly impossible to scale experimental results so as to rely on small-scale experiments to answer questions that have a practical impact on the final design. This is a crucial aspect of the deficiency of our physical understanding of these flows that frequently leads to unsatisfactory yield or product quality and costly attempts to rectify the problem. These difficulties are so acute that in some cases it has been necessary to build extremely expensive full-scale models prior to the final design. Even so, in the absence of a good understanding on these processes, it is difficult or impossible to predict plant

244

performance during severe off-design conditions, e.g., in case of accident. Fundamental research in this area is therefore essential to develop reliable design methods, to predict performance, and to enhance safety.

Several environmental processes also involve multi-phase flow phenomena and progress in this area would be beneficial for their understanding and control. For example, the injection of atmospheric gases in the ocean by breaking waves and their transport by turbulence and Langmuir circulation plays an important role in the food chain and very likely in the accumulation of carbon dioxide in the atmosphere. In smaller water bodies air entrainment caused by rain may be of comparable importance. Many processes for the treatment of sewage and industrial liquid waste rely on the action of aerobic micro-organisms the action of which is dependent on an adequate oxygen supply.

Since other sections in this report deal with solid-fluid multi-phase flows (Jenkins, Homsy, Davis) we focus here on gas-liquid systems.

GAS-LIQUID MULTI-PHASE FLOW

Contrary to solid-fluid multi-phase flows, a unique feature of gas-liquid systems is that the configuration of the interface between the phases is not known a priori. As a matter of fact, the so-called topology of the possible flow regimes exhibits a bewildering variety of forms. Some examples for the simplest situation – nominally steady flow in a horizontal or vertical tube – are shown in Figs. 1 and 2 respectively. All these flow regimes differ widely in their response to driving forces, e.g., the pressure gradient provided by gravity or pumps, and in their ability to transport thermal energy. For example slug flow, in which large gas bubbles are separated by liquid "slugs," tends to induce strong vibrations that may cause fatigue failure, and must therefore be avoided. A further possible topology in annular flow, the so-called inverted-annular regime in which a liquid core is separated from the wall by a vapor layer, exhibits very poor heat transfer properties and must be avoided when heat removal from the tube wall is of primary concern as in nuclear reactors.

It is therefore vital to be able to predict the flow regime prevailing in different situations and the occurrence of flow regime transitions.

This is the most pressing problem in this area, but many decades of fruitless efforts have made clear that it cannot be solved without a fundamental understanding of the mechanics of the flow. As in most other fluid mechanic situations, turbulence also plays an important role the understanding of which is even more limited than in single-phase flows.

Widespread as it may be, steady flow in a tube is hardly the only situation of interest. For example, when tubes are joined in T or Y arrangements, great uncertainties exist on the proportion of each phase that will flow in

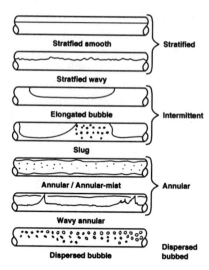

Figure 1: Some common flow patterns in horizontal liquid-gas two-phase flow (from A.E. Dukler and Y. Taitel, "Flow pattern transitions in gas-liquid systems: Measurement and modeling," in *Multiphase Science and Technology*, Vol. 2, G.F.Hewitt, J.M. Delhaye, and N. Zuber eds., Hemisphere, Washington, 1986, pp. 1-94; reproduced with permission).

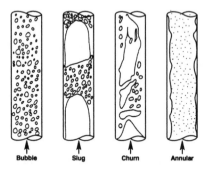

Figure 2: Some common flow patterns in vertical liquid-gas two-phase flow (from Y. Taitel, D. Barnea, and A.E. Dukler, "Modeling flow pattern transitions for steady upward gas-liquid flow in vertical tubes," *AIChE J.* **26**, 345, 1980; reproduced with permission).

each leg. A very common arrangement in heat exchangers involves the flow of the multi-phase mixture across a bank of tubes. The flow path in the core of a nuclear reactor is frequently complex and intermittently obstructed. A wealth of similar examples could be cited.

A particular mention is deserved by the thermal processes occurring in multi-phase flows. This appears to be particularly appropriate at a time when nuclear power is expected to make a strong comeback, but the practical implications of this topic widely transcend any particular industry. A striking example is the occurrence of yet-unexplained catastrophic rapid condensation events that can occur when vapor contacts cold liquid. A recent instance is the explosion in lower Manhattan in 1989 which caused casualties and significant damage to several city blocks.

Boiling is another process of immense industrial importance. The vast majority of electric power is produced by the flow of vapor through turbines, and this vapor is of course the result of boiling processes. The cooling of refrigeration systems relies on the boiling of the coolant. Boiling is being actively considered for the cooling of the next generation of super-computers and for heat rejection in space. It is still difficult to predict and control this process in many cases. For example, with some liquids, boiling incipience is retarded so that the temperature of the surface that needs to be cooled may rise to unsafe levels. Surface fouling and contamination and liquid quality have a major impact on heat transfer rates.

Under normal plant operation, the processes described occur in nominally steady conditions. Transient conditions are however also of great practical importance, ranging from the start-up of a plant to the sequence of events occurring during an accident. Major industrial safety problems – e.g., in the nuclear industry – are related to the transient behavior of multi-phase flows. Related phenomena are system-wide instabilities specifically rooted in multi-phase flow phenomena that plague steam generators, evaporators, various chemical process units, and others. These instabilities are difficult to control and induce vibrations, fluctuations, and even divergent evolution of the flow with mechanical damage and a violation of the thermal limits.

Research Needs

The research needs in multi-phase flow are too wide to be articulated in terms of *problems*; rather, it is appropriate to list *areas*:

1. Mechanics of flow regimes;

2. Mechanisms of flow-regime transitions;

3. Flow in complex geometries;

4. Heat transfer, boiling, and condensation;

5. Unsteady effects.

CAVITATION

In discussing the topic of Cavitation it is necessary to distinguish between *Flow Cavitation* and *Acoustic Cavitation.*

Flow Cavitation

The occurrence of vapor cavities in a high-speed liquid flow causes several detrimental effects such as damage of solid surfaces, vibration, noise, and loss of performance. First encountered in the naval industry in the second half of the XIX century, cavitation still remains a significant problem in commercial and military naval engineering, in the hydroelectric industry, and more generally wherever large-scale pumping of liquid is required.

Just as multi-phase flow, cavitation is another example of a phenomenon that has stubbornly resisted scaling efforts for several decades. Here, however, the problem of scaling is often compounded by the variability associated with water quality (chiefly nuclei content and gas saturation) and characteristics of the exposed solid surfaces (e.g., roughness) that have not yet been fully identified nor characterized. Even the basic physics of the so-called cavitation nuclei (the "weak spots" of the liquid at which cavities form) remains unclear. This critical area transcends mechanics and its clarification probably requires a significant input from disciplines such as physical chemistry and surface science. Experimentally, the present resolution capabilities are in the 10 μm range, at least one, but more likely two, orders of magnitude too large.

Aside from the problem of nucleation, the occurrence of cavitation is strongly intertwined with turbulence and complex transient three-dimensional flows. For example, cavitation on a propeller is often triggered by the interaction with vortices generated by solid surfaces upstream of the propeller. This process is further complicated by the effect of turbulence, maneuvering, and other sources of flow unsteadiness. The current rapid developments in our ability to resolve complex flows open up new opportunities for cavitation research.

Cavitation damage is still an important problem. For example, in 1983, the spillways of Glen Canyon dam in Arizona were severely damaged by cavitation strong enough to erode through three-feet thick concrete walls. Protection against cavitation occurrence requires that the minimum pressure in the flow be kept above certain (poorly known) levels. This objective is achieved by conservative design to limit the velocity of the liquid relative to the exposed solid surfaces and by ventilation, the deliberate introduction of gas bubbles. Both practices affect performance and the second one is also a significant noise source.

Since, as noted before, a multi-phase flow can behave very differently from a single-phase one, the occurrence of cavitation can significantly degrade the performance of hydraulic machinery. In the impossibility of accurately predicting this event, it is necessary to use "defensive" design practices which ultimately result in increased costs.

Cavitation noise is of particular concern to the U.S. Navy because, due to its distinctive high-frequency content, it is quite easy to detect and to track. For this reason, for example, the development of each new class of submarines involves a substantial effort in the study of cavitation and means by which it is to be avoided. Cavitation noise is however also of concern in commercial shipping as it may cause a high level of discomfort for crew and passengers.

To further illustrate the wide-ranging import of cavitation phenomena one might cite the recent example of cavitation-induced failure of artificial heart valves.

Acoustic Cavitation

The pressure reduction responsible for the growth of vapor bubbles can be caused by sound waves as well as by flow. The resulting broad range of phenomena, referred to as Acoustic Cavitation, has a significant impact on a number of areas.

In the first place, ultrasound is widely used in Medicine, both for diagnostic and therapeutic purposes. While it is known that cavitation can occur *in vivo* at intensities normally used in these applications, it is not clear whether this possibility presents health hazards, nor what the safety limits are. The detection of cavitation nuclei *in vivo* is an even more difficult problem than in flow cavitation.

A recent medical development in which acoustic cavitation plays a very important role is extra-corporeal lithotripsy, the comminution of kidney stones by strong focussed acoustic pulses. Extension of this approach (which necessarily will involve cavitation) to soft-tissue surgical procedures are currently being actively investigated.

Ultrasonic cleaning is an important industrial practice used in industries ranging from jet engines to electronics. One may expect increased importance of this process with the development of the new micromachines currently under intense study. Issues of optimization and physical mechanisms still remain unresolved. The process of ultrasonic atomization, which can be useful, e.g., for the production of microparticles from melts, seems to involve acoustic cavitation but is poorly understood.

An area that has recently been gaining in importance is that of sonochemistry, where acoustic cavitation gives rise to unique conditions of temperature and pressure and exotic compounds. The related phenomenon of sonoluminescence, with light emissions of picosecond duration, is a scientific mystery whose practical potential extends all the way to nuclear fusion.

Research Needs

1. Nucleation;

2. Interaction between turbulence and cavitation;

3. Unsteady cavitation processes;

4. Silencing of cavitation noise;

5. *In vivo* cavitation and its biological significance;

6. Biomedical applications of cavitation: lithotripsy, treatment of fracture and wounds, angioplasty, and others;

7. Sonochemistry: mechanisms, effects, control;

8. Sonoluminescence.

BUBBLES

Gas and vapor bubbles of course feature very prominently in the problem areas of multi-phase flows and cavitation discussed previously, but there are significant scientific and technological questions which depend on the detailed understanding of bubble phenomena at the "local" level of individual bubbles. Some examples follow.

- Although cavitation damage has been conclusively linked to the occurrence of vapor bubbles in the flow, its detailed mechanics and possibly physico-chemical aspects are not completely understood;

- Central to the mechanics of several regimes of gas-liquid flows are the processes of bubble coalescence and splitting. Little is known about both, but particularly about the former. The role of surface forces and contamination appears to be essential and here we encounter the area of surface rheology which, while of great importance in tertiary oil recovery, the spreading of liquids on solids, emulsion and foam formation and stability, and many other situations, is in its scientific infancy. In this respect it may be noted that bubbles also constitute ideal systems for surface rheological studies and therefore appear here in the twofold role of subject of, and instrument for, research;

- Sonoluminescence and sonochemistry have already been mentioned in connection with acoustic cavitation. Their fundamental dynamics can only be understood at the level of individual bubbles. An effective control of these processes also presupposes such an understanding;

- A similar comment applies to most other cavitation effects, such as the bubble-produced "micro-scouring" action of ultrasonic cleaning;

- A last aspect worth noting is the dynamics of bubbles in connection with micromachines. Bubbles are highly energetic systems that can be accurately controlled and employed, e.g., to power micro-motors, micro-pumps, etc., and also possibly in bio-medical applications.

ACKNOWLEDGMENTS

The author is grateful to A. Acrivos, R.E. Apfel, C. Brennen, L.A. Crum, P. Griffith, T.J. Hanratty, T.T. Huang, J. Katz, R.T. Lahey, O. Manley, K. Stebe, and G.B. Wallis for several suggestions.

In the preparation of this appendix use has been made in part of the document "Proposal for a Program for Basic Studies of Flow Mechanics and Transport involving High-Reynolds Number Multiphase Flows" (1986) prepared for the National Science Foundation by a group led by Prof. T.J. Hanratty.

Support from the National Science Foundation and the Department of Energy under grants CBT-8918144 and DE-FG02-89ER14043 is gratefully acknowledged.

SUGGESTED READING

The literature on multiphase flow and cavitation is very large with thousands of articles published every year. Here we indicate the principal journal and volume series, some recent books, and a handful of representative articles.

International Journal of Multiphase Flow, Pergamon, Exeter UK; first volume (1973).

Annual Reviews in Multiphase Flow, issued as a supplement to International Journal of Multiphase Flow, Pergamon, Exeter UK; first issue, (1994).

G.F. Hewitt, J.M. Delhaye, and N. Zuber eds. (1982) *Multiphase Science and Technology*, Hemisphere, Washington; first volume.

Volumes with various titles containing the papers presented at the Summer and Winter Meetings of the American Society of Mechanical Engineers; for each meeting there are several volumes devoted to multiphase and cavitating flows.

G. Hetsroni editor, (1982) *Handbook of Multiphase Systems*, Hemisphere, Washington.

M.C. Roco editor, (1993) *Particulate Two-Phase Flow*, Butterworth-Heinemann, Boston.

J.F. Davidson, R. Clift, and D. Harrison editors, (1985) *Fluidization*, Academic, New York.

P. Basu and J.F. Large editors, (1988) *Circulating Fluidized Bed Technology*, Pergamon, Exeter.

R.C. Clift, J.R. Grace, and M.E. Weber, (1979) *Bubbles, Drops, and Particles*, Academic, New York.

S. Morioka and L. van Wijngaarden editors, (1995) *Waves in Liquid/Gas and Liquid/Vapour Two-Phase Systems*, Kluwer, Dordrecht.

J.R. Blake, J.M. Boulton-Stone, and N.H. Thomas editors (1994) *Bubble Dynamics and Interface Phenomena*, Kluwer, Dordrecht.

C.E. Brennen (1994) *Cavitation and Bubble Dynamics*, Oxford U.P.

N. Zuber (1964) "On the dispersed two-phase flow in the laminar regime" *Chem. Eng. Sci.* **19**, 897-917.

G.K. Batchelor (1972) "Sedimentation in a dilute dispersion of spheres" *J. Fluid Mech.* **52**, 245-268.

L. van Wijngaarden (1972) "One-dimensional flow of liquids containing small gas bubbles" *Ann. Rev. Fluid Mech.* **4**, 369-396.

S.B. Savage (1984) "The mechanics of rapid granular flows" *Adv. Appl. Mech.* **24**, 289-366.

D. Leighton and A. Acrivos (1987) "The shear-induced migration of particles in concentrated suspensions" *J. Fluid Mech.* **181**, 415-439.

J.F. Brady and G. Bossis (1988) "Stokesian dynamics" *Ann. Rev. Fluid Mech.* **20**, 111-157.

T.J. Hanratty (1991) "Separated flow modelling and interfacial transport phenomena" *Appl. Sc. Res.* **48**, 353-390.

A.S. Sangani and A.K. Didwania (1993) "Dynamic simulations of flows of bubbly liquids at large Reynolds numbers" *J. Fluid Mech.* bf 250, 307-337.

D. Kaftori, G. Hetsroni, and S. Banerjee (1995) "Particle behavior in the turbulent boundary layer" *Phys. Fluids* **7**, Part I: 1095-1106; Part II: 1107-1121.

CONTROL OF TURBULENT FLOWS

W. C. Reynolds
Department of Mechanical Engineering
Stanford University
Stanford, CA 94305

INTRODUCTION

Turbulent flows occur in or around most important systems, including aircraft and ships, electric generating stations, materials processing plants, and the human cardiovascular system. The ability to control turbulent flows so as to enhance mixing or reduce drag would have many important consequences for U.S. leadership in technology.

Turbulence is the result of inertial instabilities in the fluid flow. The parameter characterizing the ratio of inertial to viscous effects is the Reynolds number (honoring 19th century scientist Osborne Reynolds) $Re = VL/\nu$ where V is the flow velocity, L is the physical scale of the flow, and ν is the kinematic viscosity of the fluid. At low Reynolds numbers instabilities are suppressed by viscous effects, for example in very low-speed, small-scale systems (capillaries, microelectromechanical systems). But with familiar fluids (air, water) at typical device scales the Reynolds numbers are sufficiently high that instabilities grow and lead to large-scale coherent motions (eddies). The largest eddies are filled with turbulent motions of smaller scales, and the ratio of the largest to the smallest scales varies as $Re^{3/4}$ (Tennekes and Lumley 1972). These large eddies control the most important dynamics of the turbulence, and therefore understanding the processes that produce and sustain these eddies is the key to turbulence control (Cantwell 1981).

Instabilities in flows are of two different types (Huerre and Monkewitz 1985, Huerre 1990). *Convective instabilities* amplify disturbances that travel downstream without upstream effect, and occur when there is no mechanism for upstream disturbance propagation. If the disturbance is removed, the perturbation will propagate downstream and the flow will relax to an undisturbed state. For example, jets and boundary layers are convectively unstable. In contrast, *absolute instabilities* occur in situations where there is upstream propagation of the disturbance field, for example in the separated flow over a backstep where the recirculation carries disturbances upstream. Once an absolute instability has been triggered, the disturbance source can be removed and the disturbance will remain in place forever. Therefore, it is more difficult to control disturbances if the flow is absolutely unstable. A convectively unstable flow can be made absolutely unstable with the addition of acoustic or electronic feedback upstream, and this can be used to generate self-excited flow oscillations (Reisenthel *et al* 1991).

OVERVIEW OF TURBULENCE STRUCTURE AND CONTROL STRATEGIES

Considerable understanding of the large-scale structures in varous turbulent flows has been developed over the past three decades as a result of comprehensive research using modern diagnostic instrumentation and computer simulations. This work has led to several applications of *passive control*. For example, jet engine exhausts are now scalloped to reduce the scale of the coherent jet eddies and hence reduce the jet noise. If instead one wants to increase the mixing of jet fluid with ambient fluid, the discharge can be made elliptical, which produces elliptical vortex structures that exhibit large major-minor axis oscillations that significantly enhance mixing (Ho and Gutmark 1987, Husain and Hussain 1989, 1991). Using suction just outside of a round jet flow to produce a local region of backflow, an absolute instability and an associated large-amplitude self-excited oscillation can be produced that significantly enhances the mixing and spreading of the jet (Strykowski and Niccum 1991).

For applications in which the goal is to *reduce* the turbulence *e.g.* to *reduce drag*, the instability must be stabilized. In some instances this can be done passively by modifying the geometry or physical character of the system. For example, the addition of dilute amounts of polymers to liquid flows, which modifies the viscoelastic behavior of the liquid in a way that inhibits the growth of turbulence, has been widely used in applications drag reduction (Berman 1978).

In the near-wall region of a turbulent boundary layer, the dynamically important eddies are relatively long streamwise vortices that bring high-speed fluid towards the surface and gather low-speed fluid near the wall into low-speed streaks. Although they are the "large" eddies in the near-wall region, their size in the spanwise direction is increasingly small at high Re, of the order of $100\nu/u^*$ where u^* is the friction velocity (typically a few percent of the free-stream flow velocity). Left untended, these streaks meander slowly and then erupt violently to produce new boundary layer turbulence (Robinson 1991). Riblets (tiny streamwise surface grooves), which partially stabilize the low-speed streaks and inhibit bursting, have been used as a passive control to reduce the drag of sailboat keels and aircraft wings in regions where the flow direction is known (Bushnell and McGinley 1989).

A remarkable demonstration of passive control is provided by the suppression of the shedding of vortices into the wake behind a circular cylinder caused by placing a second much smaller cylinder in just the right place alongside the main cylinder well outside of the wake (Strykowski and Sreenivasan 1990). Recently Hill (1992) showed how the optimum location for the controlling geometric modification can be determined using the adjoint of the basic linear instability problem. In view of the inherently non-linearity of turbulence, it is remarkable how much insight to structure and control can be obtained through linear stability analysis, and that research using linear theory continues to produce very important results.

Some flows can be controlled by forcing the basic instability to occur in a desired way (*open-loop active control*). The round jet provides a rich example. The coherent structures (vortex rings) in the near-field of the jet are formed by a basic instability (*Kelvin-Helmholtz*) of the shear layer between the jet fluid and the ambient fluid (Ho and Huerre 1984). The instability occurs over a broad band of frequencies, and in a natural jet leads to a train of unequally spaced vortex rings. Interaction between the unequally spaced rings leads to vortex merging, the process by which structures of larger scale are generated and the layer grows in thickness. By forcing the jet acoustically, the vortex spacing can be made more uniform, delaying the formation of larger scales and thereby reducing the jet spreading rate and noise generation for some distance downstream (Hussain and Hasan 1985).

Open-loop active control is most useful for applications where the goal is to enhance the parent instability. For example, by forcing the jet with multiple frequencies and modes in a carefully-chosen systematic manner, the near-field ring vortices can be made slightly eccentric from one another. The rings then act to tilt one another and so they can be made to move in different directions. The jet can be split into two jets (bifurcating jet) or exploded into a shower of vortex rings (the blooming jet of Lee and Reynolds 1985). The spreading angle of the jet, normally about 15°, is thereby increased to as much as 70°, and the jet entrainment can be increased by as much as a factor of four (Juvet and Reynolds 1989). The phenomena are most easily controlled at low Re, but can be controlled at high Re with sufficiently strong excitation (Parekh and Reynolds 1989).

Tokumaru and Dimotakis 1991 used open-loop controlled oscillation of a cylinder to reduce large-scale shedding behind the cylinder. However, in most instances where the goal is to reduce the turbulence, some sort of *closed-loop active control* must be used in order to stabilize (fully or partially) the controlling instability. The basic idea is to introduce disturbances counter to those arising from the basic instability. This requires sensing of the disturbance introduced by the instability and feedback control of an actuator that introduces an opposing disturbance. Liepmann *et al* (1982) used feedback-controlled surface heaters to introduce cancelling disturbances into a transitioning boundary layer, thereby retarding transition to turbulence. Breuer *et al* (1989) used feedback-controlled wall motion for a similar demonstration in air. Ffowcs-Williams and Zhao (1989) used closed-loop active control for the cylinder wake flow.

Research on chaotic dynamical systems has led to a new framework for thinking about turbulence and its control. Much of the dynamical systems work deals with systems of relatively low dimension, and turbulence is known to have a relatively high dimension (Keefe *et al* 1992). However, there is evidence (Aubrey *et al* 1990) that the low-speed streaky structure in the near-wall region of a turbulent boundary layer (which is what must be controlled to control the boundary layer) can be described *locally* by a system of relatively low dimension (perhaps 10). The meandering of low-speed streaks is interpreted as hovering of the state of the flow near an unstable fixed point in the low-dimensional state space, and the bursting (the intermittent events that produce high wall stress) as a jump to a different unstable fixed point that occurs when the state has wandered too far from the first unstable fixed point.

From the dynamical systems viewpoint, the idea of control is to sense the current local state and through appropriate manipulation keep the state close to a given unstable fixed point, thereby preventing further production of turbulence. In the case of a boundary layer, preventing all bursting should lead to a relaminarized boundary layer; reducing the bursting frequency by 50% might yield a comparable reduction in the skin friction. In the case of a jet, relaminarization would lead to a quiet flow and very significant noise reduction.

The sensing and actuation must take place in regions where the flow is unstable. In the case of an aircraft jet engine exhaust this means everywhere downstream, which seems unfeasible. Therefore, it does not seem likely that closed-loop feedback control will find much application in free shear flows, except perhaps in confined flows (*e.g.* combustors) where optical sensing and actuation throughout the flow may be practical.

However, in a turbulent boundary layer the controlling instabilities are in the near-wall region. Numerical experiments indicate that sensing and actuating at the wall could lead to significant drag reductions (Choi *et al* 1992). Microelectromechanical systems (MEMS) technology offers new methods for sensing and manipulating turbulent boundary layers at the surface. Thus, there is a demonstrated possibility for controlling turbulent boundary layers and an emerging technology for doing it. Turbulent boundary layer control offers a very high potential payoff in terms of aerospace industry competitiveness and defense capability. In what follows we focus on the possible methods, challenges, and opportunities for turbulent boundary layer control.

ACTIVE TURBULENT BOUNDARY LAYER CONTROL

In control one needs a dynamical model of the "plant" to be controlled. For boundary layer control the plant model would be a low-dimensional set of coupled nonlinear ordinary differential equations arising from an approximate analysis of the local near-wall flow (Lumley 1991). The equations are essentially for the time-dependent complex amplitudes (amplitude and phase) of the basis functions used to represent the local near-wall flow. These functions are also parameterized by the local near-wall length scale and flow direction, which of course depend upon the current local boundary layer. See Berkooz et al 1993 for the formal theory of these basis functions.

The sensing would need to identify both the parameters and amplitudes in the near-wall model for effective model-based control. Decisions on the next control action would then require computations based on the model. Sensing might be based on surface pressure measurements or on the velocity gradient at the wall inferred from heat transfer measurements. Actuation would require some modification of the surface boundary condition, such as surface distortion, temperature perturbation, or transpiration.

The process described above would have to be carried out independently over a significant portion of the surface fast enough to prevent bursting. Under typical aircraft cruise conditions the wall unit scale ν/u^* is of the order of $2.6\mu m$ (Gad-el-Hak 1993). The low-speed streaks vary in size, with the smallest being approximately 20 wall units wide and 400 units long. Therefore, the spanwise spacing of the control elements must be of the order of $50\mu m$. If one could be sure of the downstream direction, the streamwise spacing could be as much as $1000\mu m$, but since the direction is likely to change significantly with flight conditions a streamwise spacing of perhaps $100\mu m$ would be required.

Therefore, the surface would be tiled with integrated sensor-actuator-controller tiles approximately $50\mu m \times 100\mu m$, corresponding to 20,000 sites/cm^2. This tile density is well within the range of current silicon fabrications in microelectonics and MEMS technology. However, communication between distant tiles would be a problem. Therefore, each tile probably would be able to share information only with its immediate neighbors, and all computing would have to be done locally by each tile in its own integrated processor. Research on systems integration of massively repeated arrays will be important to these developments.

To make one adjustment per local burst cycle, which at the conditions used above occur locally on average once every $56\mu sec$, each processor would have to do all its analysis and actuation within about $10-20\mu sec$. A $50MH_z$ integrated microprocessor could do at most about a hundred simple computations for each actutation decision. Thus, research on control methodologies requiring a much lower computational intensity is needed.

A possible alternative to model based control would be *adaptive neural net control*. A neural net is a simple computational algorithm that predicits the next state of a system as a weighted non-linear function of its current and past states. The weights are determined adaptively using the history of prediction error. Our preliminary experiments with simple neural nets in a model wall-region flow (Jacobson and Reynolds 1993) suggest that a neural net approach has promise in boundary layer control. The hope is that the underlying non-linear dynamical system hovers long enough near a fixed point that the adaptive network can get good information for adjusting the weights. In essence the neural net would replace the physics-based dynamic model of model-based control. Research on neural nets leading to new design procedures for weakly linked local networks could provide important contributions.

In the boundary layer control application, the network should be designed so that the weights would play the role of the parameters of the basis functions in model-based control, so that the weights would change on the time scale of the local flight conditions. This would allow the weights to be continually adapted over many burst cycles in each local control tile. Only a few predictions per burst should be required. A basic theory of neural nets that would enable the designer to configure the net so that the weights were determined by the basis function parameters rather than their current amplitudes and phases would be an important research contribution. An adaptive neural net can be implemented as a relatively simple ASIC (Applications Specific Integrated Circuit) once the net design has been chosen. ASIC design tools for integrated MEMS/neural net systems are needed.

Regardless of what control strategy is followed (dynamical systems, neural net, or other), sensing is a crucial problem. MEMS technology for surface pressure transducers is very advanced, but since pressure is a non-local feature of the flow it is not likely to be useful in local wall-layer control. Hot strips on the surface seem more promising, for which the heat transfer rate to the fluid is a function of the mean velocity gradient normal to the surface. Silicon and other MEMS materials have high thermal conductivities, and at small scales the heat transfer to the solid substrate may dwarf the heat transfer to the fluid unless the sensor can be thermally guarded. The thermal wake of one sensor will contaminate the signal of other sensors. Research into ways to build accurate high-density arrays of thermal wall stress sensors is needed. See Najafi (1991) for a useful overview of current MEMS sensing technology.

Actuation is also crucial. Lumley (1991) has argued that controlled insertion of a Gaussian hill of the order of 10 wall units ($26\mu m$ for the situation above) would be sufficient for wall-layer control. This sort of protuberance would produce a pair of streamwise vortices in the near-wall region having common flow towards the wall, opposing the motion of low-speed fluid away from the wall that is recognized as a precursor of a burst. Piezoelectric actuation has been suggested, but there is not at present a piezoelectric-in-silicon technology that can produce this deflection at the scales and densities indicated above. Moreover, a $26\mu m$ bump $50\mu m$ in diameter represents a very high deformation of the surface, and it is doubtful that thin silicon membranes could take this repeated stress. Research in developing highly deformable MEMS materials could be extremely important. Thermal methods may provide an attractive method for generating the high pressures required to deform the surface in a MEMS actuator (Muntz et al 1992). Research on a variety of MEMS actuation methods is encouraged.

We have experimented with 2 mm vibrating surface elements in a low-speed water flow, and have developed an actuator that injects low momentum fluid into the wall region on demand, taking the fluid in diffusely from nearby. Our program explores actuator concepts in very slow three-dimensional laminar flows where the streaky structure scale is of the same order as the width of the actuator, and emphasizes actuator concepts that could be built in arrays at much smaller scale using MEMS technology. The actuator shoots low momentum fluid into the boundary layer, forming a pair of streamwise vorticies with common flow away from the wall. The idea is to use this injected flow to oppose the motion of high-speed fluid towards the wall, the event that creates the high skin friction. However, the actuator presents the flow with a discontinuous surface, for which clogging and fragility are likely to be problems. Research on better means for actuation is clearly needed. Experiments in large-scale flows will be helpful in identifying approaches to be explored at smaller scales using MEMS technology.

The emergence of MEMS technology has opened the possibility of turbulent boundary layer control, but present MEMS is far from adequate and there are great opportunities in MEMS research. Mounted on the external surface of the aircraft, the MEMS controllers would have to withstand air speeds of 300 m/sec, temperatures of $-40°$, dust and rain, wing flexure and vibration, fuel spills, deicing, and the shoes of aircraft service personnel. This will not be possible without the development of very robust MEMS technology. Research leading to MEMS on rugged flexible substrates will be essential to the realization of turbulent boundary layer control. Current MEMS fabrication is done in the batch mode. The development of continuous processes that could turn out MEMS-tiled sheet material at low cost would have extremely high payoff in this and many other applications.

In addition to the research needs and opportunities in hardware discussed above, there is need for more and better theory. The foundations of optimal control theory as applied to the equations governing fluid flow were only recently developed (Abergel and Temam 1990). The theory for non-linear control of continuous chaotic systems is in its infancy (Keefe 1993). Neural net control theory is virtually non-existent. Basic research in the fundamental control theory that would enable turbulent boundary layer control could make a very significant contribution.

REFERENCES

Abergel, F. & Temam, R. (1990) On some control problems in fluid mechanics *Theoret. Comput. Fluid Dynamics*, **1** 303.

Aubry, N., Holmes, P., Lumley, J. L. & Stone, E. (1988) The dynamics of coherent structures in the wall region of a turbulent boundary layer *J. Fluid Mech.*, **192** 115-173.

Berkooz, G., Holmes, P. & Lumley, J. L. (1993) The proper orthogonal decomposition in the analysis of turbulent flows *Ann. Rev. Fluid Mech.*, **25** 539-575.

Berman, N. S. (1978) Drag reduction by polymers *Ann. Rev. Fluid Mech.*, **10** 47-64.

Breuer, K. S., Haritonidis, J. H. & Landahl, M. T. (1989) The control of transient disturbances in a flat plate boundary layer through active wall motion *Phys. Fluids A*, **1** 574-582.

Bushnell, D. M. & McGinley, C. B. (1989) Turbulence control in wall flows *Ann. Rev. Fluid Mech.*, **21** 1-20.

Cantwell, B. J. (1981) Organized motion in turbulent flows *Ann. Rev. Fluid Mech.*, **13** 457-515.

Choi, H., Moin, P. & Kim, J. (1992) Turbulent drag reduction: studies of feedback control and flow over riblets *Report TF-55, Thermosciences Div., Dept. of Mech. Engrg.* Stanford.

Ffowcs-Williams, J. E. & Zhao, B. C. (1989) The active control of vortex shedding *J. Fluids Struct.*, **3** 115-122.

Gad-el-Hak, M. (1993) Innovative control of turbulent flows *AYAH Paper 93-3268* AYAH Shear Flow Conference, Orlando, July 6-9.

Hill, D. C. (1992) A theoretical approach for analyzing the restabilization of wakes *AIAA Paper 92-0067* Reno.

Ho, C-M & Gutmark, E. (1987) Vortex induction and mass entrainment in a small-aspect ratio elliptic jet *J. Fluid Mech.*, **179** 385-405.

Huerre, P. (1990) Local and global instabilities in spatially developing flows *Ann. Rev. Fluid Mech.*, **22** 473-537.

Huerre, P. & Monkewitz, P. A. (1985) Absolute and convective instabilities in free shear layers *J. Fluid Mech.*, **159** 369-383.

Husain, H. S. & Hussain, F. (1989) Elliptic Jets; Part 1. Characteristics of unexcited and excited jets *J. Fluid Mech.*, **208** 257-320.

Husain, H. S. & Hussain, F. (1991) Elliptic jets; Part 2. Dynamics of coherent structures: pairing *J. Fluid Mech.*, **233** 439-482.

Hussain, A. K. M. F. & Hasan, M. A. Z. (1985) Turbulence suppression in free turbulent shear flows under controlled excitation. Part 2. Jet-noise reduction *J. Fluid Mech.*, **150** 159-168.

Jacobson, S. A. & Reynolds, W. C. (1993) Active control of boundary layer wall stress using self learning neural networks *AIAA Paper 93-3272* AIAA Shear Flow Conference, Orlando, July 6-9.

Juvet, P. & Reynolds, W. C. (1989) Entrainment control in an acoustically controlled shrouded jet *AIAA Paper 89-0969* AIAA Shear Flow Conference, Tempe.

Keefe, L. (1993) Two nonlinear control schemes contrasted on a hydrodynamiclike model *Phys. Fluids A*, **5** 931-947.

Keefe, L., Moin, P. & Kim, J. (1992) The dimension of attractors underlying periodic turbulent Poiseuille flow *J. Fluid Mech.*, **242** 1-30.

Lee, M. & Reynolds, W. C. (1985) Bifurcating and blooming jets *Report TR-22, Thermosciences Div., Dept. of Mech. Engrg*, Stanford.

Liepmann, H. W. & Nosenchuck, D. M. (1982) Active control of laminar-turbulent transition *J. Fluid Mech.*, **118** 201-204.

Lumley, J. L. (1991) Control of the wall region of a turbulent boundary layer *Turbulence Structure and Control* AFOSR Workshop, Ohio State.

Muntz, E. P., Shiflett, G. R., Erwin, D. A. & Kunc, J. A. (1992) Transient energy-release pressure-driven microdevices *J. Microelectromechanical Systems*, **1** 155-163.

Najafi, K. (1991) Smart sensors *J. Micromech. Microeng.*, **1** 86-102.

Parekh, D. & Reynolds, W. C. (1989) Forced instability modes in a round jet at high Reynolds numbers *Phys. Fluids A*, **1** p 1447.

Robinson, S. K. (1991) Coherent motions in the turbulent boundary layer *Ann. Rev. Fluid Mech.*, **23** 601-639.

Reisenthel, P., Xiong, Y. & Nagib, H. M. (1991) The preferred mode in an axisymmetric jet with and without enhanced feedback *AIAA paper 91-0315* 29th Aerospace Sciences Meeting, Reno.

Strykowski, P. J. & Niccum, D. L. (1991) The stability of countercurrent mixing layers in circular jets *J. Fluid Mech.*, **227** 309-343.

Strykowski, P. J. & Sreenivasan, K. R. (1990) On the formation and suppression of vortex 'shedding' at low Reynolds numbers *J. Fluid Mech.*, **218** 71-107.

Tennekes, H, & Lumley, J. L. (1972) A first course in turbulence *MIT Press* Cambridge, Mass.

Tokumaru, P. T. & Dimotakis, P. E. (1991) Rotary oscillation control of a cylinder wake *J. Fluid Mech.*, **224** 77-90.

FRACTAL GEOMETRY AND MULTIFRACTAL MEASURES IN FLUID MECHANICS

Katepalli R. Sreenivasan
Mason Laboratory
Yale University
New Haven, CT 06520-2159

Philosophy is written in this grand book, the universe, which stands continually open to our gaze. But the book cannot be understood unless one first learns to comprehend the language and read the letters in which it is composed. It is written in the language of mathematics, and its characters are triangles, circles and other geometric figures without which it is humanly impossible to understand a single word of it; without these, one wanders about in a dark labyrinth.

INTRODUCTION

Galileo Galilei was merely expressing the sentiment that had reigned supreme in the days of the ancient Greeks, in whose world truth and beauty were sought in terms of composites of perfect Euclidean solids. It is a ceaseless wonder that the scientific enterprise, which is based on such perfect and abstract forms, has the impact it does on human civilization. As Mandelbrot (1982) remarks, 'Clouds are not spheres, mountains are not cones, coastlines are not circles, and the bark is not smooth, nor does the lightning travel in a straight line... Nature exhibits not simply a higher degree but an altogether different level of complexity.' Modern science has embraced the view that irregularity is a legitimate subject for study; that, with this study, comes a deeper understanding of Nature; and that, with this understanding (and some ingenuity) come useful applications and, perhaps, also the betterment of human society.

Many seemingly irregular objects possess some degree of order. While the intrinsic reason for this order is not always clear, the claim (Mandelbrot 1982) is that notions of fractal geometry and multifractal measures are central to discerning the order in a large class of systems. Nature is richer in its many manifestations than can be gift-wrapped conveniently in the garb of fractals and multifractals, but there is little doubt that they provide powerful tools for describing, often understanding, aspects of empirical experience that were not quantifiable before.

The concepts basic to fractal geometry are relatively simple; in their origin, they are not all that new either. But the recognition that these simple notions form a unified language for a variety of disciplines in natural science is due to Mandelbrot (1982 and earlier).

Research into fractals and multifractals has proceeded roughly in three related directions: (i) the characterization of fractal-like objects by experiment and numerical simulation; (ii) the search for a theoretical understanding of the physical mechanisms producing the observed fractal features; (iii) the prediction of useful physical quantities on the basis of fractal geometry. To these, one must also add the effort at the level of pure mathematics.

Our objective is to assess briefly the role of fractals and multifractals in a broad class of fluids-related problems. A projection for the near future is an intended purpose of the report. However, since such an attempt would be devoid of substance without some discussion of the current status, the latter will form the bulk of the report. As applications of fractals in fluid dynamics have yet to mature, the report captures a snap-shot of the changing scene. We focus on activities that are common to both fluid dynamics and fractals, and ignore isolated topics that are likely to be covered elsewhere (such as fractal structure in chaotic mixing). We restrict attention to how fractals enter these physical problems, rather than describe classical results.

Since fractals (especially multifractals) are not likely to be familiar to all potential readers, an informal review is included (sections 2 and 3). Much of the material to be covered below can be found elsewhere (Mandelbrot 1982, Meakin 1987, Feder 1988, Avnir 1989, Vicsek 1989, Sreenivasan 1991, Schertzer and Lovejoy 1991). Other references cited are not meant to be exhaustive.

BASIC NOTIONS CONCERNING FRACTALS

Self-similarity and self-affinity

Two basic notions are central to fractals. The first one relates to self-similarity and self-affinity. In rough terms, an object is self-similar if it is composed of smaller pieces, each of which is a replica of the whole. Each small piece can be obtained from the original whole by a similarity transformation, or a contraction which reduces the original object *by the same scale factor in all coordinates*. An affine transformation is one in which different coordinates are contracted by different factors. Objects with parts that are affine copies of the whole are called self-affine. Note that both similarity and affinity are linear transformations.

Non-integer dimensions

The second important notion is that the dimension of an object need not be an integer. We intuitively understand that a point has the dimension zero, a straight line has the dimension one, a square has the dimension two, a cube has the dimension three, and so forth. A fractal curve, because of its infinite detail, is more space-filling than an Euclidean curve. It therefore seems natural to associate with it a dimension greater than unity. In general, fractal objects have dimensions which exceed their topological dimensions.

Dimensions of self-similar objects

For fractal curves, the question of what constitutes their length—and likewise, in other cases, the area or volume—is ill-posed. Length estimates for a classical curve, made for instance by walking along its length with increasingly finer resolution, will converge to a finite value—which we may call its true length. This is not so for fractal curves. In general, the estimates for the length, area or volume of fractal objects increase as one measures them with better refinement. Further, the two sets of quantities—length, area, and volume on the one hand, and the resolution on the other—do not vary arbitrarily but are related by power laws. For example, the length estimate $L(r)$ of a fractal curve varies with the resolution r as $L(r) \sim r^{1-D}$, where D is the dimension of the curve; D is clearly equal to unity for classical curves (in the limit of small r), and is larger than unity for fractal curves.

In practice, for any physical phenomena, there are cut-offs at some scales on either end, called the inner and outer cut-offs, beyond which power-law relations do not hold. The cut-off scales contain much physics, and play an important role in the interpretation of fractal scaling.

There are many definitions of dimensions (Hausdorff dimension, similarity dimension, box-dimension, compass dimension, capacity dimension, information dimension, correlation dimension, and so forth). This situation can be quite confusing for a beginner; a reference to Mandelbrot (1982), Farmer et al. (1983), Falconer (1985), Feder (1988) and Avnir (1989) is quite useful. The various dimensions are related by equalities in some cases and inequalities in others. There is no simple guide as to which dimension is the most suitable in a given context: sometimes one is easier to measure, perhaps also more appropriate, than the others. Note that the definition of dimensions is by no means restricted to self-similar objects.

For later convenience, we briefly define the box-dimension here. The box-counting method has the advantage of being easily programmed on a computer. Consider a fractal object residing in a d-dimensional Euclidean space. Divide the space into d-dimensional boxes of a certain size and count the number of boxes containing the object. Repeat the procedure by successively reducing the box size. Let $N(r)$ be the minimal number of boxes of size r which contain the object. The box-dimension is defined as $D_b = logN(r)/log(1/r)$. Although the dimension is defined in the limit of $r \to 0$, in practice, there is often a range of scales for which D_b is a constant and obtained from the (constant) slope in the log-log plot. A straight line in such log-log plots is often a sign of statistical (rather than exact) self-similarity.

Time records

Consider a trace of turbulent velocity, say, measured as a function of time. Since the two quantities (velocity and time) are independent of each other, the coordinate axes can be stretched (or contracted) independently. As we shall see presently, such objects are not candidates for self-similarity, but for self-affinity.

For illustrative purposes, consider the so-called fractional Brownian motion (fBm), which is a generalization of the standard Brownian motion (Mandelbrot & Van Ness 1968). The fractional Brownian motion $X(t)$ is a single valued stationary function of the time variable t, such that its increments $\Delta X(t) = X(t + t) - X(t)$ have a Gaussian distribution with the variance $\langle [\Delta X(t)]^2 \rangle \sim t^{2H}$. The case $H = \frac{1}{2}$ corresponds to the classical Brownian motion.

If the time axis of an fBm trace is stretched excessively, the trace looks smooth on all scales and intuition suggests (and detailed calculations confirm) that the dimension of this highly stretched trace is unity. This is called the global dimension of the trace. On the other hand, if the time axis is compressed excessively, the signal will look rough on all scales and possesses a dimension greater than unity. This dimension, called the latent dimension, can be shown to be either 2 or $\frac{1}{H}$, whichever is smaller. For a range of stretching factors such that the vertical and horizontal scales are of the "same order", one can associate a non-trivial (local) dimension D with the trace. It can be shown that $D = 2 - H$. (Since D for a graph can lie only between 1 and 2, it follows that $0 < H < 1$.)

The notion of the dimension for a self-affine graph is not as straightforward as that for self-similar objects, and its measurement requires a good understanding of various pitfalls. The lack of appreciation of this elementary fact has occasionally resulted in false claims.

A related useful note: The exponent H appears also in the analysis of time records with memory. The auto-correlation function of the increments of the classical Brownian motion is a delta-function centered at zero (that is, it has no memory). On the other hand, processes for which $H > \frac{1}{2}$ possess a positive correlation that extends indefinitely (in practice, up to some cut-off scale); such processes are persistent in that an increasing trend in the present is followed, on the average, by a tendency to increase also in the future. Hurst (1951) showed that such phenomena are preponderant (floods, rainfall, sunspot numbers, and other phenomena unrelated to fluid dynamics). When $H < \frac{1}{2}$, the tendency is for the trace to show a decreasing tendency in the future when the present trend is to increase (anti-persistence).

MULTIFRACTALS

The notion central to multifractals is that of a singular measure (e.g., Evertsz & Mandelbrot 1992). It is enough here to think of a measure as a positive definite and additive quantity distributed on an interval (or set). By singular measure, we mean that the measure's density is everywhere either undefined or zero. For the formalism of multifractals to be useful, it must however be applicable to near-singular measures, which are singularities smoothed out by some physical mechanism (such as viscosity). This is indeed the case.

As an example of near-singular measures, consider the turbulent energy dissipation rate distributed on a line at high Reynolds numbers. Figure 1 shows a typical example. The signal is highly intermittent; a few of the peaks are several hundred times larger than the mean of the signal. The higher the Reynolds number, the stronger the peaks; in the limit of infinite Reynolds number, the peaks can be thought to be infinitely large and truly singular. At all large but finite Reynolds numbers, the peaks in the trace (if resolved adequately) can be fitted by smoothed algebraic singularities. It is clear from inspection that the first few moments (such as the mean, variance, ...) will not provide adequate information about such near-singular measures; many high-order moments, or the entire probability density (PDF) itself, will be needed. Unfortunately, the $PDFs$ depend strongly on the Reynolds number. The point, however, is that the mechanism that produces these intermittent distributions (namely the energy cascade—which itself is a partial abstraction of the vortex stretching process producing small scales of turbulence) is thought to be Reynolds-number-independent (as long as the Reynolds number is 'sufficiently high'). If this is true, one needs a statistical quantity which sheds light on this Reynolds-number-independent process. Multifractals provide tools for characterizing such singular measures in Reynolds-number-independent ways.

For another intuitive means of introducing multifractals consider the following questions about figure 1. Over what fraction of the line interval does dissipation occur? What is the dimension of the set supporting the dissipation? Since some dissipation occurs everywhere, the trivial answers are, respectively, 'everywhere' and 'unity'. These are not useful answers, however, and researchers have traditionally proceeded by setting a threshold to

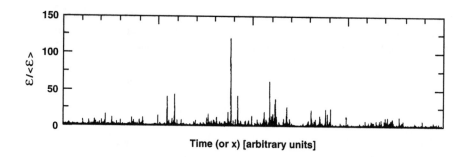

Figure 1: A typical time trace of a representative component of the energy dissipation rate $\epsilon \sim (\partial u/\partial t)^2$ normalized by its mean. It is conventional to interpret this as a spatial cut in the direction of mean motion, x, so that the abscissae may be taken as a line cut in the x-direction. The data were obtained in the atmospheric surface layer about 6 m above the ground. The so-called microscale Reynolds number was about 2500.

distinguish significant levels of dissipation from insignificant levels. Unfortunately, the choice of this threshold is quite subjective and influences answers strongly (Sreenivasan 1985). In any case, it is not enough to seek information about the oversimplified binary picture of whether or not there is dissipation, but it is essential to know something about its different magnitudes. Multifractals provide proper tools for handling these questions. Roughly, a given level of activity is characterized by a fixed value of the Hölder exponent (see below), and information is sought about the dimension of the set supporting this given level of activity.

Formally, if the measure (dissipation) in a box of size r centered at a position x on a line interval varies with r as $r^{\alpha(x)}$, the measure *density* can be written as $\mu' \sim r^{\alpha-1}$. This is singular in the sense that, as $r \to 0$, the measure density μ' diverges for all values of the Hölder exponent $\alpha < 1$. The smaller the α, the larger the strength of this singularity. Roughly, one can associate (Frisch & Parisi 1985, Halsey et al. 1986) with every iso-α region a fractal dimension $f(\alpha)$. Since α can vary continuously between α_{min} and α_{max}, so could $f(\alpha)$. The curve of $f(\alpha)$ *versus* α, called the $f(\alpha)$ curve or the multifractal spectrum, thus represents an infinity of dimensions, each of which corresponds to the set supporting singularities of a given strength α.[1]

While this description is intuitively helpful, it does not provide a convenient scheme for obtaining from experimental data a converged $f(\alpha)$ curve.

---

[1]The multifractal formalism of Frisch & Parisi (1985) does not deal with measures but with velocity increments which have the property that the sum of velocity increments over two adjacent subintervals equals that over the whole interval.

For this purpose, an equivalent description (Hentschel & Procaccia 1983) in terms of the so-called generalized dimensions, D_q, is more useful. It has been shown (Frisch & Parisi 1985, Halsey et al. 1986) that the pair of variables (τ, q), where $\tau = (q - 1)D_q$ and the real number q lies between $-\infty$ and $+\infty$, are Legendre transforms of the pair (f, α). By evaluating generalized dimensions and Legendre transforming them, one can obtain the $f(\alpha)$ curve. This aspect will not be discussed further.

One can also measure the $f(\alpha)$ curve directly (Chhabra & Jensen 1989) by noting a formal connection that exists between multifractals and thermo-dynamics of statistical mechanical systems (Feigenbaum 1987).[2]

More basic to multifractals than the $f(\alpha)$ curve is the self-similarity of the underlying multiplicative process that generates the multifractal (Mandelbrot 1989, Chhabra & Sreenivasan 1992, Evertsz & Mandelbrot 1992). To understand this, let us start with a measure uniformly distributed on unit interval, divide it into a specified number of subintervals and redistribute the measure unequally on them. The redistribution is obtained by multiplying the parent measure by 'multipliers' taken from a certain PDF ('multiplier distribution') defined on the interval. Imagine that this process continues indefinitely. If the multiplier distribution is the same at each stage of refine-ment, one obtains a statistically self-similar multifractal measure (Novikov 1990, Sreenivasan & Stolovitzky 1995).

We now return to some applications.

AGGREGATON IN PARTICLE-LADEN FLOWS

Fluid dynamics of particle-laden flows is replete with applications. If interparticle interactions are ignored, the essential problem is one of under-standing how various physical properties of the flow (such as effective vis-cosity) are altered by particle loading. Alternatively, the effect of the fluid flow on the particle motion is also of interest. Under certain circumstances, often dictated by hydrodynamics, inter-particle interactions may become im-portant and lead to the formation of aggregates (that is, structures in which particles stick together irreversibly).

There are two basic aspects to the study of aggregation: kinetics and geometry. Kinetics involve the quantitative description of the time evolution of the aggregates and their size distribution, whereas geometry is concerned with a quantitative description of the structure of the aggregates. The first aspect has been studied for a long time, whereas the second aspect, which

---

[2]The quantities f, α, τ and q are analogous, respectively, to entropy, internal energy, free energy and inverse temperature. This means that, if small scale turbulence (see section 7) is treated in some rough sense as a statistical mechanical system, the measurement of these multifractal quantities is equivalent to measuring its thermodynamic properties.

used to be the backwaters of aggregation studies until recently, has taken on a life of its own since the advent of fractals.

Aggregation can occur in a variety of ways: electrodeposition (induced by electric field), sedimentation (induced by gravity), filtration (caused by the particle motion stopped by small pores), and so forth, and can be either of the particle-cluster type or cluster-cluster type. A simple kinetic model (Witten & Sander 1981) called the diffusion-limited-aggregation (DLA) has been studied extensively and thought of (with some minor modifications) as a paradigm for a number of processes such as protein aggregation, colloid clusters of gold and silica, soot formation, viscous fingering in porous media (at least when the flow rates are high), dielectric breakdown, dendritic solidification, and so forth. The common feature among many of these phenomena is that a suitably defined potential governed by the Laplace equation can be defined. It is therefore worth examining DLA briefly.[3]

In the two-dimensional version of DLA, one considers a lattice on a plane and first chooses the origin for the cluster. A particle, the 'seed' for the aggregate, is placed at the origin. One then considers a large circle of radius R, centered at the origin, and chooses a point at random on this circle. A particle is released at a site nearest to this point and is allowed to execute a random walk on the lattice. If the random walker reaches a site nearest to the origin, it stops and stays stuck to the seed. (If the particle exits the circle without getting close to the origin, it is abandoned.) Another particle is released from a lattice point close to another randomly chosen point on the circle, and allowed to stick to the seed or the two-particle cluster (as the case may be). The process is continued until a cluster of the desired size is reached. The algorithm can be extended to any higher dimension.

The DLA aggregates are fractal structures with a dimension of about 1.7 in two dimensions and 2.5 in three dimensions (Meakin 1987, Argoul et al. 1988). The growth of the cluster is governed by the so-called harmonic measure which is the probability that a random walker approaching the cluster from the far-away circle hits the cluster in a certain infinitesimal interval along the boundary. The harmonic measure is a solution of Laplace's equation for the electrostatic potential when the cluster boundary is at zero potential and the circle is at a potential of unity. It is intermittent in appearance and amenable to multifractal analysis (Amitrano et al. 1986, Mandelbrot & Evertsz, 1990). As remarked in section 3, the basic evidence for the self-similarity in the DLA structure comes from the self-similarity of the multiplicative process (Evertsz & Mandelbrot 1992); it has also been pointed out that $\tau(q)$ does not exist for negative values of q.

---

[3]Just as the Ising model describes the essential physics of a wide variety of materials near the critical point, the hope has been expressed that DLA would describe a variety of *growth* processes.

Most fractal and multifractal characteristics of DLA have been extracted from computer simulations; to our knowledge, there are no exact analytical results.

APPLICATIONS IN POROUS MEDIA AND VISCOUS FINGERING

Porous media

The Stokes equation governing the fluid motion in porous media is linear. It can be reduced to the Darcy law (according to which the velocity is linearly proportional to the pressure drop) by using appropriate assumptions of homogeneity. Equivalently, this can be written as the Laplace equation for the pressure. This suggests analogies with DLA, except that the boundary conditions in the porous media are more difficult to assess.

The difficulty lies in describing the boundary between fluid-filled regions and solid-like regions. One procedure is to model porous media as a network of capillary tubes. This model conceives of the solid phase as being continuous with interconnected fluid-filled pores running through it. A major simplification is achieved in this way because the Poiseuille law valid for laminar flow through pipes holds for each capillary. Alternatively, one assumes the fluid phase as continuous and solid particles as obstacles for the flow.

Fractals appear in studies of flow through porous media because of the random character and the extreme variety of shapes encountered. The pore space, the solid phase and the solid-pore interface could all exhibit fractal scaling.[4] Various estimates for the fractal dimensions of these three aspects have been made (Feder 1988, Avnir 1989).

In practical applications, the one unknown is the permeability of the medium—which, in general, is a tensorial quantity. The permeability in most cases is measured by pressure drop experiments or estimated empirically. A useful goal would be to relate the permeability of a medium to its characteristic fractal dimensions. This has not yet been accomplished.

Viscous fingering in the Hele-Shaw cell

When a low viscosity fluid is pushed into a high viscosity immiscible fluid, Saffman-Taylor fingers develop (Saffman & Taylor 1958); these fingers occur singly and are broad and smooth in shape. The equation governing viscous fingering in the Hele-Shaw cell is formally the same as that for flow through porous media, except that: (a) the permeability in the former is not real but related to the gap between the plates, while that in the latter depends on

---

[4]All porous media may not have fractal pore space, but the surface of the grains is very often fractal due to long-term chemical or sintering processes.

the local volume fraction; (b) the former has a well-defined surface tension at a normal fluid-fluid interface, while the use of surface tension in the latter is rather murky. The Saffman-Taylor fingers correspond to the wavenumber with maximum instability, and their width varies as the square-root of the surface tension between the two fluids (if all other conditions are held fixed). As the surface tension is lowered, the fingers split more and more; but there is a practical limit to how low the surface tension can get. If the high-viscosity fluid is a miscible colloidal solution with shear-dependent viscosity, the interface grows to be fractal-like in appearance (Nittmann et al. 1990) even when the capillary number[5] is moderately high. It appears that the more non-Newtonian the solution, the more the tendency to fractal fingering. In spite of recent studies to model this behavior (de Gennes 1987, Wilson 1990), the basic physics of fingering in non-Newtonian fluids is not well-understood.

It has been argued (Patterson 1984) that, if interfacial tension is ignored[6], the viscous fingering problem is analogous to DLA; indeed, the viscous fingering patterns in radial Hele-Shaw cells (which do not have the anisotropic constraining effects of rectangular cells) are similar to the DLA structure and possess roughly the same fractal dimension. The growth of viscous fingers depends on the pressure gradient, which therefore plays a role analogous to the harmonic measure for DLA. Since the pressure field is not easily measured, related growth measures have been defined and characterized by multifractals (Feder 1988). If the medium in the Hele-Shaw cell is porous, one obtains fractal structure even for Newtonian fluids and the capillary number is moderately high. For a review of viscous fingering, see Homsy (1987).

Percolation and diffusion

Percolation involves the spreading of a fluid in a random medium, where the words 'fluid' and 'medium' are used in a general sense. The role of randomness in percolation is quite different from that in diffusion. In the latter, the Brownian particle executes a random walk, whereas percolation deals with randomness that is frozen into the medium. While any position in the medium can be reached by diffusion, the spreading in percolation is confined to finite regions except when the so-called 'percolation threshold' is exceeded. In model studies, one considers a square lattice whose sites are either randomly occupied (with probability p) or empty (with probability $1 - p$). Occupied sites represent parts of the medium through which the fluid can percolate while empty sites represent those parts that cannot be

---

[5]The capillary number is the ratio $\mu U/\sigma$, where μ, U and σ are, respectively, the viscosity coefficient for the driven fluid, fluid velocity and interfacial tension.

[6]In practice, this is far from being correct. The arguments postulating similarity between viscous fingering and DLA must be examined critically in spite of the resemblance of the observed fractal patterns.

invaded by the fluid. Connected sites form clusters. On an infinite sample, all clusters remain finite below the percolation threshold, $p = p_c$. For $p > p_c$, infinite clusters appear with a finite probability, and the probability of this occurrence varies with p as $(p - p_c)^\beta$. Percolation clusters are self-similar and possess, in the limit of large clusters, a fractal dimension of 1.89 (Aharony 1986). The 'hull' of the diffusion front is also a fractal with a dimension of 1.75, whereas its external perimeter has a dimension of about 1.37 (Sapoval et al. 1985).

One should also mention here the 'invasion percolation' (applicable for low flow rates) where the water displacing oil in porous rocks may trap regions of oil (Wilkinson & Willemsen 1983). Randomness encountered by the invading fluid would now also depend on the trapped regions.

Branched polymers have size distributions that are self-similar (Daoud & Martin 1989), and are therefore candidates for fractal description. Some useful analogies exist between polymers and percolation studies (Havlin & Avraham 1987). As in percolation, the fractal dimension of branched polymers can be related to other indices characteristic of the polymer size above a 'percolation threshold'.

ONSET OF CHAOS IN NEWTONIAN FLUID FLOWS

Multifractals have played a powerful role in characterizing universality at the onset of chaos in low-dimensional systems. The renormalization theory has been worked out for the onset of chaos for period-doubling (Feigenbaum 1978, 1979) and quasiperiodic cases (Feigenbaum et al. 1982, Ostlund et al. 1983). Experiments in forced Rayleigh-Benard convection (Stavans et al. 1985, Jensen et al. 1985) and the near-field of oscillating cylinders at low Reynolds numbers (Olinger & Sreenivasan 1988) strongly support the universality theory: the $f(\alpha)$ curve describing the non-uniform distribution of the invariant measure on the attractor at the onset of chaos agrees well with that calculated for one-dimensional circle maps. This is the power of universality, and illustrates an application of multifractals where powerful theory and imaginative experiments have come together satisfactorily.

As already mentioned, the $f(\alpha)$ curve provides a thermodynamic—and therefore a degenerate—description of the dynamical system. Even so, it has been possible to develop (Feigenbaum et al. 1986) a basis for extracting the multiplicative process leading to the observed multifractal state (up to a level of detail that depends on our knowledge of the system in terms of other statistical measures—such as the similarity structure exhibited by the power spectral density).

It should be noted that dynamical universality was indeed known before the multifractal formalism came to the fore. For example, the Feigenbaum number in period doubling bifurcations was experimentally observed in convection experiments (Libchaber & Maurer 1980). Furthermore, it was also known that the microscopic information about a deterministic dynamical system and its scaling properties could be characterized in detail by the so-called scaling function (Feigenmaum 1979). However, the experimental measurement of the scaling function is quite difficult, and the advent of multifractals (*albeit* statistical) made the search for universality significantly easier.

We wish to emphasize that chaos (which involves temporal complexity) is quite different from turbulence (which involves spatial as well as temporal complexity); however, transition to chaos is often relevant to early stages of transition to turbulence.

NON-REACTING AND REACTING TURBULENT FLOWS

High-Reynolds-number turbulence consists of a wide range of dynamically interacting scales. The ratio of the largest to the smallest scale increases roughly as the $\frac{3}{4}$-ths power of the large-scale Reynolds number. The conventional wisdom is that statistical similarity prevails over a range of intermediate scales; the precise form of this similarity and the scale-range over which it holds are matters of much interest. It has been thought that fractals and multifractals provide proper tools for better description, and hence better understanding, of aspects of turbulence. In the following summary statements, we indicate by C the results that have been repeated by more than one investigator (although counter examples may exist), or by P the results which come essentially from one laboratory. For references to original sources and further discussions, see Sreenivasan (1991).

Fractal scaling of flames, iso-surfaces and interfaces

The main question here is whether these objects can be treated as thin surfaces with many self-similar convolutions. It is useful to quote a few results for this paradigm problem involving fractals.

(a) For the scalar interface (i.e., outer boundary of scalar-marked regions in unbounded shear flows), fractal scaling occurs over much of the interval between the integral scale and the Kolmogorov scale. The dimension in this scale range is 2.35 ± 0.05 (C).

(b) In fully turbulent parts of shear flows, iso-scalar contours possess a fractal scaling with a dimension of 2.67 ± 0.05. The scaling range is smaller than that for (a), and the inner cut-off occurs at a multiple of the Kolmogorov scale; it can be estimated *a priori* (Constantin et al. 1992) (C).

The original heuristic explanation (Sreenivasan 1991) for these observations relied on Reynolds number similarity. In Constantin et al. (1992), by combining the so-called 'co-area formula' culled from measure theory with the convection-diffusion equation governing the scalar evolution, it has been possible to obtain the fractal dimensions of iso-scalar surfaces and interfaces in turbulent flows. The only information about the velocity field that enters the calculation is the scaling exponent of the first-order structure function. The dimensions so obtained agree well with experiment.

(c) For the range between the Batchelor scale and the Kolmogorov scale, the fractal dimension of iso-scalar surfaces as well as interfaces approaches 3 as the Schmidt number approaches infinity. The finite Schmidt number correction is logarithmic (P).

(d) The results (a) and (b) seem to hold also for vorticity interfaces and iso-vorticity contours (P).

(e) The dimension of flame surfaces depends on the ambient turbulence level, but the

flame front in both diffusion and premixed flames appears to have a fractal dimension of 2.35 *for large turbulence levels* (C). Note that, in contrast to high-Reynolds-number isothermal flows, the scale range of convolutions in high temperature flames is small, except when the turbulence levels are high. Several fractal-based closure models have been attempted in combustion (P).

Results from time series analysis

The difficulty in the determination of the fractal structure of a time series lies partly with the definition of a suitable cover, and partly with proper recognition of the cut-offs between global, local and latent dimensions. These artifacts are now moderately well-understood, and the fractal structure of a time series of velocity or temperature fluctuations in high-Reynolds-number turbulent flows has been explored (Sreenivasan & Juneja 1992). These time traces resemble fBm traces with the exponent $H \approx 0.35$, and the (local) fractal dimension $D = 2 - H \approx 1.65$. This is consistent with the classical formalism of Kolmogorov (1941). An implication is that the dimension of iso-velocity and iso-temperature surfaces in fully developed turbulence is about 2.65 — consistent with the result (b) in section 7.1. At moderate Reynolds numbers, the scaling is better for spatial aspects and ambiguous for temporal data with the exception of those taken for interfaces, for which modest scaling occurs even at moderate Reynolds numbers.

Multifractal scaling

The evidence is strong that multifractals are useful tools for describing scaling properties of structure functions (Frisch & Parisi 1985, Benzi et al. 1984), and of turbulent energy dissipation rate and scalar dissipation rate in turbulent flows (Meneveau & Sreenivasan 1987, 1991, Prasad et al. 1988); provisional evidence (Sreenivasan 1991, Chorin 1994) suggests that other positive definite quantities, such as the square of turbulent vorticity, can be described similarly. This type of work has produced the following results.

(a) Phenomenological models for small-scale intermittency, with outcomes consistent with experiment, have been constructed. These models have yielded realistic intermittency corrections in the inertial and dissipation ranges, produced a refinement of the scaling of the power spectra in the dissipation range (Castaing et al. 1990), and generated stochastic signals which do not differ from the real turbulent signals in most respects (Juneja et al. 1994).

(b) The nature of the observed near-singularities appears consistent with the mathematical result (Cafarelli et al. 1982) on the partial regularity of weak solutions of the Navier-Stokes equations.

(c) By assuming that the scalar as well as vector fields are fractal graphs with the measured dimension, a broad theoretical apparatus has been constructed (Constantin & Procaccia 1993, Procaccia & Constantin 1993) to explain the Kolmogorov scaling in the inertial range and its various modifications (Monin & Yaglom 1975, Anselmett et al. 1984).

GEOPHYSICAL PHENOMENA

Geophysical fields (such as cloud radiance, rainfall, temperature and pollution records, sea surface infrared reflectivity, sea surface geometry, lightning paths, and so forth) are a result of nonlinear processes involving different fields at widely varying scales. In each case, the statistical invariance of a wide scale range suggests that fractal concepts may be useful. One could ask, for instance, if lightning paths and cloud boundaries are fractal and, if so, measure their fractal dimensions. Among the first fractal measurements made in geophysics was the dimension of cloud boundaries (Lovejoy 1982): the dimension of fair-weather clouds as well as large clouds is about 2.35 (the same as that of scalar interfaces in turbulent flows), whereas clouds strongly affected by the mean wind shear possess smaller dimensions (see Sreenivasan 1991).

In geophysics, there are numerous potential candidates for fractal description. As already remarked in section 3, it is not enough to seek information about the binary picture of whether or not there is rainfall (for example), but one needs to know something about the different rainfall rates. The variability and intermittency of rainfall records suggest that multifractals could be quite useful. Preliminary multifractal analysis has been made for rainfall rates, cloud radiance and other geophysical fields. For a survey of articles on these topics, see (Schertzer & Lovejoy 1991). Because accurate computations of high-order statistics require extraordinary amounts of data, geophysical measurements have generally been restricted to low-order multifractal measures.

It is claimed that high-order moments may diverge in geophysical situations (Mandelbrot 1972, Schertzer & Lovejoy 1991). However, measurements in the atmospheric surface layer (Fan 1990) suggest that the apparent divergence is an artifact of relatively small data records.

APPLICATIONS FOR IMAGE COMPRESSION AND DATA INTERPOLATION

The application of fractals in the construction of natural scenery—such as mountains, clouds, lakes, and so forth is well-known (see, for example, Mandelbrot 1982, Voss 1985, 1989, Peitgen & Richter 1986). This technology has found application in diverse ventures such as movie-making, art and music, which cannot be discussed here.

Fractals find a useful application in image compression and reconstruction. Lovejoy & Mandelbrot (1985) constructed cloud images using the model that a rain field is composed of self-similar pulses; the areas of these pulses had power-law distributions and rain rates were random. The resulting pictures look quite realistic even though real clouds are stratified in the vertical direction, and the governing principle for reconstruction should be self-affinity rather than self-similarity.

Image compression by taking advantage of fractal structure is the subject of Barnsley (1988); related applications involve fractal interpolation schemes by minimizing the Hausdorff distance. However, extensive application of these ideas to turbulence data has not been made. It is worth emphasizing that image compression on the basis of self-similarity or self-affinity is not especially the domain of fractals alone; in fact, applications based on wavelet theory work quite well (Zubair et al. 1992).

A QUALITATIVE ASSESSMENT OF PAST ACCOMPLISHMENTS

It is sometimes implied that fractals embody one of the basic truths about complexity: all known truths about Nature are expressible in the form of some generalized concepts, and that fractals represent one such generalized concept. *This is the appeal of fractals.*

This philosophical appeal has occasionally produced an exaggerated response, as pointed out by Kadanoff (1987), and the 'wheat' cannot always be separated from 'chaff'. *Note, however, that other glamorous concepts such as catastrophe theory never came close, even in their heydays, to enjoying the degree of appeal that fractals possess.* (Whether popular appeal is always correlated with scientific significance is another matter.)

Fractals have been taken seriously in mainstream science for about fifteen years. We hope that the summary given above makes clear the impressive fact that *fractals have had tremendous impact on providing descriptive and incremental understanding in many fields.*

To the examples already cited, we can add a few more here. Suppose we need to model the spread of forest fires, or the seepage of ground water or radioactive substances. While exact formulation of these problems is in principle possible, it would be futile to take this approach because of their extreme complexity. Instead, simulations of the type carried out in percolation studies can be quite useful in providing an overall picture (even if incomplete). Similarly, a host of other phenomena such as rainfall rates can be modelled by multifractals. *Fractals provide tools for modeling a variety of complex systems with some realism.*

Even though the evidence is still not compelling, it is strong enough to think that fractals are well-suited for handling scaling phenomena. *There is a certain tangible benefit of unity that fractals have brought to apparently unrelated areas of science.*

Much of the work using fractals has so far occurred in the physics community (or those small pockets of fluid mechanics community with relatively strong ties to physics), and the focus has not been engineering applications. However, *for fractal-based models to solve complex problems at the level of engineering utility, it is essential for the engineering community to take sustained interest in these tools.*

Whenever fractal (and multifractal) scaling is observed in fluid flows, it is nearly always statistical. This means that they provide only partial information—even if valuable and unique. *Therefore, as with all partial information, the degree to which one can make sense of an observation depends on the ingenuity of the individual trying to extract it. This is the ultimate constraint.*

We noted earlier that fractal-related work falls into three broad classes: description, explanation and prediction. Description and measurements of fractal dimensions consumed most of the energy in the early days. This effort was not trivial because it involved an understanding of the potential of fractals when they were still not commonplace, as well as the improvisation of theoretical, experimental and computational techniques. As for the second aspect, in percolation, multifractal measures, scalar interfaces in turbulence, and in other areas, a variety of results exist in which phenomenology and rigor play complementary roles. As regards predictions, a few powerful ones have already been pointed out in section 6, even though their engineering consequence is unclear. *On the whole, the past efforts have been quite rewarding.*

FUTURE NEEDS

The question of what constitutes future opportunities in fractals may well be thought to belong to the domain of mathematics. That, however, is not very useful here; an important mathematical result was recently proved[7] (Shishikura 1992) but it has produced hardly a ripple elsewhere. On the other hand, opportunities in specific fields of fluid mechanics will have to be assessed in the context of those specific fields. That would be a Herculean task for this brief report, and it therefore seems best to restrict it to a few broad questions common to most applications of fractals.

(a) The situation *typical* of most fractal-related studies is that much of our knowledge comes from experiment and simulations. While simulations and experiments are very useful, they are often limited by approximations, finite size effects, noise, and other artifacts, and cannot supplant theoretical results. *There is at present a big backlog of observations without the backup of solid explanations.*

(b) Much of the physics in scale-similar phenomena is hidden in cut-off scales. *Greater attention should be paid to cut-off scales and cross-over phenomena.*

(c) It would be essential to know, at least for one hard problem like turbulence, the relation between dynamics on the one hand and fractal geometry and multifractal measures on the other. (As already remarked in section 6, some progress has been made for the passive scalar problem.) What aspects of the hydrodynamic equations yield the fractal structure of its solutions? How may one show, without empirical input, that the turbulence structure is fractal (if that is indeed the case), obtain dimensions of its various facets and generate a closed list of fractal dimensions to define turbulence uniquely? *Without such knowledge, it is difficult to make a case for the inevitability of fractals as the tool of choice for studying large classes of nonlinear problems.*

(d) There are practical issues that appear to be within the realm of near-term possibility. For example, it seems possible to construct a reasonable model for flame speeds in premixed flames; model the gross spread rates of jets; generate a large-eddy-simulation model based on multifractals. Similarly, one can shed light on conditions under which flame extinction occurs because of excessive local stretch; generate synthetic signals which, if used as initial conditions for direct numerical simulations for turbulence, yield rapid convergence; generate good data compression techniques. *Such efforts should be driven more by expertise in the respective fields rather than in fractals* per se. Only when fractals integrate increasingly with applications-oriented research will more useful results emerge.

---

[7]The principal result is that the Hausdorff dimension of the boundary of the Mandelbrot set is 2.

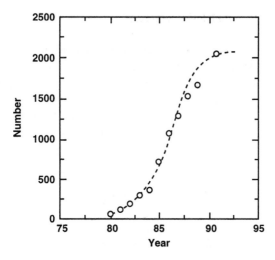

Figure 2: The number of entries in the Permutation Index on Fractals and Multifractals, as a function of time since the invention by Mandelbrot of the word 'fractal' around 1972 and the word 'multifractal' by Frisch & Parisi in 1984. Note that these entries may not all refer to independent papers, and so should be treated as representative of the number of related publications.

EPILOGUE

We may end this report with a statistical extrapolation. Figure 2 shows the number of entries in *Science Citation Index* on fractals and multifractals. The increase in interest in these topics has been nothing short of phenomenal. Various curve fits were tried for the data, and the best fit was the Landau equation describing supercritical bifurcation (Landau & Lifshitz 1959). As is well-known, the symptoms of such a bifurcation are the initial exponential growth and subsequent saturation due to nonlinear effects. In the present instance, the fact that the newly emerging 'science of complexity' is slowly subsuming fractals as its legitimate part may well explain the saturation. At the risk of being proven wrong, it appears that a continued exponential-like growth in fractals cannot be sustained without further stimulus in the form of major new ideas or some spectacularly successful application.

Acknowledgments: For comments on this manuscript and various discussions, I am thankful to Ashvin Chhabra, Boa-Teh Chu, Carl Evertsz, Walter Goldburg, Celso Grebogi, Bud Homsy, Daniel Joseph, David Keyes, Joel Koplik, Dan Lathrop, Mark Nelkin, Itamar Procaccia, Lex Smits, Gustavo Stolovitzky and Trevor Stuart.

REFERENCES

Aharony, A. (1986) Percolation. In *Directions in Condensed Matter Physics*, eds. G. Grinstein & G. Mazenko, World-Scientific pp 1-50.

Amitrano, A., Coniglio, A. & Di Liberto, F. (1986) Growth probability distribution in kinetic aggregation processes. *Phys. Rev. Lett.* **57**, 1016-19.

Anselmet, F., Gagne, Y., Hopfinger, E.J. & Antonia, R.A. (1984) High-order velocity structure functions in turbulent shear flows. *J. Fluid Mech.* **140**, 63-89.

Argoul, F., Arneodo, A., Grasseau, G. & Swinney, H.L. (1988) Self-similarity of diffusion-limited aggregates and electrodeposition clusters. *Phys. Rev. Lett.* **61**, 2558-61.

Avnir, D. (ed), (1989) *The Fractal Approach to Heterogeneous Chemistry*, Wiley.

Barnsley, M. (1988) *Fractals Everywhere*, Academic.

Benzi, R., Paladin, G., Parisi, G. & Vulpiani, A. (1984) On the multifractal nature of fully developed turbulence and chaotic systems. *J. Phys. A.* **17**, 3521-31.

Caffarelli, L., Kohn, R. & Nirenberg, L. (1982) Partial regularity of suitable weak solutions of the Navier-Stokes equations. *Commun. Pure. Appl. Math.* **35**, 771-831.

Castaing, B., Gagne, Y. & Hopfinger, E. (1990) Velocity probability density functions of high Reynolds number turbulence. *Physica D* **46**, 177.

Chhabra, A.B. & Jensen, R.V. (1989) Direct determination of the f (α) singularity spectrum. *Phys. Rev. Lett.* **62**, 1327-30.

Chhabra, A.B. & Sreenivasan, K.R. (1992) Scale-invariant multiplier distributions in turbulence. *Phys. Rev. Lett.* **68**, 2762-5.

Chorin, A.J. (1994) *Vorticity and Turbulence*, Springer.

Constantin, P. & Procaccia, I. (1993) Scaling in fluid turbulence. A geometric theory. *Phys. Rev. E* **47**, 3307.

P. Constantin, I. Procaccia & K.R. Sreenivasan, (1991) Fractal geometry of isoscalar surfaces in turbulence: Theory and experiments. *Phys. Rev. Lett.* **67**, 1739-42.

Daoud, M. & Martin, J.E. (1989) Fractal properties of polymers. In *The Fractal Approach to Heterogeneous Chemistry*, ed. D. Avnir, Wiley pp 109-130.

de Gennes, P.G. (1987) Time effects in viscoelastic fingering. *Europhys. Lett.* **3**, 195-7.

Evertsz, C.J.G. & Mandelbrot, B.B. (1992) Appendix B: Multifractal measures. In *Chaos and Fractals*, eds. H.O. Peitgen, H. Jurgens & D. Saupe, Springer, pp 921-953.

Falconer, K.J. (1985) *The Geometry of Fractal sets*, Cambridge University Press.

Fan, M.S. (1991) *Features of Vorticity in Fully Turbulent Flows*, Ph.D. thesis, Yale University.

Farmer, J.D., Ott, E. & Yorke, J.A. (1983) The dimension of chaotic attractors. *Physica D* **7**, 153-180.

Feder, J. (1988) *Fractals*, Plenum.

Feigenbaum, M.J. (1978) Quantitative universality for a class of nonlinear transformation. *J. Stat. Phys.* **19**, 25-52.

Feigenbaum, M.J. (1979) The universal metric properties of nonlinear transformations. *J. Stat. Phys.* **21**, 669-706.

Feigenbaum, M.J. (1987) Some characterizations of strange sets. *J. Stat. Phys.* **46**, 919-924.

Feigenbaum, M.J. (1987) Scaling spectra and return times of dynamical systems. *J. Stat. Phys.* **46**, 925-932.

Feigenbaum, M.J., Jensen, M.H. & Procaccia, I. (1986) Time ordering and the thermodynamics of strange sets: theory and experimental tests. *Phys. Rev. Lett.* **57**, 1503-1506.

Feigenbaum, M.J., Kadanoff, L.P. & Shenker, S.J. (1982) Quasiperiodicity in dissipative systems: a renormalization group analysis. *Physica D* **5**, 370-386.

Frisch, U. & Parisi, G. (1985) Appendix: On the singularity structure of fully developed turbulence. In *Turbulence and Predictability in Geophysical Fluid Dynamics and Climate Dynamics*, eds. M. Ghil, R. Benzi & G. Parisi, North-Holland, p. 84-88.

Galilei, G. (1623) *The Assayer* (as translated by S. Drake (1957), *Discoveries and Opinions of Galileo*, Doubleday, pp. 237-38). This quotation seems to have been invoked first by Benoit Mandelbrot.

Halsey, T.C. Jensen, M.H., Kadanoff, L.P., Procaccia, I. & Shraiman, B.I. (1986) Fractal measures and their singularities: the characterization of strange sets. *Phys. Rev. A.* **33**, 1141-1151.

Havlin, S. & Avraham, D.B. (1987) Diffusion in disordered media. *Adv. Phys.* **36**, 695-798.

Hentschel, H.G.E. & Procaccia, I. (1983) The infinite number of generalized dimensions of fractals and strange attractors. *Physica D* **8**, 435-444.

Homsy, G.M. (1987) Viscous fingering in porous media. *Annu. Rev. Fluid Mech.* **19**, 271-311.

Hurst, H.E. (1951) Long-term storage capacity of reservoirs. *Trans. Amer. Soc. Civil Engg.* **116**, 770-779.

Jensen, M.H., Kadanoff, L.P., Libchaber, A., Procaccia, I. & Stavans, J. (1985) Global universality at the onset of chaos: results of a forced Rayleigh-Bénard experiment. *Phys. Rev. Lett.* **55**, 2798-2801.

Juneja, A., Lathrop, D.P., Sreenivasan, K.R. & Stolovitzky, G. (1994) Synthetic turbulence. *Phys. Rev. E* **49**, 5179-94.

Kadanoff, L.P. (1986) Fractals: Where's the physics? *Phys. Today* **39** (2): 7-8, 6-7.

Kolmogorov, A.N. (1941) The local structure of turbulence in incompressible viscous fluid for very large Reynolds numbers. *Dokl. Akad. Nauk SSSR* **30**, 301-305.

Landau, L.D. & Lifshitz, E.M. (1959) *Fluid Mechanics*, Pergamon.

Libchaber, A. & Maurer, J. (1980) Une expérience de Rayleigh-Bénard de géometrie réduite; multiplication, accrochage et demultplication de fréquences. *J. de Physique* **41**,(supplement, fasc. 4, Colloque 3) C3-51 - C3-56.

Lovejoy, S. (1982) Area-perimeter relations for rain and cloud areas. *Science*, **216**, 185-187.

Lovejoy, S. & Mandelbrot, B.B. (1985) Fractal properties of rain, and a fractal model. *Tellus* **37A**, 209-232.

Mandelbrot, B.B. (1974) Intermittent turbulence in self-similar cascades: divergence of high moments and dimension of the carrier. *J. Fluid Mech.* **62**, 331-358.

Mandelbrot, B.B. (1982) *The Fractal Geometry of Nature*, Freeman.

Mandelbrot, B.B. (1989) Multifractal measures, especially for the geophysicist. *Pure and Appl. Geophys.* **131**, 5-42.

Mandelbrot, B.B. & Evertsz, C.J.G. (1990) The potential distribution around growing fractal clusters. *Nature* **348**, 143-45.

Mandelbrot, B.B. & Van Ness, J.W. (1968) Fractal Brownian motions, fractional noises and applications. *SIAM Review* **10**, 422-37.

Meakin, P. (1988) The growth of fractal aggregates and their fractal measures. In *Phase Transitions and Critical Phenomena*, eds. C. Domb & J.L. Lebowitz, Academic, 335-489.

Meneveau, C. & Sreenivasan, K.R. (1987) Simple multifractal cascade model for fully developed turbulence. *Phys. Rev. Lett.* **59**, 1424-7.

Meneveau, C. & Sreenivasan, K.R. (1991) Multifractal nature of turbulent energy dissipation. *J. Fluid Mech.* **224**, 429-84.

Monin, A.S. & Yaglom, A.M. (1975) *Statistical Fluid Mechanics II*, M.I.T. Press.

Nittmann, J., Daccord, G. & H.E. Stanley, (1985) Fractal growth of viscous fingers: quantitative characterization of a fluid instability phenomenon. *Nature* **314**, 141-44.

Novikov, E.A. (1990) The effects of intermittency on statistical characteristics of turbulence and scale similarity of breakdown coefficients. *Phys. Fluids A.* **2**, 814-20.

Olinger, D.J. & Sreenivasan, K.R. (1988) Nonlinear dynamics of the wake of an oscillating cylinder. *Phys. Rev. Lett.* **60**, 797-800.

Ostlund, S., Rand, D., Sethna, J. & Siggia, E. (1983) Universal properties of the transition from quasi-periodicity to chaos in dissipative systems. *Physica D* **8**, 303-42.

Paterson, L. (1984) Diffusion-limited aggregation and two-fluid displacements in porous media. *Phys. Rev. Lett.* **52**, 1621-24.

Peitgen, H.-O. & Richter, P.H. (1986) *The Beauty of Fractals*, Springer.

Prasad, R.R., Meneveau, C. & Sreenivasan, K.R. (1988) Multifractal nature of the dissipation field of passive scalars in fully turbulent flows. *Phys. Rev. Lett.* **61**, 74-7.

Procaccia, I. & Constantin, P. (1993) Non-Kolmogorov scaling exponents and the geometry of high Reynolds number turbulence. *Phys. Rev. Lett.* **70**, 3416.

Saffman, P.G. & Taylor, G.I. (1958) The penetration of a fluid into a porous medium or Hele-Shaw cell containing a more viscous liquid. *Proc. Roy. Soc. Lond.* **245A**, 312-29.

Sapoval, B., Rosso, M. & Gouyet, J.F. (1985) The fractal nature of a diffusion front and the relation to percolation. *J. Phys. Lett.* **46**, L149-L156.

Schertzer, D. & Lovejoy, S. (eds.) (1991) *Nonlinear Variability in Geophysics*, Kluwer.

Shishikura, M. (1992) *Proc. Complex Analytic Methods in Dynamical Systems*, IMPA, January.

Sreenivasan, K.R. (1985) On the fine-scale intermittency of turbulence. *J. Fluid Mech.* **151**, 81-103.

Sreenivasan, K.R. (1993) Fractals and multifractals in fluid turbulence. *Annu. Rev. Fluid Mech.* **23**, 539-600.

Sreenivasan, K.R. & Juneja, A. (1992), Fractal dimension of time series in turbulent flows. preprint.

Sreenivasan, K.R. & Stolovitzky, G. (1995) Turbulent cascades. *J. Stat. Phys.* **78**, 311-33.

Stavans, J., Heslot, F. & Libchaber, A. (1985) Fixed winding number and the quasiperiodic route to chaos in a convective fluid. *Phys. Rev. Lett.* **55**, 596-99.

Vicsek, T. (1989) *Fractal Growth Phenomena*, World Scientific.

Voss, R.F. (1985) Random fractal forgeries. In *Fundamental Algorithms for Computer Graphics*, ed. P. Earnshaw, Springer, p. 805-35.

Voss, R.F. (1989) Random fractals: self-affinity in noise, music, mountains, and clouds. *Physica D* **38**, 362-71.

Wilkinson, D. & Willemsen, J.F. (1983) Invasion percolation: a new form of percolation theory. *J. Phys. A.* **16**, 3365-76.

Wilson, S.D.R. (1990) Taylor-Saffman problem for a non-Newtonian liquid. *J. Fluid Mech.* **220**, 413-25.

Witten, T.A. & Sander, L.M. (1981) Diffusion-limited aggregation, a kinetic critical phenomenon. *Phys. Rev. Lett.* **47**, 1400-03.

Zubair, L., Sreenivasan, K.R. & Vickerhauser, V. (1992) Characterization and compression of turbulent signals and images using wavelet-packets. In *Studies in Turbulence*, eds. T.B. Gatski, S. Sarkar & C.G. Speziale, Springer, 489-513.

CAPILLARY AND INTERFACIAL PHENOMENA, WETTING AND SPREADING

Paul H. Steen
School of Chemical Engineering
Cornell University
Ithaca, NY 14853

INTRODUCTION

Life on this planet is concentrated at the earth's surface where air meets soil, as is common knowledge. It is also common experience that life forms of the greatest diversity and abundance are found along our ocean shores, river banks and pond rims, that is, wherever air, water and earth meet. In much the same way, an abundance of rich physical and physiochemical phenomena occurs near the contact of gas, solid and liquid matter. All objects of our world are bounded by interfaces. They may be classified as either solid/solid, gas/solid, liquid/solid, gas/liquid or liquid/liquid.

The two most far-reaching technological developments of the 20th century may be traced to the understanding of interfacial phenomena. Flight controlled by man is based on a knowledge of the boundary layer, the region of air flowing adjacent to the flying body. Here, understanding the fluid dynamics of the gas near the gas/solid interface was central to progress toward reliable and efficient flight. On a smaller scale, the operation of the transistor depends on the interface between two different solids, one of p-type and the other of n-type material. Here, it is the flow of electrons across the solid/solid interface that is central to the phenomenon. We expect that understanding interfacial effects will play no less significant a role in technological developments of the next century.

SCOPE

We focus on gas/liquid and liquid/liquid interfaces which may be grouped together as fluid/fluid boundaries. We shall consider gas/liquid pairs with soluble or insoluble gas and volatile or nonvolatile liquid and liquid/liquid pairs consisting of miscible or immiscible liquids. The concept of a sharp interface (i.e. a mathematical surface) is an idealization of a thin region across which properties vary rapidly. This idealization requires the introduction of properties such as the surface tension that are attributed to the

286

two-dimensional surface Gibbs (1948), Padday (1969). If the fluid/fluid interface intersects a solid/fluid interfacial region then a three-phase region exists.
In this case, each two-phase boundary is idealized as a two-dimensional surface as above and their intersection, the three-phase region, is idealized as
a one-dimensional continuum, called the contact-line (also common line or
triple line). We shall restrict our attention to problems where fluid/fluid
interfaces and contact-lines occur under circumstances where surface tension
and continuum-scale fluid motions are important. Examples of problems of
practical concern that fall within this scope come from a wide variety of
disciplines; refer to the Table.

RESEARCH DIFFICULTIES

Difficulties that currently hinder solution of problems in this area can be
split into those for which a knowledge of underlying physics or chemistry is
lacking and those for which all the basics are understood but the complexity
of the problem limits progress. A useful simplification is that the former
difficulties call for new knowledge and the latter for new methods. In cases
where difficulties arise from complexity, the complexity may be due to either
a broad range of length scales or of physiochemical effects.

An Illustrative Example. Packed bed contactors are used in the chemical
processing industry to bring liquid and gas into intimate contact in order to
separate chemical species. For example, in gas absorption of ammonia, the
object is to remove the water-soluble ammonia from the gas stream. A typical
industrial-scale packed bed consists of a tall cylindrical vessel (meters high)
packed with pieces of inert solid (millimeter to centimeter in size). The liquid
and gas are introduced at the ends of the vessel (same end for cocurrent or
opposite ends for countercurrent operation) and are pumped simultaneously
through the solid packing. The aim is to generate large gas/liquid interfacial
areas in order to enhance mass transfer by diffusion from one phase to the
other while not sacrificing the connectedness of either phase. The two streams
are gathered at the ends of the apparatus opposite to their introduction. If,
for example, the liquid streams are dispersed into droplets, the separate
collection of gas and liquid is hindered; dispersions are to be avoided. A
detailed discussion of the mechanics of packed-bed contactors is found in de
Santos et al. (1991).

Several difficulties may be distinguished in this example. On the scale
of millimeters, there will be wetting and spreading. There will be locations
both where liquid displaces gas along the solid surfaces (advancing contact
lines) and other places where gas displaces liquid (receding contact lines).
The basic physics of this process, even in the absence of mass transfer and in
a simplified geometry, is poorly understood. To illustrate, consider a rivulet
or thin stream of liquid flowing down a solid inclined plane in a gaseous

environment. Prediction of the meander of the rivulet (typical) is beyond
current capabilities for several reasons. First, the bulk motion of the rivulet
(mm scale) is strongly influenced by molecular interactions at the contact
line ($10^9 mscale$) and working molecular-based theories for the prediction of
the constitutive behavior of contact-line dynamics are not available Dussan
(1979) and de Gennes (1985). Constitutive equations are needed for the
boundary conditions which are required in order to solve the fluid mechanics
problem of bulk motion Dussan (1983). Second, lacking theories, empirical
correlations that characterize the contact-line behavior are needed, but are
available for very few systems de Gennes (1985). The magnitude of the experimental task of collecting data from material triplets (gas, liquid, solid) is
compounded by the apparent sensitivity of contact-line behavior to solid surface conditions (e.g. cleanliness and roughness). Third, even in the few cases
where an adequate contact line description is known, there are difficulties
in applying traditional stability methods due to the nonunique dependence
of the contact-angle on velocity Davis (1983); Dussan and Davis (1986). In
summary, the challenge here is to understand and describe the sensitivity of
the contact-line motion to the molecular-level interactions and to incorporate
that information into boundary conditions for the bulk continuum behavior.
We shall refer to this as the "microscale" problem.

Suppose now that the physiochemistry of the contact-line dynamics is
understood and that a good constitutive description is available. Then, on
the scale of the packing, referred to as the "mesoscale", one could write
down the equations and boundary conditions governing the temperature,
the fluid velocities, and the concentrations of the species of interest. The
difficulty then comes in solving these equations. The position and shape
of the contact-lines and the interface are unknown a priori. This is a free
surface and a free boundary problem. Geometry is complicated due to the
irregular shape of the packing elements and their haphazard arrangement.
Temperature and concentrations differences at the interface may influence the
surface tension and thereby introduce an additional coupling between energy
and momentum (the so-called Marangoni effect). In any case, the nonlinear
governing equations will be strongly coupled. Furthermore, even if a solution
to a model mesoscale problem could be found, the very nature of free surface
problems make the stability of that solution an issue. In summary, even
simplified mesoscale problems may be intractable even though solvable in
principle.

Now suppose the microscale behavior is known and the mesoscale problem
has been solved. It remains then to infer the behavior on the length-scale
of the apparatus (the macroscale) from knowledge of the mesoscale. How
does the mechanics local to one or a few packing elements (centimeter scale)
contribute to the cooperative effect of a large number of elements (10 meter

scale)? How does the small scale influence the performance of the overall unit? Here, the difficulty is one of disparate length scales. The complexity is of a different nature to that faced by the mesoscale problem.

RESEARCH OPPORTUNITIES

Research difficulties are a guide to research opportunities. The Table gives a sense of the breadth of discipline-specific opportunities. We attempt to identify some common difficulties of important unsolved problems from across these disciplines. They are grouped into three categories: *physical description, length-scale disparity, and physiochemical diversity.*

Physical Description. These opportunities arise from gaps in our ability to understand and describe some basic physics. First, there are problems in description on the kinematic level. For example, fluid motions adjacent to a moving triple line have multi-valued velocity fields if no-slip is imposed at the solid surface, Dussan (1983). This difficulty has been circumvented through the introduction of various slip models that give single-valued fields and that relieve the stress singularity that otherwise occurs. Slip models are effective without capturing details in the immediate vicinity of the contact line, Ngan and Dussan (1989). Another example where kinematics are the source of difficulty, discussed below, occurs (even in the absence of triple lines) with the fracturing or pinching of a material body. In contrast to these kinematical issues, it is often force-flux or dynamical descriptions that are lacking, as in the case of partial wetting in the contactor application mentioned above. There, constitutive relations between velocity and contact angle of a triple line are needed to predict its motion. Critical reviews of dynamic contact-line phenomena give a more detailed account of research needs in this area, de Gennes (1985). We now turn to opportunities somewhat less well-publicized.

A long standing and important engineering problem is the prediction of when and how an object will break; that is, undergo a change of state from one to two or more pieces. Examples of practical importance within our scope include the dryout of patches on a film that initially wet a solid (loss of simply-connectedness), the turning inside-out of an emulsion (exchange of topologies of the two phases) and, on the macroscale, the degradation of the liquid stream in a trickle-bed reactor into a dispersion (loss of connectedness) mentioned above. Although, depending on the case, various 'sufficient' and sometimes 'necessary' conditions for a change in connectedness are available, we are unaware of any context where the instant of change in topology of a continuum and its behavior at nearby times (before and after) have been predicted. Indeed, even to describe the trajectory that leads from connected to disconnected states faces kinematical difficulties since there is at least one instant (in the usual description) where at least one body point in the object is mapped into two spatial points, i.e. the kinematical mapping breaks down, Dussan (1976). A simple example where capillary-driven topological change has been closely observed is the collapse of bridges (liquid bridges, Peregrine et al. (1990) and soap-film bridges, Cryer and Steen (1992)). In the soap-film case, the multi-valuedness leads to a space-time singularity in the velocity field with finite-time blow-up, Steen (1992). Even these relatively simple problems of collapse await solution. Treatments of break-up problems in the literature typically excise from consideration the events immediately preceding and following the instant of collapse, i.e. they do not confront collapse.

Length-scale Disparity. In particulate systems (e.g. suspensions, porous media, fixed beds), rheological properties on the macroscale, given a knowledge of the microscale, are often required. Length scales may differ by several orders of magnitude. The contactor application above and many of the problems listed in the Table (e.g. the flow of foams, dispersions and of bubbly liquids) have similar difficulties. For opportunities in this area, we refer to reviews of the particulate systems, Drew (1983), of the bubbly systems, Hsieh (1988) and Miksis and Ting (1991) and of multiphase systems, Adler and Brenner (1988). Successful approaches, like the difficulties, are likely to parallel one another.

An important distinction between systems with and without free surfaces in the presence of length-scale disparity should be pointed out. The susceptibility of a free surface to instabilities means that additional 'internal' length scales beside those set by the geometry may develop. Furthermore, knowing if and when such scales develop must be part of the solution. A simple example is the difference between single-phase and two-phase flow (immiscible liquids, say) through a porous bed. Suppose the response of flow rate to pressure drop over the scale of the bed is sought. In the case of single-phase flow, averaging techniques (e.g. multiple scales, homogenization or statistical mechanical approaches) have been successful in bridging the micro- and macro-problems and the equations so-obtained are relatively well-accepted (e.g. Darcy-Brinkman equation). In contrast, when two phases of different viscosities are present in the bed, fingering instabilities of various lengthscales can develop and whether or not the fingers penetrate the bed strongly influences the resultant flow rate. If surface tension is important on the microscale, behavior can be even more complicated. Reliable predictions of flow regimes and other macroscopic characteristics for these flows are still unavailable despite the 30 years or so of sustained attention, Homsey (1987) and Pearson (1991). In summary, the degrees of freedom introduced by free surfaces and contact-lines in the microproblem exacerbate difficulties associated with length-scale disparity.

In simple geometries, characterized by a single length scale, and in the absence of flow, free surface instabilities develop on the capillary length scale. In contrast, in the presence of flow, viscous and inertial forces (depending on the regime) can strongly influence the lengthscales that evolve through instability. There are still significant gaps in our understanding of how flow influences free surface instabilities, even in relatively simple contexts, and especially regarding evolution into nonlinear structures. For example, the various regimes of two-phase pipe flow are manifestations of the nonlinear evolution of interfacial instabilities; aspects of these flows still defy reliable prediction, Preziosi et al. (1989).

Physiochemical Diversity. The difficulty here is due to the large number of competing physiochemical effects. The equations are known; their reliable and efficient solution is the issue. The competing effects may be organized into the hierarchy below where the simplest cases concern mechanical systems where thermal and multicomponent effects are absent. The next step in diversity adds thermal effects, examples of which are listed. In a similar way, the presence of two or more chemical species in a mechanical system raises the complexity. Most complicated are those problems involving mechanical, thermal and species effects with interfaces susceptible to instability. As in the foregoing the references provided below are examples of treatments of the topics; they are are not exhaustive lists.

I. Mechanical (single component, immiscible, and isothermal)

    A. Containment (contact-lines - fixed or absent)

        1. passive (stable - drops, bubbles, bridges) Michael, (1981), Rallison (1984)

        2. active (unstable - films, jets) Deryagin and Levi (1964); Bogy (1979)

    B. Wetting and Spreading (contact-lines - free)

        1. static (drops, bubbles, bridges, foams) Dussan and Chow (1983)

        2. dynamic (films, fingers, foams, rivulets) Ruschak (1985), Bankoff and Davis (1987), Kraynik (1988) Payatakes (1982), Young and Davis (1987)

II. Thermomechanical (single component, immiscible or two-phase)

    A. Thermally-generated surface-tension gradients Levich and Krylow (1969), Davis (1987) and Subramanian (1991)

    B. Condensation/evaporation Burelback et al. (1988)

III. Solutomechanical (multicomponent, immiscible or miscible)

    A. Compositionally-generated surface-tension gradients Levich and Krylow (1969)

    B. Mass-transfer influences Johnson and Sadhal (1985), Davies (1963)

  IV. Thermo- and solutomechanical

    * All of the above, Miller (1973).

Progress in situations with diversity often comes from a careful formulation (i.e. narrowing) of the question(s) being asked of the system along with a judicious mix of classical analysis (asymptotics, perturbation, etc.), computation, and experiment. Here the call is for closer collaboration of the practitioners of these three vital arts.

REFERENCES

Gibbs, J.W. (1948) "On the Equilibrium of Heterogeneous Substances", *The Collected Works of Willard Gibbs*. vol. 1, 55-353.

Padday, J.F. (1969) Theory of surface tension in *Surface and Colloid Science*, (ed. E. Matijevic), New York, Wiley 1:39-251.

de Santos, J.M., T.R. Melli and L.E. Scriven (1991) Mechanics of Gas-Liquid Flow in Packed-Bed Contactors. *Ann. Rev. Fluid Mech.* 23, 233-60.

Dussan V., E.B. (1979) On the Spreading of Liquids on Solid Surface: Static and Dynamic Contact Lines. *Ann. Rev. Fluid Mech.* 11, 371-400.

de Gennes, P.G., Wetting: statics and dynamics. (1985) *Rev. Mod. Phys.* 57, 827.

Dussan V., E.B. (1983) The Moving Contact Line. In *Waves on Fluid Interfaces* (ed. R.E. Meyer) Proc. Symp. Math. Res. Center, Univ. of Wisc., Academic Press, 303-324.

Davis, S.H. (1983) Contact-line Problems in Fluid Mechanics. *J. Applied Mechanics.* 50, 977-982.

Dussan V., E.B. and S.H. Davis (1986) Stability in systems with moving contact lines. *J. Fluid Mech.* 173, 115-130.

Ngan, C.G. and Dussan V., E.B. (1989) On the Dynamics of Liquid Spreading on Solid Surface, *J. Fluid Mech.* 209, 191-226.

Dussan V., E.B. (1976) On the difference between a bounding surface and a material surface, *J. Fluid Mech.* 75(4), 602-623.

Peregrine, D.H., G. Shoker and A. Symon, (1990) The bifurcation of Liquid Bridges, *J. Fluid Mech.* 212, 25-39.

Cryer, S.A. and Steen, P.H. (1992), Collapse of the Soap-film Bridge: Quasistatic Description. J. Coll. Int. Sci.

Steen, P.H. (1993), Capillary Containment and Collapse in Low Gravity: Dynamics of Fluid Bridges and Columns. In *Emerging applications in free boundary problems* (J. Chadam and H. Rasmussen, eds.) Pitman Research Notes in Mathematics v.280, Longman Scientific and Technical, Essex, U.K.

Drew, D.A. (1983) Mathematical Modeling of Two-Phase Flow. *Ann. Rev. Fluid Mech.* 15, 261-91.

Hsieh, D.Y. (1988) On dynamics of bubbly liquids. Advances in Applied Mechanics 26, 63-133.

Miksis, M.J. and L. Ting, (1991) Effective Equations for Multiphase Flows-Waves in a Bubbly Liquid. In Advances in Applied Mechanics 28, 142-256.

Adler, P.M. and Brenner, H. (1988) Multiphase flow in Porus Media. *Ann. Rev. Fluid Mech.* 20, 35-59.

Homsy, G.M. (1987) Viscous Fingering in Porous Media. *Ann. Rev. Fluid Mech.* 19, 271-311.

Pearson, J.R.A. (1991) Fluid Mechanics and Transport Phenomena Advances in Chemical Engineering, v. 16, 97-107.

Preziosi, L., Chen, K. and Joseph, D.D. (1989) Lubricated Pipelining: stability of core-annular flow, *J. Fluid. Mech.* 201, 323-356. 21. Michael, D.H. (1981) Meniscus Stability. *Ann. Rev. Fluid Mech.* 13, 189-215.

Rallison, J.M. (1984) The Deformation of Small Viscous Drops and Bubbles in Shear Flows. *Ann. Rev. Fluid Mech.* 16, 45-66.

Deryagin, B.V. and Levi, S.M. (1964) Film Coating Theory, Focal Point Press, N.Y.

Bogy, D.B. (1979) Drop Formation in a Circular Liquid Jet. *Ann. Rev. Fluid Mech.* 11, 207-28.

Dussan V., E.B. and Chow, R.T.-P. (1983) On the ability of drops or bubbles to stick to non-horizontal surfaces of solids. *J. Fluid Mech.* 137, pp. 1-29.

Ruschak, K.J. (1985) Coating Flows. *Ann. Rev. Fluid Mech.* 17, 65-89.

Bankoff, S.G. and S.H. Davis (1987) Stability of Thin Films. PhysioChem. Hydrodyn. 9, 5-7.

Kraynik, A.M. (1988) Foam Flows. *Ann. Rev. Fluid Mech.* 20, 325-57.

Payatakes, A.C. (1982) Dynamics of Oil Ganglia During Immiscible Displacement in Water-Wet Porous Media. *Ann. Rev. Fluid Mech.* 14, 365-93.

Young, G.W. and S.H. Davis (1987) Rivulet Instabilities. *J. Fluid Mech.* 176, 1-31.

Levich, V.G. and V.S. Krylov (1969) Surface-Tension-Driven Phenomena. *Ann. Rev. Fluid Mech.* 1, 293-316.

Davis, S.H. (1987) Thermocapillary Instabilities. *Ann. Rev. Fluid Mech.* 19, 403-35.

Subramanian, R.S. (1991) The Motion of Bubbles and Drops in Reduced Gravity. In Transport Processes in Bubbles, Drops and Particles, Hemisphere Publishing Corporation, New York.

Burelbach, J.P., Bankoff, S.G. and Davis, S.H. (1988). Nonlinear stability of evaporating/condensing liquid films, *J. Fluid Mech.* 195, 463-494.

Johnson, R.E. and S.S. Sadhal (1985) Fluid Mechanics of Compound Multiphase Drops and Bubbles. *Ann. Rev. Fluid Mech.* 17, 289-320.

Davies, J.T. (1963) Mass-Transfer and Interfacial Phenomena, Advances in Chemical Engineering, v. 4, 3-49.

Miller, C.A. (1973) Stability of Moving Surfaces in fluid systems with heat and mass transport. *AIChE J.* 19, 909-915.

TABLE I

Engineering Discipline	Areas of Impact
Agricultural	sprays, emulsions, etc.
Biomedical	aveolar collapse, tear-film wetting, etc.
Chemical	liquid extraction, emulsions, dispersions, mixing, two-phase flow
Electronic	chemical vapor deposition (thin films), semiconductor packaging, etc.
Environmental	oil-spill recovery, groundwater/contaminant flow, flue gas scrubbing, spray technology, etc.
Manufacturing	film coating, injection molding, blow molding, melt-spinning, etc.
Materials	crystal growth, synthetic cellular solids, etc.
Mechanical	direct-contact heat exchangers, combustion, etc.
Metallurgical	spray-atomization solidification, sparging, molten metal foams
Petroleum	two-phase flow(porous media), oil recovery, foam fracturing, etc.

BIOFLUID DYNAMICS

Sheldon Weinbaum
Department of Mechanical Engineering
The City College of the City University of New York
New York, NY 10031

Since the internal and external environments of humans, animals and plants are air, water and solutes, biofluid dynamics touches on nearly every aspect of their function both at the cellular and whole organ level. Biofluid mechanicians have contributed significantly to such diverse problems as the transport of LDL cholesterol in arteries and the role of fluid stresses in atherosclerosis, the development of noninvasive methods for the diagnosis of cardiovascular disease, the function of the heart and its valves and the design of prosthetic devices, the role of surfactants in the lungs and an understanding of the mechanics of airway closure in pulmonary disease, the lubrication of cartilage and the mechanisms for bone healing and growth, an understanding of the flight of birds and insects and the swimming of fish and microorganisms. Future research has been enhanced by the very positive climate for mutual collaboration that has developed between medical and biological researchers and investigators in the physical sciences and engineering. This should lead to a great broadening of the scope and depth in which biofluid mechanicians can contribute to the biological sciences and the development of new fields of collaboration, particularly at the cellular level, where fluid dynamic forces and transport processes are coupled to a vast spectrum of biochemical and immunological responses.

INTRODUCTION

Since the internal and external environments of humans, animals and plants are air, water and solutes, biofluid dynamics touches on nearly every aspect of their function both at the cellular and whole organ level. An excellent overview of the more classical areas of biofluid dynamics involving the flow of blood in the large arteries, the fluid dynamic function of the heart and its valves, blood rheology and its flow in the microcirculation and the ventilation and perfusion of the lung is given in Fung (1984). Similarly, the mathematical modeling of the flight of birds and insects, the swimming of fish and mammals and the locomotion of microorganisms is described in Lighthill (1975). In the past two decades a very positive climate for mutual collaboration has developed between medical and biological researchers and investigators in the physical sciences and engineering. The result has been a great broadening of the scope and depth in which biofluid mechanicians can contribute to these classical areas and the development of new fields of collaboration in which fluid dynamic forces and transport processes are coupled to biochemical and immunological responses, particularly at the cellular level.

In this chapter we shall highlight a few of the more important new developments and research needs in the more classical areas where this cross-fertilization has led to significant new understanding, the role of fluid dynamic forces in arterial disease, the computational modeling of the heart and its valves, two-phase flow aspects of microvascular networks and recent studies in animal locomotion. In addition, several newer areas of research have been singled out. These include: biological porous media flow [flow of fluid and nutrients through the vascular interface, cartilage, bone and soft (connective) tissue]; flow instability and the role of surfactants in the lungs and non-invasive cardiovascular flow measurement.

FLUID DYNAMIC ASPECTS OF ATHEROGENESIS

Atherosclerosis is the major cause of death in the industrialized nations. In the past decade extensive studies have been performed to measure the local fluid shear stress in the vicinity of arterial bifurcations using vascular casts and non-invasive flow measurement methods. These studies have finally confirmed the hypothesis, Caro et al. (1969), that the topological pattern of early lesion formation (atherogenesis) in the major arteries is correlated with regions of low shear near vessel bifurcations and wall curvature where there is flow reversal during the cardiac cycle and or flow separation. It has also been demonstrated that these regions with predilection for lesion formation have endothelial cells which are rounded in shape, have an increased permeability to macromolecules, including low density lipoproteins LDL, and higher than normal rates of endothelial cell turnover. The case for the fluid flow pattern being an important risk factor in atherogenesis, the initiation of the disease process, is summarized in Yoshida et al. (1988). The local adaptive response of the vessel diameter to wall shear has also been demonstrated experimentally by Giddens et al. (1990) confirming the hypothesis by Kamiya and Togawa (1980) that the vessel lumen adjusts so as to maintain nearly constant average wall shear.

The critical unanswered questions in the localization of the disease process relate largely to (i) the mechanisms for the LDL entry into the artery wall and the focal increases in LDL permeability, (ii) the relationship of fluid shear stress to biochemical processes in the endothelial cells and the artery wall more generally and (iii) how the fluid transport processes in the subendothelial intima (region where lesion forms) are related to the biochemical events in the early stages of lesion formation. In Weinbaum et al. (1985) a new hypothesis and transport model for the artery wall were proposed which suggested that LDL entered the wall via leaky junctions surrounding the tiny fraction of cells that were either in mitosis or the processes of dying. These cellular level LDL leakage sites have now been confirmed experimentally. In Yuan et al. (1991) a transport model for the intima has been developed to predict the growth of the leakage spots and a new hypothesis advanced which suggests that there is a colocalization of the leakage sites and the formation of subendothelial lipid liposomes, the earliest event in atherogenesis.

The need to understand the role of fluid shear stress in endothelial cell function has led to a major new area of research in which endothelial mono-layers are grown in tissue cultures and then put in flow chambers where the different biochemical responses to fluid shear can be examined under carefully controlled conditions. Since the first studies by Dewey et al. (1981) examining the realignment of endothelial cells and their actin filaments due to fluid shear, a host of other responses have been investigated including the intra-cellular release of Ca and a variety of second messengers and the activation of various membrane receptor proteins to explain the entry of blood borne monocytes into the intima. This approach has now been utilized in many other areas where the cellular response to shear is believed to be important, Nerem and Girard (1990).

Research Needs

a. Improved non-invasive methods for measuring the fluid shear stress on arterial endothelium in vivo.

b. A coupled transport and kinetic model to explain the growth of suben-dothelial lipid liposomes and experiments to prove the colocalization hypothesis for the leakage sites.

c. Further tissue culture studies to explain the role of fluid shear stress on endothelial cell turnover and the mechanisms of monocyte attachment and entry into the artery wall.

d. Theory and experiments to elucidate the cellular mechanisms involved in the adaptive response of vessel diameter to fluid shear stress.

HEART AND VALVE FUNCTION, COMPUTATIONAL BIOFLUID DYNAMICS

Charles Peskin, Courant Institute of Mathematical Sciences, NYU
Biofluid dynamicists are often concerned with the non-linear interaction between a viscous fluid and an active or passive elastic tissue. An important

example is the heart, where blood interacts with the muscular heart walls and flexible heart valve leaflets. A special numerical technique, known as the *immersed boundary method*, has been developed by Peskin (1972) to treat this type of interaction.

In the case of the heart, the immersed boundary method has been used to elucidate the normal function of the heart and its valves as well as their malfunction during certain diseases, such as the prolapse of the mitral valve. The method has also been applied in the design of prosthetic cardiac valves to minimize valve thrombosis and optimize hemodynamic performance. The initial application of the immersed boundary method was a two-dimensional model for the contraction of the left ventrical. The methodology has since been extended to a fully three-dimensional model in which all four heart chambers and their valves are considered as well as all the major arteries and veins leaving and entering the heart, Peskin and McQueen (1989). These large vessels are equipped with sources and sinks which are used to connect the model heart to a simulated circulation. The fiber architecture of the heart and its valves is treated as a system of active and passive elements whose behavior is based on anatomical principles. A major challenge that has still not been satisfied is to express these anatomical principles in mathematical form and to derive the fiber architecture by solving the system of governing equations for the model and comparing the predictions with experimental observation.

The immersed boundary method, has also been applied to a wide variety of other active and passive fluid-elastic interaction problems. These include platelet aggregation during clotting, aquatic animal locomotion and, most recently, wave propagation in the inner ear, Beyer (1992).

Research Needs

a. More adequate access to existing supercomputers.

b. Further development of high-performance parallel computers.

c. Development of numerical algorithms for parallel computers that can be effectively applied to fluid flows with flexible boundaries.

CARDIOVASCULAR FLOW MEASUREMENT

(Don P. Giddens, School of Aerospace Engineering, Georgia Institute of Technology)

Measurement of fluid dynamic variables under disturbed conditions in human subjects is essential for understanding both normal and pathologic cardiovascular function. While *in vitro* and animal models yield great insight into various mechanistic phenomena due to their ability to offer control of parameters, ultimately this knowledge needs to be applied in the context of an intact human system. Thus, there is a critical need for non-invasive and minimally invasive *in vivo* measurements of variables such as flow, velocity, wall shear, cell and macromolecule motion. These measurements have the potential to improve the diagnosis of cardiovascular disease, to uncover the fluid dynamic factors which are involved in disease initiation and progression, and to evaluate the efficacy of therapeutic interventions.

Among the greatest contributions fluid dynamics has made to medicine are the marked advances in the diagnosis of heart and artery diseases. Within the past 15 years several techniques using dyes, heated fluids and radioisotopes have been developed for evaluating tissue and organ perfusion. These techniques have made it possible to diagnose pathological conditions which produce only minor symptoms. Ultrasound, used noninvasively in both imaging and Doppler modes, is now very effective in evaluating localized atherosclerotic plaque formation in peripheral arteries at stages where drug, diet and exercise therapies are possible, Breslau et al. (1982). Color Doppler ultrasound, while presently limited by spatial, temporal and velocity resolution, offers significant potential as a noninvasive flow visualization method, Nanda (1989). The technology associated with the magnetic resonance imaging of tissues and blood flow has advanced rapidly, but is still limited in its ability to measure velocity profiles accurately, Caprihan et al. (1986).

As new therapies are developed, the need to develop efficacy will become more important. Thus the ability to discriminate and interpret small changes in the blood flow pattern resulting from changes in plaque geometry will not only be valuable in evaluating the advance of a disease process but also its regression. Insight into the mechanism of disease progression, will depend in part on our ability to measure macromolecular transport and near wall flow phenomena. While currently the spatial resolution is marginal, detection of the relative concentration of radiolabelled LDL along the artery wall, coupled with noninvasive measurement of wall shear, should permit *in vivo* testing of hypotheses relating fluid dynamics and atherogenesis in human subjects. Similar studies utilizing platelets will prove useful in understanding the influence of fluid mechanics upon thrombosis.

Research Needs

a. Improvements in spatial resolution of Doppler ultrasound and new methods to visualize, present and interpret blood velocity data obtained *in vivo*.

b. Higher resolution of velocity profiles obtained *in vivo* by magnetic resonance imaging especially near vessel walls.

c. Techniques for measuring various cell and macromolecular interactions with human and animal vessel surfaces.

d. Automated particle tracking methods for *in vitro* study of cell motion.

e. Coupling of unsteady three-dimensional computational methods with vessel imaging and flow data to improve the interpretation of experimental measurements.

MICROCIRCULATORY FLUID DYNAMICS

It has been well appreciated since the classical studies of Fahraeus and Krogh, sixty years ago, that the hematocrit distribution (volume fraction of red cells) and the elastic properties of the red cell membrane, were critical factors in determining the rheological properties of blood in the microcirculation and its fluid dynamic behavior. Many of these properties were studied by examining the behavior of red cell suspensions in small glass tubes (see summary in Chien et al. (1984)), and numerous studies have been undertaken to elucidate the rheology of the red cell membrane (see Special Issue of ASME J. Biomech. Eng. (1990)). Increasingly sophisticated models have been developed to describe novel aspects of the flow of red cells in capillaries including a large deformation theory for the single file motion of red cells in capillaries, the tank treading motion of the red cell membrane and a simplified two-dimensional model for the transition from single to double file motion as the diameter of the capillaries increases. Many of these models are described in Skalak et al. (1989).

Since the vessel hematocrit is critical in determining the local delivery of oxygen to the tissue, many of the physiological studies in recent years have focused on network studies of the hematocrit distribution in different tissue preparations where it has been observed that the hematocrit can be far lower than predicted by the Fahraeus effect, the decrease in tube hematocrit due to the red cells traveling near the axis of the vessel where the velocity is higher than the average velocity of the plasma. This hematocrit defect, termed the network Fahraeus effect by Pries et al. (1986), has been attributed to two phase separation mechanisms, plasma skimming (preferential capture of cell free plasma near the wall of the parent vessel) and cell screening (hydrodynamic interaction of the cells with the vessel entrance geometry) at vessel branchings. An approximate three-dimensional theory for plasma skimming has been developed in Yan et al. (1991) and a theory for cell screening developed by these same investigators for dilute cell suspensions of rigid spheres. The effect of network plugging by the far less numerous, but larger white cells, has been examined by T. C. Skalak and coworkers. An important unanswered question in all existing studies is the concentration distribution of the red cells across the vessels. The recent scaling laws derived in Leighton and Acrivos (1987) for the shear induced diffusion of particles in concentrated suspensions has yet to be applied to microcirculatory flow.

Research Needs

a. Theory and experiments for the shear induced diffusion of red and white cells and platelets in different size microvessels as a function of vessel hematocrit, surface adhesion and velocity.

b. Theory and experiment for the screening of red cells at microvascular bifurcations for concentrated suspensions.

c. Three-dimensional models of the asymmetric motion of red cells in microvessels in multifile motion.

d. Further network studies on the effect of white cell plugging, cell screening and plasma skimming.

PULMONARY BIOFLUID MECHANICS & FLOW IN COLLAPSIBLE VESSELS

R. Kamm, Department of Mechanical Engineering, M. I. T.
    Until recently most studies in pulmonary fluid mechanics have focused on steady and quasisteady flow of air in the bronchial tree, the pressure-flow relation and collapsible tube flow. However, many flows are intrinsically unstable (coughing, forced expirations, high frequency ventilation) and give rise to flow-induced oscillations (wheezes). Gavriely et al. (1987) have observed that flow limitation and airway flutter often occur simultaneously. Flow induced wall flutter has been investigated for some time, but Bertram et al. (1990) have recently shown that the flutter mechanism is far more complicated than previously recognized. Our present understanding of the effect of maximal flow in the presence of flutter is incomplete and requires further study. Other recent efforts have examined the stability, mobility and regulation of the airway liquid lining, raising a variety of intriguing fluid dynamic problems. During normal breathing the thickness of the liquid layer can become unstable at low lung volumes by a mechanism that is analgous to the Rayleigh instability of a liquid jet and can form a meniscus that blocks the airway (Johnson et al. 1991). Gas absorbtion peripheral to the meniscus produces regional atelectasis and the loss of gas exchange capability. Excess liquid and regional atelectasis are common symtoms of respiratory disease. To avoid atelectasis the airway must reopen on inspiration. This raises the question as to the mechanism for the movement and elimination of the meniscus. The distribution of liquid between the vasculature, interstitium, airway

wall and lumen is the primary determinant of airway closure, but also determines airway narrowing, a critical factor in asthma. Fai-Look (1988) have examined the liquid exchange between the vascular and air compartments. This process involves some of the same issues raised in the next section.

Pulmonary surfactant is found in varying amounts throughout the airway tree. For normal conditions the surfactant produces a surface tension gradient that has the capability of pumping liquid from the periphery toward the mouth of the pathway (Marangoni convection). The impaired production of surfactant can lead to common pulmonary disorders such as neonatal or adult respiratory distress syndrome (RSD). Endogenous surfactant is a well accepted treatment for neonatal RSD, but there is a significant group in which this therapy is ineffective. Since the efficacy of this treatment depends on how effectively the surfactant is transported to the lung periphery, studies have been undertaken to examine how exogenous surfactant spreads in the lung, either from individual aerosol droplets, Gaver and Grotberg (1990), or from an instilled bolus. This spread of exogenous surfactant raises the exciting possibility that surfactants can be used to transport other medications to the periphery where they can be readily absorbed.

Research Needs

a. Further studies of the interaction between flutter in a collapsible tube and (mean) maximal flow to improve our understanding of forced expiration and labored breathing.

b. Theory and experiments to explain the formation, movement and reopening of liquid menisci in the air pathways.

c. Studies of the mechanisms regulating the distribution of liquid between vascular, interstitial and airway compartments.

d. Further studies on the use of endogenous and exogenous surfactants to produce surface tension gradients for the transport of airway liquid and the potential use of instilled surfactant as a drug delivery vehicle.

POROUS MEDIA BIOFLUID MECHANICS

Nearly all the tissues in the body are subject to a filtration flow of water and solutes that are driven by either the blood pressure, osmotic forces or stress due to body motion. The vital lubricating function of interstitial fluid in cartilage subject to loading has until recently been described by biphasic consolidation theory. However, the ability of a charged proteoglycan matrix to maintain an osmotic swelling stress has led to a reformulation of this theory as a triphasic mixture in which ionic solutes can be trapped by fixed charge in the matrix and exert an osmotic force, Lai et al. (1991). A similar situation exists in bone tissue where both cortical and trabecular bone are filled with an interconnected system of pores, which when subjected to cyclic loading generate a streaming potential. In Salzstein and Pollack (1987) consolidation theory is combined with electrokinetic theory to provide a rational explanation for the frequency response and amplitude of the measured streaming potential. An important unanswered question in bone is the cellular transduction mechanism by which mechanical loading stimulates bone growth. Weinbaum and Cowin have, recently, proposed a new theory which suggests that fluid shearing stresses acting on the membranes of the osteocytic processes in the fluid filled bone pores are the triggering mechanism for this cellular growth response.

Proteoglycan matrix is present not only in cartilage but at the endothelial surface of cells and in nearly all connective tissues. In addition, in many biological tissues there are fluid filled channels or pores that are partially or completely filled with different types of proteoglycans or gel like substances. In Tsay and Weinbaum (1991) a theory has been developed for fiber filled channels which shows that if the fibers are long and slender a Darcy-Brinkman equation can be used to describe these bounded porous media flows. This theory has been applied to the filtration flow in the intercellular clefts between endothelial cells, the aqueous drainage channels in the eye and the thin intimal layer in the artery wall. Porous media flow has also been involved in treating the internal active motion of cells, Dembo (1986), in which movement of fluid through polymerized actin gel is an essential aspect of the cell locomotion.

Blood perfusion in tissue is critical in delivering nutrients, removing wastes and in tissue energy exchange. Gases are lipid soluble and thus can easily cross the membranes of the cells lining the vasculature. In contrast, the structural organization of the pathways that allow for the passage of water and solutes and provide the molecular sieve for different size solutes is still not fully understood and has been part of a continuous search that started in the 1950's. Neither pore or fiber matrix models can explain the large body of available data on perfused individual capillaries and recently a combined junction-pore-matrix theory has been proposed for capillary endothelial clefts. In the past decade it has been realized that heat exchange, unlike gaseous exchange, does not occur in capillaries but primarily in 100 to 500 m diameter arteries and veins that are closely spaced countercurrent pairs in muscle tissue. A new bioheat equation, Weinbaum and Jiji (1985), has been derived to describe the blood tissue heat transfer from these paired vessels, but the theory breaks down at high blood flow rates or in the larger countercurrent vessels of the microcirculation where the thermal equilibration between the vessels is much slower.

Research Needs

a. The recently developed triphasic theory has widespread potential application for many connective tissues whose filtration behavior is poorly understood.

b. Theory and experiment are required to identify the cellular level mechanosensory mechanism by which mechanically induced fluid motion in bone pores may lead to bone growth.

c. New approaches are required to elucidate the structural pathways for the movement of water and solute across capillary endothelium and the artery wall.

d. Theory and experiments for the intracellular flow through porous cellular components involved in active cellular motion.

e. A more comprehensive bioheat transfer theory is needed to describe highly perfused tissue and deep muscle tissue where the assumptions in the Weinbaum-Jiji equation do not apply.

ANIMAL LOCOMOTION, SWIMMING AND FLYING

Tim Pedley, Department of Applied Mathematics, University of Leeds

Animal locomotion is determined by the forces and torques acting on the animal's external surfaces. The force and torque balances on the whole organism are used to calculate its speed and energy expenditure, Lighthill (1975). The former can be compared with experimental observation and the latter is estimated from considerations of muscle physiology. Many phenomena are well understood, such as flagellar locomotion of microorganisms in still water, steady and burst swimming of fish and fast (quasi steady) forward flight of birds and insects. Discovery of the clap-and-fling mechanism for instantaneous lift was a major triumph. Many phenomena are still not fully understood. These include the swimming of microorganisms in shear flow to explain rheotaxis and gyrotaxis motions, the unsteady aerodynamics of slow forward flight or hovering, Ellington (1984), and the swimming of small organisms in the Reynolds number range 0.1 to 10. For many of these phenomena in vivo flow visualization and force measurement data are lacking.

Similar to the heart in Section 2, a full understanding of animal locomotion requires an integration of its external hydrodynamics with the physiology of muscular contraction. One wishes to determine how a body deforms in response to its external forces and torques and to relate this loading to the electrochemical signals that initiate muscular contraction. Predictions of the animal's speed of locomotion are a by product of this calculation. The immersed boundary method developed to model the contraction of the heart has also been applied to animal locomotion, Fauci and Peskin (1988), but the physiological data required to model the muscular contraction of the animal's anatomical structures is an important unknown that needs to be determined.

The collective behavior of large populations of organisms is important in ecology, biotechnology and gravitational and evolutionary biology, Pedley and Kessler (1992). Best understood are the bioconvection patterns that arise in suspensions of swimming microorganisms. The response to gravity and chemical gradients is understood, but the response to light still needs to be elucidated. The time evolution of a well-mixed suspension of microorganisms towards a stratified state, the convective instability of that steady state and the nonlinear interactions that lead to the observed patterns are all challenging problems that require attention. Related problems with larger animals, such as the swarming of locusts, are even more important, but less amenable to analysis because these animals respond to non-physical stimuli as well as to physical ones.

Research Needs

a. *In vivo* data on muscle activity, forces exerted and flow visualization for both swimming and flying.

b. Three-dimensional Navier-Stokes codes for periodically moving boundaries of time varying shape in non-uniform flow at low, intermediate and high Reynolds numbers.

c. Iterative coupling of the codes in b. with models for muscular contraction and elastic behavior to determine unknown boundary motions.

d. Mathematical modeling of nonlinear, pattern-forming motions of animal populations.

ACKNOWLEDGMENTS

In preparing this report I have utilized the overviews provided by the following references:

Fung, Y. C. (1984) *Biodynamics: Circulation.* New York, Springer-Verlag.

Lighthill, M. J. (1975) *Mathematical Biofluid Dynamics.* Philadelphia, Soc. Ind. & Appl. Math.

Skalak, R. and Chien, S. ed. (1987) *Handbook of Bioengineering.* New York, Mcgraw-Hill.

Skalak, R., Ozkaya, N., and Skalak, T. (1989) "Biofluid Mechanics", *Ann. Rev. Fluid Mech.* **21**, 167-204.

ADDITIONAL REFERENCES

Bertram, C. D., Raymond, C. J., and Pedley, T. J. (1990) Mapping of instabilities for flow through collapsed tubes of differing length. *J. Fluids & Struct.* **4**, 125-153.

Beyer, R. P. (1992) A computational model of the cochlea using the immersed boundary method. *J. Comput. Phys.* **98**, 145-162.

Breslau, P. J., Knox, R. A., Phillips, D. J., Beach, K. W., Chikos, P. M., Thiele, B. L., and Strandness, D. E., Jr. (1982) *Vascular Diag. & Therapy* **3**, 17-22.

Caro, C. G., Fitz-Gerald, J. M., and Schroter, R. C. (1969) Arterial wall shear and distirbution of early atheroma in man. *Nature* **223**, 1159-1161.

Caprihan, A., Davis, J. G., Altobelli, S. A., and Fukushima, E. (1986) A new method for flow velocity measurement: frequency encoded NMR. *Magn. Reson. Med.* **3**, 352-362.

Chien, S., Usami, S., and Skalak, R. (1984) Blood flow in small tubes. In *Handbook of Physiol.- The Cardiovascular System IV,* Ch. 6, 217-249, Bethesda, Md: Am. Physiol. Soc.

Dembo, M. (1986) The mechanics of motility in dissociated cytoplasm *Biophys. J.* **50**, 1165-1183.

Dewey, C. F., Bussolari, S. R., Gimbrone, M. A., Jr., and Davies, P. F. (1981) Dynamic response of vascular endothelial cells to fluid shear stress. *ASME J. Biomech. Eng.* **103**, 177-185.

Ellington, C. P. (1984) The aerodynamics of hovering insect flight. *Phil. Trans. R. Soc. Lond.* **B 305**, 1-181.

Fauci, L. J. and Peskin, C. S. (1988) A computational model of aquatic locomotion. *J. Comput. Phys.* **77**, 85-108.

Gaver, D. P. III and Grotberg, J. B. (1990) Dynamics of a localized surfactant on a thin film. *J. Fluid. Mech.* **213**, 127-148.

Gavriely, N., Kelly, K. B., Grotberg, J. B., and Loring, S. H. (1987) Forced expiratory wheezes are a manifestation of airway flow limitation. *J. Appl. Physiol.* **62**, 2398-2403.

Giddens, D. J., Zarins, C. K., and Glagov, S. (1990) Response of arteries to near-wall fluid dynamic behavior. *Appl. Mech. Rev.* v. 43 **5**, S98-S102.

Johnson, M., Kamm, R. D., Lo, L. W. Shapiro, A. H., and Pedley, T. J. (1991) Nonlinear growth of surface-tension-driven instabilities of a thin annular film. *J. Fluid Mech.* **233**, 141-156.

Kamiya, A. and Togawa, T. (1980) Adaptive regulation of wall shear stress to flow change in the canine carotid artery. *Am. J. Physiol.* **239**, H14-H21.

Lai, W. M., Hou, J. S., and Mow, V. C. (1991) Triphasic theory for the swelling and deformation behaviours of articular cartilage. *ASME J. Biomech. Eng.* **113**, 245-258.

Lai-Fook, S. J. (1988) Pressure-flow behavior of pulmonary interstitium. *J. Appl. Physiol.* **64**, 2372-2380.

Leighton, D. and Acrivos, A. (1987) The shear-induced migration of particles in concentrated suspensions. *J. Fluid Mech.* **181**, 415-439.

Nanda, N. C. (1989) *Textbook of Color Doppler Echocardiography* Lea & Febiger, Philadelphia.

Nerem, R. M. and Girard, P. R. (1990) Hemodynamic influences on vascular endothelial biology. *Toxicologic Pathol.* **18**, 572-582.

Pedley, T. J. and Kessler, J. O. (1992) Hydrodynamic phenomena in suspensions of swimming microorganisms. *Ann. Rev. Fluid Mech.* **24**, 313-358.

Peskin, C. S. (1972) Flow patterns around heart valves: a numerical method. *J. Comput. Phys.* **10**, 252-271.

Peskin, C. S. and McQueen, D. M. (1989) A three-dimensional computational method for blood flow in the heart. I. Immersed elastic fibers in a viscous incompressible fluid. *J. Comput. Phys.* **81**, 372-405.

Pries, A. R., Ley, K., and Gaeghtens, P. (1986) Generalization of the Fahraeus for microvessel networks. *Am. J. Physiol.* **251**, H1324-H1332.

Salzstein, R. A.and Pollack, S. R. (1987) Electromechanical potentials in cortical bone - I. A continuum approach. *J. Biomech.* **20**, 261-270.

Special Issue on Cell Biomechanics. (1990) *ASME J. Biomech. Eng.* **112**, 233-368.

Tsay, R-Y. and Weinbaum, S. (1991) Viscous flow in a channel with periodic cross-bridging fibres. Exact solutions and Brinkman approximation. *J. Fluid Mech.* **226**, 125-148.

Weinbaum, S. and Jiji, L. M. (1985) New simplified bioheat equation for the effect of blood flow on local average tissue temperature. *ASME J. Biomech. Eng.* **107**, 131-139.

Weinbaum, S., Tzeghai, G., Ganatos, P., Pfeffer, R., and Chien, S. (1985) Effect of cell turnover and leaky junctions on arterial macromolecular transport. *Am. J. Physiol.* **248**, H945-H960.

Yan, Z-Y., Acrivos, A., and Weinbaum, S. (1991) Fluid skimming and particle entrainment into a small circular side pore. *J. Fluid Mech.* **229**, 1-27.

Yoshida, Y., Yamaguchi, T., Caro, C. G., Glagov, S., and Nerem, R. M., Eds. (1988) *Role of Blood Flow in Atherogenesis,* Springer-Verlag, Tokyo.

Yuan, F., Chien, S., and Weinbaum, S. (1991) New view of convective-diffusive transport processes in the arterial intima. *ASME J. Biomech. Eng.* **113**, 314-329.

GEOPHYSICAL FLUID DYNAMICS

John A. Whitehead
Woods Hole Oceanographic Institution
Department of Physical Oceanography
Woods Hole, MA 02543

INTRODUCTION

Geophysical Fluid Dynamics includes theoretical, experimental and numerical studies that operate in conjunction with data acquisition and analysis for the geophysical sciences of oceanography, meteorology, solid earth science and planetary physics. The ultimate aim of these branches of natural science is understanding the behavior of the Earth systems. From this perspective, geophysical fluid dynamics is sometimes considered to only be a service discipline to these larger natural sciences. Unfortunately this view is shortsighted since numerous studies of geophysical fluid dynamics operate at the forefront of understanding in continuum physics itself. For instance important fundamental contributions have been made in such basic fluid mechanics problems as turbulence, bifurcation studies, chaos, fractal characterizations, boundary layers, critical layers and critical control. Ultimate applications have encompassed not only the natural sciences but engineering, biology, chemistry, ecology, and nonlinear optics. Some studies have also made profound contributions to the understanding of basic predictability of our macroscopic world.

In spite of the intellectual stimulus of cross disciplinary applications, over the past decade geophysical fluid dynamics has continued to branch into specialized segments aligned with corresponding applications in geophysical sciences. Thus for example oceanography has branched into dynamics of microstructure, general circulation, coastal dynamics, and equatorial studies; meteorology has branched into studies of climate, upper atmosphere, and mesoscale dynamics; geophysics has branched into hydrology, mantle convection, volcanology, and dynamo studies; and planetary dynamics has branched into inner and outer planet dynamics. Many segments contain specialized numerical modelling efforts and some include almost no laboratory studies but instead use surrogate data sets obtained by direct field observation. Most segments have the common bond of including effect of frame rotation and/or density variation, and many share similar kinds of waves, eddies, boundary layers, and instability mechanisms. As in most problems in fluid mechanics, nonlinearity is important. Linear superposition cannot be invoked if additional physical mechanisms are added, and the general tools of nonlinear dynamics have to be used in studying the increasing complexity that arises.

To give a view of some of the present challenges, eight specific examples of studies will be presented. These describe progress in understanding the circulation in the thermocline of the ocean, buoyancy-driven flows in rotating systems, geological fluid mechanics, coastal oceanography, two dimensional turbulence, waves and wave-mean flow interaction, mixed boundary conditions, and the dynamo problem. Enough detail is introduced in the first four to allow one to picture the interleaving between theoretical considerations, the actual earth or planetary problems and some of the fundamental properties of rotating or stratified fluids. The next four are condensed for brevity. This is not a comprehensive collection of all present studies. A truly representative collection would involve more than forty such examples. This section ends with a brief summary.

THE STRUCTURE OF THE OCEANIC GENERAL CIRCULATION

The circulation of the ocean presents a problem in fluid mechanics of fundamental importance in understanding the Earth's environment and climate. Over the past decade great advances have been made in understanding the fluid dynamics of the oceanic shell of our planet.

The problem can be characterized as a boundary value problem in which the spatial structure of the oceans' velocity, temperature, salinity and pressure fields are sought in response to specified surface forcing produced by the atmosphere. In reality, the ocean and atmosphere and are coupled, but as a first approximation the circulation of the oceans can be imagined to be forced by specified wind stresses and heat and fresh water fluxes at the surface between air and sea. The circulation problem is further complicated by the nonlinear interaction of a wide range of scales of motion in a fluid that contains turbulent eddies of scales of hundreds of kilometers embedded within circulatory patterns, the gyres, of thousands of kilometers in extent.

A central problem in oceanography has been the need to explain the overall spatial structure of the wind driven circulatory gyres in mid latitudes which extend in depth to at least two kilometers below the sea surface, the region of the so-called oceanic thermocline. The key element of the fluid mechanics has been the application of two fundamental principles. First, that over much of the ocean the principal internal force acting on the fluid is the pressure gradient. Because of the overwhelming importance of the Earth's rotation at the time and space scales characteristic of the circulation this produces a Coriolis acceleration. Second, and also because of the dominating effect of the Earth's rotation on the dynamics, the vorticity of the circulation (the elementary rotation of each fluid particle) becomes the principal controlling dynamical quantity of interest. More precisely, the structure of the circulation and its explanation depend on the treatment of Ertel's scalar *potential vorticity* or q. This is the vorticity component perpendicular to density surfaces divided by the local vertical spacing between these surfaces. The distribution of q over the oceans determines the pathways of the fluid flow but the distribution of q is itself determined by the flow. The problem is inherently nonlinear.

Recent fluid mechanical models of the thermocline circulation have made progress by ignoring to a first approximation the effects of turbulent dissipation over most of the ocean. The potential vorticity is then a conserved quantity, and the use of the potential vorticity conservation theorems then allows the potential vorticity and the complete flow field to be determined. Two mechanisms which determine q are described in the recent theories. In one, the ventilation of the thermocline by surface waters, forced down by the action of the winds sets the potential vorticity of the subducted fluid at the sea surface. The value of q subsequently preserved, determines the path and structure of the ventilated water. For water at deeper levels which is not exposed to the sea surface within the great circulating gyres the slow but inexorable diffusion of q within closed q circuits builds large plateaus of horizonally uniform q whose known value allows a determination of the circulation. Nonlinear coupling of these two models for different regions provides an explanation for the observed oceanic structure.

The physics of the models still has strong controversial elements. Simple applications of theorems from instability theory as well as direct observations show the existence, alluded to earlier, of an embedded turbulent eddy field. It remains to be seen to what extent the ideal fluid theories developed in the past decade must be altered to account for the cooperative effects of the eddies in redistributing the potential vorticity. Similarly, the theories are still fluid mechanically deficient in describing one vital branch of the circulation pattern namely the flow through the intense western boundary currents where the simplifications in the force and vorticity balance change dramatically. The consequences of the deficiency are as yet unknown.

Continued fluid mechanical research will be required to elucidate the dynamics of this important component of our planet's climate system.

Research Needs

a. Understanding of the interaction between the wind-driven (horizontal) and the thermo-haline (vertical) circulation.

b. Coupled ocean-atmosphere variability on interannual, interdecadal and secular scales.

c. A clearer understanding of the role of friction and mixing in the dynamics.

d. Coupled boundary layer — interior circulation studies.

BUOYANCY DRIVEN FLOWS IN ROTATING SYSTEMS

The dynamics of planetary and stellar atmospheres is governed by the interaction of thermal convection and effects of rotation. The tendency of the latter to induce two-dimensional fluid motions gives rise to a surprising high degree of regularity of flows even in the range of asymptotically high Reynold numbers. The band structures and the regular arrangement of cyclonic vortices on Jupiter are particularly striking examples. Even with the most simple application of fluid dynamical principles it has been possible to understand the restriction of the bands to low latitudes and to predict the extent towards much higher latitudes on Saturn, a feature which was confirmed by the Voyager observations.

More recent research has clarified the close connection between Rossby-wave like thermal convection modes and the generation of huge zonal jets in the deep atmospheres of the major planets. Although they are probably not directly connected with buoyancy driven motions, it is worth mentioning that the large anticyclonic oval eddies and the Great Red Spot, in particular, have been understood to a considerable extent on the basis of ingenious laboratory experiments and numerical simulations. Further research is likely to lead to a more profound understanding of the regularly spaced small cyclonic eddies.

Laboratory experiments and associated theoretical studies of buoyancy driven flows in rotating systems have not only led to deep insights into processes in planetary atmospheres and in stars, but have provided some valuable models for hydrodynamic instability and the transition to chaotic fluid motion. The rich variety of features seen in the baroclinic annulus experiment, the special chaotic state induced by the Küpers-Lortz instability in a convection layer rotating about a vertical axis, and the various transitions of centrifugally driven convection columns in the narrow gap annulus are good examples for the interesting nonlinear dynamics offered by buoyancy driven flows in rotating systems.

As easily accessible nonlinear systems from the theoretical as well as the experimental point of view and as keys for the understanding for dynamical processes in planetary and stellar atmospheres, the various cases of buoyancy driven flows in rotating fluids will continue to attract the attention of fluid dynamicists and physicists.

Research Needs

a. The application of modern laboratory techniques for the measurement of flows and temperature fields is likely to provide new insights into the dynamics and to stimulate theoretical research. Future progress in the field will depend strongly on increased experimental activities.

b. Analytical and numerical studies are needed to help understand and complement the experimental work.

GEOLOGICAL FLUID MECHANICS

Buoyancy driven flows in fluids that experience great compositional and physical changes are being studied in response to geological and geophysical questions. The questions range from basic considerations of mechanics of volcanos, mantle convection, understanding the flows associated with plate tectonics to earthquake and fracture behavior. Such flows exhibit new classes of instabilities and unanticipated structures compared to flows in fluid with more uniform properties. For example, cooling in magma chambers results in modification of the boundaries of the chamber. Certain minerals preferentially crystallize and the interior fluid may become compositionally layered. Vast layers of strata in volcanic provinces have long been documented by geologists. In another example convection cells with very large viscosity variation produce flows with completely different structures from convection with uniform viscosity fluid. Hot low viscosity mantle material can travel with unusual effectiveness through vertical conduits that take the form of lubricated pipes. These can support solitary waves that may convey large parcels of material from deep in the mantle to the surface of the earth over geological time. To initiate these pipes, large spherical or mushroom shaped plume heads may develop that may have initiated epochs of gigantic volcanism in earth history. Interaction between fluid flow and fracturing or accreting solid boundaries also leads to novel problems. Such areas as earthquake mechanics, dike propagation, hydrothermal alteration and eruption mechanics are now being explored. The results are achieving the level where they can be compared with geological observations and are leading to new understanding of our earth and its evolution.

Research Needs

a. Studies of non-Newtonian fluid dynamics with various rheologies.

b. Understanding of problems with interaction between boundary conditions and interior circulation.

c. Fluid dynamics with very large variation in viscosity and other properties.

d. Fluid dynamics of suspensions and multi-phase flow.

COASTAL OCEANOGRAPHY

The coastal ocean is physically distinct from the remainder of the ocean because of the boundary processes which inevitably occur and because of the pervasive presence of the continental margin, which means that topographic influences often dominate the physics of the flow. Over the last 10–20 years, a good deal has been accomplished in terms of understanding coastal currents, both observationally and theoretically. With the advent of modern oceanographic instruments, major moored array experiments have been conducted in many different shelf environments around the world and a lowest order description of many important coastal current systems has emerged. From a theoretical standpoint, perhaps the greatest development is the generation of a simple set of models of wind-forced coastal-trapped waves. These models have genuine predictive skill for alongshore currents and sea level in the coastal ocean on time scales of typically about 2–20 days. The models build on geophysical fluid dynamics ideas about topographic vorticity waves and exploit the free-mode properties of the waves to expand the complete forced solution. Since there are very few areas in oceanography where simple models have advanced to the level of having predictive skill, their significance should not be underestimated.

While these models are important, and represent an important first step, they have little or no skill in terms of predicting the cross-shelf currents. While the stronger alongshore currents contribute to alongshore advection, it is the cross-shelf currents which lead to much of the thermal and nutrient variability on the continental margins, and it is these currents which transport pollutants and sediments away from the coastline towards the deep-ocean reservoir. Despite our considerable success with alongshore currents, cross-shelf currents have not yet yielded to theoretical understanding except in special cases (such as within the near-surface turbulent boundary layer). Difficulties with the cross-shelf flow appear to be due in part to poorly characterized forcings (such as offshore eddies and small scale, $0(10–100$ km), wind variations), and to the probable increased importance of nonlinear processes and turbulence.

From the standpoint of the coastal oceanographer, the main challenges for the next decade lie in developing a better understanding of a) the role of boundary layer turbulent mixing b) the nature of cross-shelf flows. One particularly interesting problem will be the role of finite-amplitude frontal instabilities. Fronts of various sorts are nearly ubiquitous in the coastal ocean on a wide range of scales, and their instabilities may often dominate cross-shelf transport processes. It seems likely, based on observational results, that frontal instabilities and eddy shedding may be key processes in cross-shelf exchange. The dynamical problem is not simple because of wind forcing, bottom topography and the lack of validity of some standard simplifying assumptions. Even if only numerical approaches to the problem are employed, advances in computational schemes will be needed in order to resolve simultaneously the front and the larger scale flow environment which may create and maintain the front. In any case, it seems likely that the next decade of fluid dynamical problems in coastal oceanography will see a growing emphasis on smaller scale, more nonlinear processes.

Research Needs

a. A basic understanding of rotating flows with density variations and bottom topography.

b. Role of boundary layer and turbulent mixing.

c. The nature of cross-shelf flows.

d. Nature of frontal instabilities in rotating fluids with bottom topography.

d. Understanding of eddy shedding by fronts, coastal features, and bottom topography.

TURBULENCE IN GEOPHYSICAL FLUIDS

Turbulence in geophysical fluid dynamics has many manifestations that differ from fully developed three dimensional turbulence. Little is known about the interaction of turbulence with background for properties such as ambient stratification, rotation, or large scale shear. Only the crudest suggestions have been advanced, and few have been tested by experiments or field data.

Two dimensional turbulence is thought to exist in some form in both the atmosphere and ocean because in many regions the two components of velocity at right angles to the gravity vector are much greater than vertical velocity. Numerical studies have been conducted which have the resolution to compute the evolution of two dimensional turbulence for times that are long enough for comparison with statistical theories. The results seem to be mixed. Some features of the computed fields such as the cascade to long length scales reflect the predictions of the statistical theories, but other features such as patchiness and coherence of the eddies in the latter stages seem to differ. It is not known how closely either one agrees with "real" two dimensional turbulence. Virtually no laboratory experiments have yet been conducted, nor are there regions of the ocean or atmosphere sufficiently uncontaminated by other flows available to test the theories and calculations. In no case can the theories or numerical models yet be imbedded in larger flows so that stress or temperature can be reliably transported by two dimensional turbulence in ocean models.

Research Needs

a. Further two-dimensional turbulence studies by high-resolution numerical models.

b. Comparison of numerical models of two-dimensional turbulence with laboratory studies is feasible and would be extremely useful.

c. The interaction of turbulence with stratification, with shear, with planetary rotation, and with large density changes needs to be studied. Numerical models can now begin to do this with supercomputers and results could be compared with laboratory results.

WAVE MODES IN GEOPHYSICAL FLUIDS

For geophysical fluid dynamics purposes the ocean, atmosphere, core of the earth and atmospheres of planets can be considered to be rotating spherical shells of fluid whose density varies with depth and location and which have very complicated bottom configurations. These possess an enormous number of wave modes! (Acoustic and seismic modes are not being included here since they apply to solid as well as fluid processes.) Well over half of these are now understood in a linear sense and many have specific names. Frequencies of these waves in the ocean and atmosphere of the earth range from 10^{-8} hertz for the largest internal Rossby waves to ten hertz for surface capillary waves on the air-water interface. Some are trapped in certain locations such as the bottoms or sides of the ocean, or at the equator, and others cannot penetrate to certain latitudes or heights because they reach some critical regions where the energy is turned back. The manner in which the waves interact with larger scale flows is usually poorly understood but progress is being made. For example the wind generation of gravity waves is explained by a number of different theories and no rigorous test with either ocean observation or controlled experiments has resolved which theory is correct. In another example some equatorially trapped waves evidently interact with zonal wind fields to produce climatic changes associated with El Niño and ENSO, but again present theories and numerical models are not tested by either controlled experiment or unambiguous direct observations. In a

318 Geophysical Fluid Dynamics

third example gravity waves propagate upward in the lower atmosphere and break in the stratosphere where they deposit angular momentum and energy. The process is well understood and theory agrees with observations. These examples show that the interaction of larger flows and waves are gradually becoming understood in a number of key areas.

Research Needs

a. Numerical studies have been extremely useful because the equations can be solved by calculation in geometries that are like the actual geometries. They need to continue using the best computers.

b. Experiments are needed since few laboratory tests have been made of the theories. Some results agree with data from the geophysical systems but usually some adjustable parameters are needed to insure agreement.

c. Wave-mean flow interaction is still poorly understood.

d. The interaction of wave modes with friction or mixing is poorly understood.

FLUID MECHANICS WITH MIXED OR NONLINEAR BOUNDARY CONDITIONS

Another growing body of present research involves the response of interior flows to mixed boundary conditions. Geophysical fluids always contain more than one component. For example the atmosphere contains air, water vapor, and other gases. The ocean contains both water and salt. The interior of the earth contains many substances. Planetary atmospheres are similarly composed of many gases. Often one component has flux conditions into and out of the boundaries that differs from the flux condition of the other component even though both contribute to a dynamically important property such as density. In addition the velocity or stresses require boundary conditions. For instance, the example of thermocline circulation described above had both a vorticity, and density boundary condition. The mixed boundary conditions can lead to subtle but important processes, such as oscillations, bifurcation, and cataclysmic transition from one state to another. Important applications such as climate or earth history require thorough understanding.

Research Needs

a. Mathematical understanding of these problems needs to be improved.

b. Experimental work is needed as mathematical problems become clarified.

c. A hierarchy of models, from simple box models to coupled general circulation models needs to be used to clarify the physics of the situation.

THE DYNAMO PROBLEM

How does a flowing fluid that can conduct electricity produce a magnetic field with no external forcing? Dynamos exist on numerous planets and stars yet the fluid mechanics is still a matter of much research. Great success has been achieved in this century at the simplest level where one specifies a flow field and calculates whether a magnetic field grows. At the next level, where one specifies only boundary conditions, so the fluid is free to flow as it will, solutions have only been found in the past decade. Halfway between these two are classes of dynamos driven by turbulence model equations where the eddies organize themselves to produce a large external mean field. Numerous theoretical solutions have been found for these over the past ten years.

Research Needs

a. Most recent advances have been made by applied mathematicians. Support for these studies should continue.

b. There is little experimental work on the dynamo. The few laboratory attempts to produce a dynamo have been largely unsuccessful. Facilities needed apparently are a container meters in size or larger and velocities in excess of tens of meters a second. The required fluid is liquid sodium or sodium-potassium mixture (NaK). All are within the capability of modern engineering practice.

c. Significant advances have been made using large computers. These efforts should continue.

SUMMARY

In summary, there are numerous challenging subjects in geophysical fluid dynamics that are presently being investigated. Although they are usually pictured as being in support of one of the geophysical disciplines, the actual concepts and methodology transcend any one discipline. There are a number of institutional constraints that hamper full development of this field. The mathematical aspects of many of these problems are conducted by applied mathematicians and communication with earth scientists is sometimes a problem. Numerical methods have made significant contributions by enabling more detailed and accurate calculations to be done using more general nonlinear equations. The numerical work is sometimes done by groups that view simulation rather than understanding as their objective so that the results are not easily tested against experiment. Laboratory experiments have been relatively rare in the past few years and more are needed. They have continued to be done principally in engineering or physics departments and again communication with earth scientists is sometimes a problem. Finally, fluid dynamicists in the individual disciplines are sometimes only regarded as providing as service work to the natural science. This attitude unfortunately causes some scientist to lose perspective of the total value of the work. This is reinforced by the fact that many geophysical fluid dynamicists work at specialized organizations. Thus their contributions to fluid dynamics as a unified body of understanding is fragmented and lost.

These obstacles are partly of institutional and partly of economic nature. They can be overcome by improved communication and funding of geophysical fluid dynamics in its own right rather than as a service discipline to data gathering enterprises such as expeditions, cruises, field studies or satellite missions. On the positive side, geophysical fluid dynamics is finding a renewed intellectual and applications-oriented vitality in the central role that it plays in dynamic meteorology and physical oceanography. These two sister disciplines are at the core of efforts to understand, model, and predict global change due to humanity's increasing impact on its environment. The dynamics of the coupled ocean-atmosphere system is both the physical centerpiece and the intellectual paradigm for the broader Earth system, encompassing chemical and biochemical processes, as well as other subsystems, like snow and ice or the biosphere. Thus the rapidly developing study of the entire Earth system and of its global change has much to learn from the relatively advanced meteorological and oceanographic disciplines whose theoretical essence is distilled in geophysical fluid dynamics.

GENERAL REFERENCES

Many topics in Geophysical Fluid Dynamics are reviewed in the Annual Review of Fluid Mechanics. They are listed in the "Chapter Titles" index under the heading "Geophysical Fluid Dynamics".

CSEDI Science Plan, a Science Plan for Cooperative Studies of the Earth's Deep Interior U. S. SEDI Coordinating Committee, (1993).

Ridge Science Plan 1993-1997, available from RIDGE office, Woods Hole Oceanographic Institution Department of Geology and Geophysics Woods Hole, MA 02543.

WOCE Discussions of Physical Processes. U. S. WOCE Planning Report number 5, 143 pp. U. S. Planning Office for WOCE, College Station, TX.

Scientific Design for the Common Module of the Global Ocean Observing System and the Global Climate Observing System: An Ocean Observing System for Climate, Final Report of The Ocean Observing System Development Panel, Department of Oceanography, Texas A&M University, College Station, Texas 265 pp.

Coastal Ocean Processes: A Science Prospectus, K. H. Brink, J. M. Bane, T. M. Church, C. W. Fairall, G. L. Geernaert, D. E. Hammond, S. M. Henrichs, C. S. Martens, C. A. Nittrouer, D. P. Rogers, M. R. Roman, J. D. Roughgarden, R. L. Smith, L. D. Wright, and J. A. Yoder, *Woods Hole Oceanographic Institution Technical Report* WHOI-92-18. 88 pp.

Transport Processes and the Hydrological Cycle, Alessandro Marani and Andrea Rinaldo, Eds. Instituto Veneto Di Scienze, Lettere ed Arti, Venice, (1992).

Climate Change, Eds. J. T. Houghton, G. J. Jenkins, and J. J. Ephraums, Report prepared for IPCC Working Group 1, Published by the Press Syndicate of the University of Cambridge, The Pitt Building, Trumpington Street, Cambridge, CB2 1RP England.

Proceeding of the International Conference on Monsoon Variability and Prediction, Volume 1. International Centre for Theoretical Physics, Trieste, Italy, (1994). WCRP-84, WMO/TD- No. 619

Cloud-Radiation Interactions and their Parameterizations in Climate Models, Report of international workshop, National Oceanic and Atmospheric Administrations Science Center, Camp Springs Maryland, U. S. A. 18-20 October, (1994). WCRP-86, WMO/TD-No. 648

Cirricula in the Atmospheric, Oceanic, Hydrologic, and Related Sciences, American Meteorological Society, 45 Beacon Street, Boston, (1994) MA 02108-3693.

VORTEX DOMINATED FLOWS

Norman J. Zabusky
Department of Mechanical and Aerospace Engineering
Rutgers University
Piscataway, NJ 08855-0909

INTRODUCTION

Many practical and geophysical flows are vortex dominated as observed in vortices shed from steady and maneuvering aircraft and high-energy events like hurricanes and tornadoes. The generation, interaction and dispersal or mixing of vorticity plays a profound role in a wide class of applied and fundamental fluid flows.

A flow is vortex dominated if the dissipation is low (high Reynolds number) or there is transient bursting (or intermittent) event. The flow is then dominated by localized concentrations of vorticity (or coherent vortex structures) embedded in a nearly passive-incoherent background (or sea) of intermixed and distributed vorticity.

A better ability to predict and control flows will arise from a deep understanding of the processes leading to the formation (cyclogenesis), evolution and persistence of coherent structures while interacting with weakly correlated vortex processes. These aspects introduce the ideas of data assimilation and signal feedback for control of aircraft, ship and chemical process performance. Imagine forecasting meteorological or oceanographic events in which local environmental measurements and remote (e.g. satellite) observations are fedback into local space-time regions in the computer simulation code to reduce errors and improve the reliability of predictions. Similarly in the latter, we have sensors located within the flow which provide feedback signals to force the flow in a hopefully stable manner.

Finally, the incompressible and compressible Euler equations provide a unique test environment for the entire gamut of mathematical physics questions associated with nonlinear fields and their evolving topologies

RECENT HISTORY

Prior to the early-seventies there was little understanding of vorticity dynamics beyond some general theorems, linear stability analyses (Drazin and Howard) and point vortex simulations in 2D and a (few poorly understood) Biot-Savart single filament simulations in 3D. In turbulence, we knew about

upward (low wave number) cascade of energy in 2D and had a vague understanding of physical space mechanisms of intermittency in 3D. Large strides
have been made since in understanding realistic vortex dominated flows. This
has arisen because of the synergetic use of large-scale computation (with
visualizations and quantifications) and well-diagnosed laboratory experiments.
This is illustrated by examining the style and content of the contributions to
Annual Reviews of Fluid Mechanics since 1973. (Some references are given
below.)

In the last decade, we have begun to appreciate the processes associated
with coherent manifestations of vorticity. For example, in 2D and multilayer
2D, we now appreciate the role of merger (or pairing) of like-signed vorticity, binding or entrainment of opposite-signed vorticity and axisymmetrization and "stripping" of isolated regions in strain fields. In 3D the diverse
space-time configurations associated with breakdown of isolated tubes through
the occurrence of stagnation points and reconnection of tube-like regions has
been viewed as a scattering process. It would be valuable to associate these
physical-space processes with the wave-number cascade processes and non-
gaussian turbulent statistics.

APPLICATIONS

Applications of vortex dominated flow studies are being made to: geo-
physically motivated 2D barotropic and multilayerflows and 3D flows in the
presence of fixed and moving solid and porous boundaries, free surfaces, stratification and rotation. In 2D and multilayer-2D, high resolutions have been
obtained with pseudospectral, piecewise-parabolic method (ppm), and particularly the lagrangian contour dynamical/surgery codes. In laboratory experiments, laser sheets, particle image velocimetry and multi-element probes
have enhanced our quantitative understanding of processes and phenomena.
Many of the topics discussed below will overlap with contributions to other
sections in this volume, since vorticity is omnipresent in realistic and turbulent fluid motions.

Research Needs

MATHEMATICAL AND COMPUTATIONAL METHODS

The mathematical theory of 2D incompressible flow is well-developed but
far from complete. The modeling of turbulence and the effect on particle
dispersion remains a valuable goal. In 3D, singularity and near-singularity
development for inviscid and weakly dissipative flows (sheet-like structures
and 3D reconnection environments), and questions concerning uniqueness of
solutions to inviscid initial value problems after the singularity formation

time, have not been satisafactorily resolved. Small viscosity (asymptotically motivated) mechanisms for selecting the correct branch for continuation of solutions, effects of strain, axial flow, etc.

Explain generic processes on the basis of the evolution of the topology of critical points and separatrix lines and surfaces and the rate of strain tensor in 2D and 3D. Include filamentation, curvature singularities on vortex sheets, axisymmetrization, stripping and gradient-intensification, merger of like signed vortex domains, binding of opposite-signed vortex domains, survivability (robustness) in strain-shear environments; global "blocking highs" (which often appear as dipoles in the vorticity field) in the earth's atmosphere; stretching-intensification, bridging and reconnection of vortex tubes and vortex breakdown. It would be valuable to associate these physical-space processes with the wave-number cascade processes and non-gaussian turbulent statistics.

Low-order models

The ideas of low degree-of-freedom chaotic nonlinear systems has enriched our understanding of complex and higher dimensional phenomena, including aspects of mixing and predictability. The aim is to use the conservation properties to select the smallest number of variables to represent generic process, such as vortex merger or stripping in 2D and 3D; and axial variations including filamentation, helical symmetry and axial flows and breakdown, local binding and pancake formation during reconnection in 3D. This will help us interpret experiments (e.g. merger in stratified 2D and multilayer 2D rotating environmaents and vortex tube reconnections in wind tunnels) by juxtaposing data sets from large-scale direct numerical simulations.

We should develop rigorous (asymptotically-motivated and verifiable) model representations like:

- the "moment", "elliptical" or "conformal dynamics" models in 2D and multilayer 2D;

- the self-and-mutual stretching and core deformation models for 3D filamentary regions.

A possible model in 3D is the representation of cores by several intersecting vortex ribbons (narrow sheets).

Computational/numerical methods and algorithms

These include the deterministic and random vortex method, the vortex in cell method, boundary integral methods and contour dynamics/contour surgery. New work is needed in describing boundary layer phenomena by discrete vortex methods.

New work in formulating two-fluid representations where the large-scale phenomena are represented by coherent entites in physical space and the weak turbulent background by a fluid or kinetic set of continuum equations. An important issue is the validation of the correspondence of Lagrangian (e.g. CD/CS) scheme to the regularized Navier-Stokes (hyperviscosity filtering, sub-grid scale or LES models) for fundamental coherent-structure processes at long times.

PREDICTION AND CONTROL

The identification of sensitive, relatively unpredictable processes to aid in choosing optimal data assimilation methods. Test data assimilation methods in the context of simple models first, and then, by identifying sensitive processes, perhaps with the help of simplified models like conformal dynamics, one can proceed to more complex, realistic models approaching those used currently in atmospheric and oceanic modeling.

How can we alter the parameters associated with vortex shedding from solid objects (for example, bluff bodies, wing and propeller tips) by airfoil design or feedback control to enhance or mitigate stability and "breakdown" of downstream motions for example, the vortex breakdown on low aspect ratio wings.

In geophysical fluid problems we are plagued with the predictability issue and data assimilation. There is a need to make use of other algorithmic approaches including contour dynamics/contour surgery method in modeling atmospheric and oceanic flows, particularly to examine the effects of lack of adequate resolution in conventional numerical methods at long times.

The extensive list of generalizations of contour dynamics/contour surgery, such as complex rigid and dynamics external boundaries (e.g. coastlines and islands), 3D stratification, bottom topography, surface temperature and wind-stress forcing, Ekman pumping and idealized radiative damping, as well as the numerical advance of a moment-accelerated algorithm will enable a deep, rational quantification of many basic processes in GFD.

Experimental data could be obtained from studies of helical-straight vortex tube interactions in a wind or water tunnel, where the helical vortex tube is produced by a rotating propeller and the straight vortex tube by a horizontal-fixed wing. Many other interacting configurations can be arranged easily, including varying circulations and cores (laminar and turbulent) of tubes and different propeller frequencies to study chaotic particle dynamics and (bursting) intermittent phenomena. Also vortex-fixed surface interactions can be arranged. Synchronization of events is readily resolved by using phase-locking with the frequency of the propeller.

STRATIFIED, ROTATING, COMPRESSIBLE AND WAVE-LIKE FLOWS

(Including acoustic, gravity, & Rossby wave interactions)

Vortex-Wave Processes - Direct and Inverse

By this we mean low-to-intermediate Mach number phenomena—namely the direct wave radiation from coherent vortex events like merger, breakdown and reconnection and interaction with solid structures and topography. Also, the inverse effect of waves impinging on (scattering off) vortex layers and localized vortex structures, (cores) etc.

What are the properties of compressible vortices with transitional and high mach number, etc.? How can the steady states of compressible subsonic and supersonic flows with confined vorticity be used to understand the dynamics of important practical problems such as the compressible mixing layer?

Theoretical studies of the shallow water equations have indicated emission and subsequent back reaction of gravity (acoustic) waves. Hence, develop mathematically justifiable equations which ignore the fast-time scale effects of gravity waves. Also, "balanced equation models" in oceanographic fluid dynamics often work surprisingly well, but need rigorous derivation.

Understanding vortex component of natural flows: tornadoes, hurricanes, the polar vortex and ozone hole, the interaction of small-scale, low-level potential vorticity anomalies near the polar vortex edge, atmospheric blocking, planetary-scale vortices (e.g. the red spot of Jupiter), the gulf stream and its gyres, transport by ocean eddies, and explosive cyclogenesis.

ACKNOWLEDGMENTS

The author has benefited from substantial inputs from: R. Adrian, H. Aref, and D.G. Dritschel. In addition the author has benefited from conversations and correspondence with: R. Caflisch, L. Campbell, V. Gryanik, E. Hopfinger, T. Kambe, S Lele, D. Manin, S. Nazarenko, L. Polvani, D. Pullin and G. Reznik.

BIBLIOGRAPHY

References from Annual Reviews of Fluid Mechanics

Adrian, R.J., Particle-imaging techniques for experimental fluid mechanics. *Annu Rev Fluid Mech.* **23**:261-304.

Baker, G.R. and Saffman, P.G., (1980) Vortex interactions. *Annu Rev Fluid Mech.* **11**:95-122.

Browand, F.K. and Maxworthy, T. Experiments in rotating and stratified flows: oceanographic application. *Annu Rev Fluid Mech.* :273-305.

Chong, M.S.and Perry, A.E. A description of eddying motions and flow patterns using critical-point concepts. *Annu Rev Fluid Mech.* **19**:123-55.

Dickinson, R.E. Rossby waves-long-period oscillations of oceans and atmospheres. *Annu Rev Fluid Mech.* **10**:159-95.

Hall, M.G. Vortex breakdown. *Annu Rev Fluid Mech.* **4**:195-218.

Klemp, J.B. Dynamics of tornadic thunderstorms. *Annu Rev Fluid Mech.* **19**:369-402.

Leibovich, S. The structure of vortex breakdown. *Annu Rev Fluid Mech.* **10**:221-46.

Leonard, A. Computing three-dimensional incompressible flows with vortex elements. *Annu Rev Fluid Mech.* **17**:523-59.

Leonard, A. and Shariff, K., Vortex rings. *Annu Rev Fluid Mech.* 1992. **24**: 235-79.

Liu, J.T.C. Coherent structures in transitional and turbulent free shear flows. *Annu Rev Fluid Mech.* **21**:285-315.

Lugt, H.J. Autorotation. *Annu Rev Fluid Mech.* **15**:123-47.

Oertel, Jr., H. Wakes behind blunt bodies. *Annu Rev Fluid Mech.* **22**:539-64.

Pullin, D.I., (1992) Contour dynamics methods. *Annu Rev Fluid Mech.* **24**: 89-115.

Rhines, P.B. (1986) Vorticity dynamics of the oceanic general circulation. *Annu Rev Fluid Mech.* ch. **18**:433-97.

Robinson, S.K. (1991) Coherent motions in the turbulent boundary layer. *Annu Rev Fluid Mech.* **23**:601-39.

Smith, J.H.B. (1986) Vortex flows in aerodynamics. *Annu Rev Fluid Mech.* **18**:221-42.

Tobak, M. and Peake, D.J., (1982) Topology of three dimensional separated flows. *Annu Rev Fluid Mech.* **14**:61-85.

Werle, H. (1973) Hydronamic flow visualization. *Annu Rev Fluid Mech.* **5**:361-82.

Widnall, S.E. (1975; b) The structure and dynamics of vortex filaments. *Annu Rev Fluid Mech.* **7**:141- 65.Vol 7.

Zabusky, N.J. (1993) Visualization and Quantification of Vortex Dominated Flows. *Annu Rev Fluid Mech.*, 29, (to be published)

Other References

Lugt, H.J. (1983) Vorticity flow in nature and technology. New York, Wiley, ENGR QC 159 L95.

Saffman, P.G. (1981). Dynamics of vorticity. *J. Fluid Mech.* **106**: 49-58.

Saffman, P.G. (1992) *Vortex Dynamics*, Cambrige University Press

Van Dyke, M. (1982) *An Album of Fluid Motion*, Parabolic. Press, Stanford, Calif.

Proceedings of recent meetings

(1989) Workshop on Mathematical Aspects of Vortex Dynamics (Leesburg, VA) *Mathematical Aspects of Vortex Dynamics*, Ed. R. E. Caflisch. SIAM.

(1992) *Topological Aspects of the Dynamics of Fluids and Plasmas*, Ed K. Moffat et al, Kluwer.

(1991) *Vortex Dynamics and Vortex Methods*, Eds. C. R Anderson and C. R. Greengard. Lectures in Applied Mathematics Vol 28, American Math Society.

(1988) *Vortex Motion*, Proceedings of the IUTAM Symposium on Fundamental Aspects of Fluid Motion, Aug-Sept 1987, Tokyo, Japan. Eds, H. Hasimoto and T. Kambe. North-Holland.